6

MATHEMATICS MANUAL

MATHEMATICS MANUAL

Methods and Principles of the Various Branches of Mathematics for Reference, Problem Solving, and Review

FREDERICK S. MERRITT

Senior Editor, *Engineering News-Record*

McGRAW-HILL BOOK COMPANY, INC.

New York   Toronto   London

1962

MATHEMATICS MANUAL

To my wife Elsie, whose patience and encouragement
made this book possible

PREFACE

The vine of mathematics, rooted in logic, shoots intertwining branches upward. Each leaf is a tool potentially useful in solving practical problems. In classical study, teachers trim this vine to manageable size and pursue one branch at a time from root to tip. Textbooks reflect this course. But there is no need for a reference work to follow the same path. In fact, there are many advantages in applied mathematics in considering the vine as a whole; this book reflects that viewpoint.

This manual has one aim: to make it easy for you to solve the mathematical problems you encounter in your work and hobbies.

This objective makes broad coverage essential; therefore the scope of the book ranges from simple arithmetic through higher mathematics, including matrices, tensors, probabilities, and statistics. In addition, the book is designed to provide useful information speedily.

In this single volume you will find the important definitions, principles, theorems, corollaries, relationships, and methods of the most commonly used branches of mathematics. Illustrative examples accompany nearly every topic to clarify the meaning of formulas, to show how to use them, and to demonstrate the techniques under discussion. References to textbooks, other reference works, and mathematical tables are noted at the end of each chapter to guide you to sources of additional information.

Each subject is treated concisely. Explanations are given, but they are kept short and simple. This treatment makes for speedier extraction of information. It also permits inclusion of more topics than would otherwise be possible.

Furthermore, history, derivations, proofs, and obsolete techniques have been deliberately omitted. This also enables the manual to cover more topics. In addition, the elimination of extraneous material prevents distraction from the main objective of extracting problem-solving information.

The author hopes you will be pleased with the contents and with the arrangement, which differs from that used in most textbooks. The manual places related topics close together, even though they may belong to different branches of mathematics. This arrangement permits continuity of treatment and should make it easy for you to find the information you seek.   It groups alternative methods and allows you to use any of the mathematical tools available for solution of a problem without the need for waiting, as with textbooks, for the tool to be derived.   For example, Chap. 3, Algebra, contains the important methods of solving equations, including solutions involving derivatives, which usually are treated in differential calculus.

The manual's novel treatment of plane and solid geometry in Chap. 4 demonstrates another advantage of the departure from conventional arrangement.   To make proofs and constructions easier, each article comprises theorems and corollaries applicable to a specific purpose.   For example, one article presents the theorems and corollaries for proving lines equal, another gives those for proving lines parallel, and another lists those for proving lines perpendicular.   And incidentally, if you can use some suggestions on how to prove a geometric statement or carry out a geometric construction, you should find helpful the articles at the beginning of Chap. 4.

Throughout the book is material that will be useful with high-speed electronic computers: nondecimal number systems, matrices, determinants, and numerical solutions of algebraic, trigonometric, and differential equations.   Frequently, alternative methods of solving a problem are grouped to enable you to select the one most suitable to your purpose or equipment, be it manual computation, slide rule, desk calculator, or electronic computer.

For speedy location of specific formulas or methods, you can use either the detailed table of contents at the front of the book or the extensive index at the rear.   Generally, you will find the index faster.   But if you can't find what you are seeking immediately, try the table of contents. This will guide you to the articles covering the general subject, and a quick examination of the text may turn up the information you need.

A few of the formulas and methods in the manual originated with the author.   But for the rest the author is indebted to many excellent textbooks and reference works.   Space considerations preclude listing all the sources, but the most important are included in the bibliographies at the end of each chapter.

Frederick S. Merritt

CONTENTS

Chapter 1

BASIC DATA

1-1. Common Mathematical Symbols

$=$	Equals
$\neq$	Is not equal to
$\equiv$	Is identical to
$+$	Plus
$-$	Minus
$\pm$	Plus or minus
$\times$ or $\cdot$	Multiplied by
$\div$ or $/$	Divided by
$>$	Greater than
$\geqq$ or $\geq$	Equal to or greater than
$<$	Less than
$\leqq$ or $\leq$	Equal to or less than
$\sqrt[n]{}$	nth root
$\to$ or $\doteq$	Approaches as a limit
$\lvert a \rvert$	Absolute value of a

∞	Infinity
$\propto$	Varies as
$\approx$	Approximates
$\sim$	Is similar to
$n!$ or $\lfloor n$	Factorial $n =$
	$n(n-1)(n-2)\cdots 1$
$:$	Ratio sign, is to
$\displaystyle\sum_{b}^{a}$	Sum from b to a
$\displaystyle\int_{b}^{a}$	Integrate from b to a
$\angle$	Angle
$\perp$	Perpendicular
$\parallel$	Parallel
$\%$	Per cent

1-2. Greek Alphabet

	Capital	Lower Case		Capital	Lower Case
Alpha	A	α	Nu	N	ν
Beta	B	β	Xi	Ξ	ξ
Gamma	Γ	γ	Omicron	O	o
Delta	Δ	$\delta,\ \partial$	Pi	Π	π
Epsilon	E	ϵ	Rho	P	ρ
Zeta	Z	ζ	Sigma	Σ	$\sigma,\ s$
Eta	H	η	Tau	T	τ
Theta	Θ	$\theta,\ \vartheta$	Upsilon	Υ	υ
Iota	I	ι	Phi	Φ	$\phi,\ \varphi$
Kappa	K	$\kappa,\ \varkappa$	Chi	X	χ
Lambda	Λ	λ	Psi	Ψ	ψ
Mu	M	μ	Omega	Ω	ω

1

1-3. Conversion of Inches to Feet. When converting fractions of an inch to decimals of a foot, it is convenient to remember that $\frac{1}{8}$ in. is approximately 0.01 ft. Hence, $\frac{5}{8}$ in. = 0.05 ft and $\frac{7}{8}$ in. = 0.07 ft.

In.	Ft	In.	Ft	In.	Ft
1	0.0833	5	0.4167	9	0.7500
2	0.1667	6	0.5000	10	0.8333
3	0.2500	7	0.5833	11	0.9167
4	0.3333	8	0.6667	12	1

• How many feet to the nearest hundredth in $55\frac{5}{8}$ in.?

$$55\frac{5}{8} \text{ in.} = 48 \text{ in.} + 7\frac{5}{8} \text{ in.} = 4 \text{ ft } 7\frac{5}{8} \text{ in.}$$
$$4 \text{ ft } 7\frac{5}{8} \text{ in.} = 4 + 0.58 + 5 \times 0.01 = 4.63 \text{ ft.}$$

1-4. Common (English) Measurement System

Length

1 foot (ft) = 12 inches (in.)
1 yard (yd) = 3 ft
1 rod = $5\frac{1}{2}$ yd = $16\frac{1}{2}$ ft
1 statute mile = 320 rods = 5,280 ft = 1,760 yd
1 league = 3 miles
1 furlong = 220 yd
1 chain = 66 ft

Area

1 sq ft = 144 sq in.
1 sq yd = 9 sq ft
1 sq rod = $30\frac{1}{4}$ sq yd
1 acre = 160 sq rods = 43,560 sq ft
1 sq mile = 640 acres
1 circular inch (cir in.) = 0.7854 sq in.
1 circular mil (cir mil) = area of circle 0.001 in. in diameter

Volume

1 cu ft = 1,728 cu in.
1 cu yd = 27 cu ft
1 cord = 128 cu ft

Liquid Measure

1 pint (pt) = 4 gills = 16 ounces (oz)
1 quart (qt) = 2 pt
1 gallon (gal) (U.S.) = 4 qt = 231 cu in.
1 cu ft = 7.4805 gal
1 Imperial gal (British) = 1.20094 U.S. gal

Dry Measure

1 quart (qt) = 2 pints (pt)
1 peck = 8 qt
1 U.S. bushel (bu) = 4 pecks = 2,150.42 cu in.

Weight (Avoirdupois)

1 pound (lb) = 7,000 grains = 16 ounces (oz)
1 stone = 14 lb
1 ton = 2,000 lb = 0.8929 long tons
1 long ton = 2,240 lb = 1.12 tons

Nautical Units

1 nautical mile = 6,076.10333 ft
1 fathom = 6 ft
1 knot = 6,076.10333 ft per hour
1 knot = 1.1516 statute (ordinary) miles per hour

1-5. Metric System. This is a decimal system in which prefixes are applied to the basic units of length, weight, etc., to denote fractional or multiple units. Thus, a centimeter (cm) is 0.01 meter and a kilogram is 1,000 grams.

Prefixes

pico = 10^{-12} tera = 10^{12}
nano = 10^{-9} giga = 10^{9}
micro = 10^{-6} mega = 10^{6}
milli = 10^{-3} kilo = 10^{3}
centi = 10^{-2} hecto = 10^{2}
deci = 10^{-1} deka = 10

The meter is the unit of length. Another length unit is the Angstrom (A), which equals 0.0001 micron, the micron being 10^{-6} meter. The are is the unit of area; 1 are = 100 square meters (sq m).

The liter is the unit of volume or capacity. It equals 1,000 cu cm.

The gram (g) is the unit of weight. Other units are the carat, which equals 200 milligrams (mg), and the metric ton, which equals 1,000 kilograms (kg).

1-6. Conversion Factors, English-Metric

Length

1 in. = 2.54001 cm   1 m = 39.37 in. = 3.281 ft
1 ft = 30.4801 cm   1 cm = 0.3937 in.
1 mile = 1.60935 km   1 km = 0.6214 mile

Area

1 sq ft = 0.09290 sq m   1 sq m = 10.7639 sq ft
1 acre = 4,046.86 sq m   1 hectare (ha) = 2.4710 acres
1 acre = 0.4047 hectare   1 hectare (ha) = 0.003861 sq mile
1 sq mile = 258.9998 hectares 1 hectare (ha) = 107,638.7 sq ft

Volume

1 cu in. = 0.01638716 liter	1 liter = 1,000 cu cm = 61.03 cu in.
1 cu in. = 16.38716 cu cm	1 liter = 0.03532 cu ft
1 U.S. gal = 3.785 liters	1 liter = 0.2642 gal
1 qt (liquid) = 0.9463 liter	1 liter = 1.057 qt (liquid)
1 cu ft = 28.317 liters	

Weight

1 lb (avoir.) = 0.45359 kg	1 kg = 2.205 lb (avoir.)
1 ton = 0.9072 metric ton	1 metric ton = 1.1023 tons
1 ton = 907.2 kg	1 metric ton = 0.9842 long ton
	1 metric ton = 2,204.62 lb

Pressure

1 psi = 0.0703 kg per sq cm
1 psf = 4.882 kg per sq m
1 kg per sq cm = 14.22 psi

Energy or Work

1 joule = 0.10197 kg-m = 0.7376 ft-lb
1 joule = 0.0009478 Btu = 0.2388 calories
1 ft-lb = 1.356 joules = 0.1383 kg-m
1 kw-hr = 3.600×10^6 joules = 859,900 calories = 3,412 Btu
1 calorie = 4.187 joules = 3,968 Btu
1 Btu = 0.000252 calorie = 1,055 joules = 778.2 ft-lb
1 kg-m = 7.233 ft-lb

Power

1 hp = 0.7457 kw = 550 ft-lb per sec
1 hp = 1.014 metric hp = 0.7068 Btu per sec
1 kw = 1.341 hp = 0.9478 Btu per sec
1 metric hp = 0.9863 hp = 75 kg-m per sec

Temperature

$$°F = \tfrac{9}{5}C + 32$$
$$°C = \tfrac{5}{9}(F - 32)$$
$$°K = C + 273.160$$

1-7. Fundamental Constants

$$\pi = 3.141592653589793$$
$$e = 2.718281828459045$$
$$\log_{10} e = 0.434294481903252$$
$$\log_e 10 = 2.302585092994046$$
$$1 \text{ degree } (°) = 0.01745329 \text{ radian}$$
$$1 \text{ radian} = 57.2957795°$$

Chapter 2

ARITHMETIC

2-1. Integer. A whole number, or an aggregate of units (1,2,3,4,5, 6,7, . . .).

2-2. Even Number. An integer exactly divisible by 2.

2-3. Odd Number. An integer not exactly divisible by 2.

2-4. Prime Number. An integer divisible only by itself and 1. To test a number for primeness, divide it successively by all primes smaller than its square root; if each of these divisions has a remainder, the number is a prime.

2-5. Composite Number. An integer with factors other than itself and 1. It can be represented as the product of two or more integers. Also, it can be factored into a product of primes.

• What are the prime factors of 84?

$84 = 2^2 \cdot 3 \cdot 7$, so 2, 3, and 7 are the prime factors of 84.

2-6. Reciprocal of a Number. Unity divided by the number.

2-7. Existence Law for Arithmetic. If a and b are numbers, the sum of a and b is a unique number c. $a + b = c$.

2-8. Commutative Law. Numbers can be added or multiplied in any order. $a + b \equiv b + a$; $ab \equiv ba$.

2-9. Associative Law. The sum or product of three or more numbers does not depend on their grouping. $(a + b) + c \equiv a + (b + c)$; $(ab)c \equiv a(bc)$.

2-10. Distributive Law. $a(b + c) = ab + ac$.

2-11. Order Relation. If a and b are two numbers, then either $a < b$, $a = b$, or $a > b$. If c is a third number, and if $a < b$ and $b < c$, then $a < c$.

2-12. Decimal System. Representation of a number as a power of 10 (Art. 3-35).

A decimal point divides a number into an integer portion and a fraction. The successive digits to the left of the decimal point represent, in order, the first digit times unity (10^0), the second digit times ten (10^1), the third digit times one hundred (10^2), the fourth digit times one thousand (10^3), etc.

The digits to the right of the decimal point represent, in order, the first digit times one-tenth (10^{-1}), the second digit times one-hundredth (10^{-2}), etc.

$$3{,}246.18 \equiv 3 \times 1{,}000 + 2 \times 100 + 4 \times 10 + 6 \times 1 + 1 \times \tfrac{1}{10}$$
$$+ 8 \times \tfrac{1}{100}$$
$$\equiv 3 \times 10^3 + 2 \times 10^2 + 4 \times 10^1 + 6 \times 10^0 + 1 \times 10^{-1}$$
$$+ 8 \times 10^{-2}$$

2-13. Nondecimal Number Systems. Representation of a number as the multipliers of powers of a number (base) other than 10 (Art. 2-12).

$$(N)_n \equiv a_0 n^0 + a_1 n^1 + a_2 n^2 + a_3 n^3 + \cdots + a_m n^m$$

where $(N)_n$ is a given number to the base n and $a_0, a_1, a_2, \ldots, a_m$ are digits less than n.

If 2 is the base, the system is called binary.

$$(29)_{10} \equiv 1 \times 2^4 + 1 \times 2^3 + 1 \times 2^2 + 0 \times 2^1 + 1 \times 2^0 \equiv (11101)_2$$

The binary number is composed of the digits by which the powers of 2 are multiplied.

If 12 is the base, the system is called duodecimal. For this system, two new symbols are needed to represent 10 and 11.

$$(1321)_{10} \equiv 9 \times 12^2 + 2 \times 12^1 + 1 \times 12^0 \equiv (921)_{12}$$

Here the duodecimal number is composed of the digits by which the powers of 12 are multiplied.

2-14. Exponential Notation. Very large or very small numbers often may be conveniently expressed with powers of 10 (Arts. 2-12 and 3-35).

$$3{,}650{,}000{,}000 \equiv 3.65 \times 10^9$$
$$0.000\ 000\ 365 \equiv 3.65 \times 10^{-7}$$

To find the exponent of 10 to be used, count the number of decimal places that the decimal point must be shifted to be just to the right of the integer portion of the coefficient (the number by which the power of 10 is to be multiplied). Decimal movements to the right are negative.

To multiply numbers expressed in exponential notation, multiply the coefficients and add the exponents of 10. To divide, divide the coefficients and subtract the exponent of the divisor from that of the dividend.

$$\frac{4{,}000{,}000 \times 800{,}000 \times 0.015}{20{,}000 \times 0.25 \times 200} = \frac{4 \times 10^6 \times 8 \times 10^5 \times 15 \times 10^{-3}}{2 \times 10^4 \times 25 \times 10^{-2} \times 2 \times 10^2}$$
$$= \frac{4 \times 8 \times 15 \times 10^{(6+5-3)}}{2 \times 25 \times 2 \times 10^{(4-2+2)}}$$
$$= \frac{24 \times 10^8}{5 \times 10^4} = 4.8 \times 10^4 = 48{,}000$$

Numbers in exponential notation can be added and subtracted only when their exponents of 10 are identical; then the coefficients are added or subtracted and the exponents remain unchanged.

2-15. Divisibility by Small Integers. A number is divisible by 2 if its right-hand digit is zero or divisible by 2.

A number is divisible by 3 if the sum of its digits is divisible by 3; it is divisible by 4 if its last two digits are divisible by 4 or are two zeros; it is divisible by 5 if its right-hand digit is zero or 5.

A number is divisible by 6 if it is an even number and the sum of its digits is divisible by 3; it is divisible by 8 if its last three digits are divisible by 8; it is divisible by 9 if the sum of its digits is divisible by 9.

2-16. Factor (Divisor). An integer that exactly divides another integer. 3 is a factor of 12, since $12/3 = 4$.

2-17. Common Divisor. An integer that exactly divides each of two or more integers. 3 is a common divisor of 12 and 48, since $12/3 = 4$ and $48/3 = 16$.

2-18. Greatest Common Divisor. The largest factor that exactly divides two or more integers. 12 is the greatest common divisor of 36 and 48, since $36/12 = 3$ and $48/12 = 4$, and 36 and 48 have no common divisor greater than 12.

If d is the greatest common divisor of two integers a and b, the quotients a/d and b/d are relatively prime (have no common divisor other than 1).

The greatest common divisor of any set of numbers $a,b,c, \ . \ . \ .$ also is the greatest common divisor of any one of them, a, and of d, the greatest common divisor of the remaining numbers of the set.

2-19. Euclid's Algorithm for Greatest Common Divisor. A procedure in which each operation depends on the results of the preceding operation.

To find the greatest common divisor of two numbers a and b $(a > b)$, divide a by b. If there is no remainder, b is the greatest common divisor. If the remainder is b_1, divide b by b_1. If there is no remainder, b_1 is the greatest common divisor of a and b. If the remainder of the second division is b_2, divide b_1 by b_2. Continue until a value b_n is found that is exactly divisible into b_{n-1}. b_n is then the greatest common divisor of a and b.

• Find the greatest common divisor of 26,232 and 6,444.

Step 1: $26,232/6,444 = 4 + 456/6,444$
Step 2: $6,444/456 = 14 + 60/456$
Step 3: $456/60 = 7 + 36/60$
Step 4: $60/36 = 1 + 24/36$
Step 5: $36/24 = 1 + 12/24$
Step 6: $24/12 = 2$

The last division leaves no remainder; therefore, 12 is the greatest common divisor of 26,232 and 6,444.

NOTE: The number of steps in the procedure can be reduced by selecting for the quotient the next larger integer and obtaining a negative remainder, if the absolute value of that remainder is smaller than the remainder obtained in ordinary division.

• In the last example, step 3 could have been

$$456/60 = 8 - 24/60$$

Subsequent steps, then, would have been:

Step 4: $60/24 = 2 + 12/24$
Step 5: $24/12 = 2$

As before, the answer is 12, but it is obtained in one less step.

2-20. Multiple of an Integer. An integer exactly divisible by the given integer.   12 is a multiple of 4, since $12/4 = 3$.

2-21. Common Multiple. An integer divisible by two or more given integers is their common multiple.   The common multiples of 3 and 4 are 12, 24, 36, etc., since these numbers are exactly divisible by both 3 and 4.

2-22. Least Common Multiple (lcm). The smallest integer divisible by two or more given integers.

The least common multiple of two or more numbers relatively prime to each other is their product.   If two numbers are not relatively prime, their least common multiple equals their product divided by their greatest common divisor (Arts. 2-18 and 2-19).

The least common multiple of three numbers not relatively prime can readily be determined when the least common multiple of any two of them is known.   If m is the least common multiple of a and b, the least common multiple of a, b, and c is the least common multiple of m and c.

• Find the least common multiple of 6, 14, and 18.
First determine the least common multiple of 6 and 14.   Their product is 84 and their greatest common divisor is 2, so their lcm is $84/2 = 42$.
Then compute the lcm of 42 and 18.   Their greatest common divisor is 6; hence their lcm is $42 \times 18/6 = 126$, which also is the lcm of 6, 14, and 18.

2-23. Fraction. An expression representing division of one number by another, written a/b, $\dfrac{a}{b}$, as a decimal, or in exponential form (Arts. 2-12 to 2-14). $3/4$, $9/2$, and $20/21$ are fractions.

The numerator is the part of the fraction above (or left of) the line; it is equivalent to the dividend in division.   3 is the numerator in $3/4$.

The denominator is the part of the fraction below (or right of) the line; it is equivalent to the divisor in division.   4 is the denominator in $\frac{3}{4}$.

When the numerator is less than the denominator, the fraction is termed proper. $\frac{1}{2}$, $\frac{1}{4}$, and $\frac{7}{8}$ are proper fractions.   When the numerator is equal to or larger than the denominator, the fraction is improper. $\frac{8}{7}$, $2\frac{1}{4}$, and $\frac{3}{2}$ are improper fractions.

2-24. Mixed Number. Sum of an integer and a fraction.   It is usually written without the plus sign.   Thus, $1 + \frac{5}{16}$, usually written $1\frac{5}{16}$, is a mixed number.

2-25. Similar Fractions. Fractions having the same denominator.

To convert a fraction from one denominator to another, divide the new denominator by the given denominator and multiply the given numerator by the quotient to obtain the new numerator.

• Convert two-thirds to eighteenths.

$$\frac{2}{3} = \frac{2 \times {}^{18}\!/_3}{18} = \frac{12}{18}$$

• Convert $\frac{3}{7}$ and $\frac{5}{6}$ to similar fractions.
Choose $6 \times 7 = 42$ as the common denominator.   Then

$$\frac{3}{7} = \frac{3 \times {}^{42}\!/_7}{42} = \frac{18}{42}$$

$$\frac{5}{6} = \frac{5 \times {}^{42}\!/_6}{42} = \frac{35}{42}$$

2-26. Least Common Denominator. The least common denominator of two or more fractions is the least common multiple of the denominators (Art. 2-22).

• Find the least common denominator of $\frac{7}{30}$, $\frac{11}{45}$, $\frac{17}{135}$, and $\frac{14}{25}$.

The least common multiple of the four denominators can be found by first computing the least common multiple of successive pairs, then determining the least common multiple of the lcm's.   Begin with 30 and 45.

The greatest common divisor of 30 and 45 is 15; the lcm is $30 \times {}^{45}\!/_{15} = 90$.
The greatest common divisor of 135 and 25 is 5; the lcm is $135 \times {}^{25}\!/_5 = 675$.
Finally, the least common denominator is the lcm of 90 and 675, or

$$90 \times {}^{675}\!/_{45} = 1{,}350$$

2-27. Equality of Fractions. If two or more fractions yield the same quotient when the indicated division is carried out, the fractions are equal. $\frac{3}{4} = \frac{6}{8} = \frac{15}{20} = 0.75$.

Two fractions are equal if multiplication or division of both numerator and denominator of either fraction by the same number yields the numerator and denominator, respectively, of the other fraction. $\frac{3}{4} = \frac{6}{8}$, since $(3 \times 2)/(4 \times 2) = \frac{6}{8}$.

2-28. Order of Fractions. If two fractions have equal denominators, the one with the larger numerator is larger. If the numerators are equal, the fraction with the smaller denominator is larger.

- Which is larger, $^{19}\!/_{32}$ or $^{38}\!/_{67}$?

Since $\dfrac{19}{32} = \dfrac{19 \times 2}{32 \times 2} = \dfrac{38}{64}$, and $\dfrac{38}{64} > \dfrac{38}{67}$, then $\dfrac{19}{32}$ is larger.

- Which is larger, $^{5}\!/_{17}$ or $^{14}\!/_{51}$?

Since $\dfrac{5}{17} = \dfrac{5 \times 3}{17 \times 3} = \dfrac{15}{51}$, and $\dfrac{15}{51} > \dfrac{14}{51}$, then $\dfrac{5}{17}$ is larger.

2-29. Basic Operations on Fractions. Multiplication or division of both numerator and denominator of a fraction does not change its value. $\tfrac{3}{4} = (3 \times 2)/(4 \times 2) = \tfrac{6}{8}$.

To multiply a fraction by a number, multiply the numerator or divide the denominator by that number. $4 \times \tfrac{3}{2} = 1\tfrac{2}{2}$; $6 \times \tfrac{3}{12} = \tfrac{3}{2}$.

To divide a fraction by a number, divide the numerator or multiply the denominator by that number. $1\tfrac{2}{2} \div 4 = \tfrac{3}{2}$; $1\tfrac{2}{5} \div 6 = 1\tfrac{2}{30}$.

2-30. Cancellation. Division of numerator and denominator of a fraction by the same number to simplify it.

- Evaluate $\dfrac{6 \times 80 \times 810}{9 \times 16 \times 300}$.

Divide numerator and denominator by common factors 8, 9, 3, and 2 and by 10 twice:

$$\frac{6 \times 80 \times 810}{9 \times 16 \times 300} = \frac{\overset{2}{\cancel{6}} \times (\overset{1}{\cancel{8}} \times \overset{1}{\cancel{10}}) \times (\overset{1}{\cancel{9}} \times 9 \times \overset{1}{\cancel{10}})}{\underset{1}{\cancel{9}} \times \underset{2}{\cancel{16}} \times (\underset{1}{\cancel{3}} \times \underset{1}{\cancel{10}} \times \underset{1}{\cancel{10}})} = 9$$

2-31. Interchangeability of Integers and Fractions. Unity can be represented by a fraction with equal numerator and denominator. Hence, any integer can be represented as a fraction by converting 1 to a fraction with a desired denominator and multiplying the numerator by the integer. $5 = 5 \times \tfrac{8}{8} = \tfrac{40}{8}$; $4 = 4 \times \tfrac{12}{12} = \tfrac{12}{3}$.

To convert a mixed number to a fraction, change the integer part to a fraction with the same denominator as the fraction part and add the numerators.

$$2\tfrac{5}{16} = 2 + \tfrac{5}{16} = 2 \times \tfrac{16}{16} + \tfrac{5}{16} = \tfrac{32}{16} + \tfrac{5}{16} = \tfrac{37}{16}$$

To convert an improper fraction to a mixed number, divide the numerator by the denominator. If there is no remainder, the fraction is equal to an integer. If there is a remainder, the result is a mixed number; the integer part is the largest integer by which the denominator can be multiplied without exceeding the numerator, and the fraction part is the remainder divided by the denominator. $1\frac{5}{2} = 7 + \frac{1}{2} = 7\frac{1}{2}$; $5\frac{3}{8} = 6 + \frac{5}{8} = 6\frac{5}{8}$.

2-32. Addition and Subtraction of Fractions. Convert to similar fractions (Arts. 2-25 and 2-26) and add or subtract numerators.

• Add $\frac{5}{8}$ and $\frac{3}{7}$.

First convert both to similar fractions with the least common denominator $8 \times 7 = 56$: $\frac{5}{8} = (5 \times 7)/(8 \times 7) = \frac{35}{56}$; $\frac{3}{7} = (3 \times 8)/(7 \times 8) = \frac{24}{56}$.

$$\frac{5}{8} + \frac{3}{7} = \frac{35}{56} + \frac{24}{56} = \frac{59}{56} = 1\frac{3}{56}$$

$$\frac{5}{8} - \frac{3}{7} = \frac{35}{56} - \frac{24}{56} = \frac{11}{56}$$

2-33. Addition and Subtraction of Mixed Numbers. Add or subtract the fraction parts, then the integer parts.

If, in a subtraction of fractions, the minuend (number from which another number is to be subtracted) is smaller than the subtrahend (number to be subtracted), add 1 to the minuend and reduce the integers by 1. If, in an addition, the sum of the fractions exceeds 1, convert the sum to a mixed number and add its integer to the sum of the integers.

• Subtract $2\frac{9}{32}$ from $5\frac{3}{16}$.

$$5\frac{3}{16} = 5\frac{6}{32} = \quad 4\frac{38}{32}$$

$$-2\frac{9}{32} = -2\frac{9}{32} = -2\frac{9}{32}$$

$$\text{Difference} = \quad 2\frac{29}{32}$$

• Add $5\frac{3}{16}$ and $2\frac{31}{32}$.

$$5\frac{3}{16} = \quad 5\frac{6}{32}$$

$$+2\frac{31}{32} = +2\frac{31}{32}$$

$$\text{Sum} = \quad 7\frac{37}{32} = 8\frac{5}{32}$$

2-34. Multiplication of Fractions. Multiply numerators, then multiply denominators. Cancel where possible to reduce the amount of multiplication (Art. 2-30).

$$\frac{8}{13} \times \frac{9}{16} \times \frac{26}{27} = \frac{\overset{1}{8} \times \overset{1}{9} \times \overset{2}{26}}{\underset{1}{13} \times \underset{2}{16} \times \underset{3}{27}} = \frac{1}{3}$$

2-35. Multiplication of Mixed Numbers. Convert the mixed numbers to improper fractions (Art. 2-31) and multiply numerators, then denominators. Cancel where possible to reduce the amount of multiplication.

$$3\tfrac{1}{3} \times 8\tfrac{1}{10} = {}^{10}\!\!/\!_3 \times 8\tfrac{1}{10} = 8\tfrac{1}{3} = 27$$

2-36. Reciprocal of a Fraction. Interchange numerator and denominator (equivalent to dividing 1 by the fraction). $\tfrac{3}{4}$ is the reciprocal of $\tfrac{4}{3}$; 4 is the reciprocal of $\tfrac{1}{4}$.

2-37. Division by Fractions. Multiply the dividend by the reciprocal of the divisor (Art. 2-36).

• Divide $\tfrac{9}{16}$ by $\tfrac{3}{8}$.
The reciprocal of $\tfrac{3}{8}$ is $\tfrac{8}{3}$; so $\tfrac{9}{16} \div \tfrac{3}{8} = (\tfrac{9}{16})(\tfrac{8}{3}) = \tfrac{3}{2}$.

2-38. Division by Mixed Numbers. Convert to improper fractions, then multiply the dividend by the reciprocal of the divisor.

• Divide $4\tfrac{5}{7}$ by $2\tfrac{1}{5}$.

$4\tfrac{5}{7} = {}^{33}\!\!/\!_7$, $2\tfrac{1}{5} = {}^{11}\!\!/\!_5$, and the reciprocal of ${}^{11}\!\!/\!_5$ is $\tfrac{5}{11}$
$4\tfrac{5}{7} \div 2\tfrac{1}{5} = {}^{33}\!\!/\!_7 \times \tfrac{5}{11} = {}^{15}\!\!/\!_7 = 2\tfrac{1}{7}$

2-39. Simultaneous Multiplication and Division of Mixed Numbers. If both the numerator and denominator of a fraction consist of mixed numbers, first convert to improper fractions. Then multiply the fractions in the numerator and the reciprocals of the fractions in the denominator. Cancel where possible.

• Evaluate $\dfrac{3\tfrac{1}{3} \times 8\tfrac{1}{10}}{4\tfrac{5}{7} \times \tfrac{5}{11}}$.

$$\frac{3\tfrac{1}{3} \times 8\tfrac{1}{10}}{4\tfrac{5}{7} \times \tfrac{5}{11}} = \frac{{}^{10}\!\!/\!_3 \times 8\tfrac{1}{10}}{{}^{33}\!\!/\!_7 \times \tfrac{5}{11}} = \left(\frac{10}{3} \times \frac{81}{10}\right)\left(\frac{7}{33} \times \frac{11}{5}\right) = \frac{63}{5}$$

2-40. Conversion of Fractions to Decimals. Divide the numerator by the denominator and express the result in decimal form (Art. 2-12).

Convert $1\tfrac{1}{9}$ to a four-place decimal.

$$9\,\underline{)\,11.0000}$$
$$1.2222$$

2-41. Conversion of Decimals to Fractions. To obtain the numerator, shift the decimal point to the right to eliminate the decimal. The denominator is 1 followed by as many zeros as the decimal point was shifted for the numerator.

• Express, 2.3125 as an ordinary fraction.

$$2.3125 = \frac{23,125}{10,000} = 2\,\frac{3,125}{10,000} = 2\,\frac{125}{400} = 2\,\frac{5}{16}$$

The numerator and the denominator were divided twice by 25.

• Evaluate 0.11111 · · ·

Let $n = 0.11111 \cdots$ Note that $0.01111 \cdots = 0.1n$. Then, $n = 0.1 + 0.1n$; from which $0.9n = 0.1$ and $n = 0.1/0.9 = \frac{1}{9}$.

2-42. Rounding Off Approximate Numbers. Discarding insignificant digits in a number (digits on the right of a number) or replacing them with zeros. When a number is being rounded off, the last digit of the retained portion may have to be increased by 1 to compensate for the discard. Leave *unchanged* the last digit of the retained portion if the first digit of the discard is smaller than 5. *Increase* the last digit of the retained portion by 1 if the first digit of the discard is greater than 5. If the first digit of the discard is exactly 5, add 1 to the last digit to be retained if it is an odd number; leave it unchanged if it is even.

Five figures	Four figures	Three figures	Two figures
1.8927	1.893	1.89	1.9
12.735	12.74	12.7	13
468.45	468.4	468	470

2-43. Addition of Decimals. Write each number, one below the other, with the decimal points aligned. Then add each column; place the decimal point in the sum in line with the other decimal points.

• Add 256.1, 32.03, and 1.296.

$$
\begin{array}{r}
256.1 \\
32.03 \\
1.296 \\
\hline
\text{Sum} = 289.426
\end{array}
$$

To check the addition, add each column from the bottom up; then add from the top down. The sum should be the same in both additions.

2-44. Subtraction of Decimals. Write the numbers, one below the other, with the decimal points aligned. Subtract the numbers in each column; place the decimal point in the remainder in line with the other decimal points.

• Subtract 4.287 from 5.106.

$$
\begin{array}{r}
\text{Minuend} = 5.106 \\
\text{Subtrahend} = 4.287 \\
\hline
\text{Remainder} = 0.819
\end{array}
$$

To check, add the remainder to the subtrahend. The result should be the minuend.

2-45. Absolute Error. The approximate value of a number minus its true value is the absolute error. It may be positive or negative.

• If the distance between two points should be exactly 512 ft, and a survey yields a measurement of 512.01 ft, what is the error?

The absolute error is $512.01 - 512 = +0.01$ ft.

The absolute error of a sum is the sum of the absolute errors of the numbers being added. The absolute error of a difference is the algebraic difference between the absolute errors of minuend and subtrahend.

• If five numbers, each accurate to 0.01, are to be added, what is the maximum possible error in the sum?

The maximum possible error in each number is ± 0.005. Hence, the maximum possible error in the sum is $\pm 0.005 \times 5 = \pm 0.025$.

2-46. Relative Error. Ratio of the absolute error in a number to the exact value of the number (often expressed as a percentage).

• If the distance between two points should be exactly 24 in., but a measurement gives a reading of 24.03 in., what are the absolute and relative errors?

The absolute error is $24.03 - 24 = +0.03$ in.

The relative error is $+0.03 \times 100/24 = +0.125$ per cent.

The relative error of a product is approximately the algebraic sum of the relative errors of the factors.

• What is the per cent error in an area when one side is off by -0.2 per cent and the other by $+0.3$ per cent?

The relative error in area, the product of the two sides, is

$$-0.2 \text{ per cent} + 0.3 \text{ per cent} = +0.1 \text{ per cent}$$

The relative error of a quotient is approximately the relative error of the dividend minus the relative error of the divisor.

• What is the per cent error in the computed unit strain in a bar when its total deformation is measured with an error of $+0.2$ per cent and its length with an error of -0.1 per cent?

Since unit strain is determined by dividing the total deformation by the length of the bar, the relative error in unit strain is

$$+0.2 \text{ per cent} - (-0.1 \text{ per cent}) = +0.3 \text{ per cent}$$

2-47. Addition and Subtraction of Approximate Numbers. Keep no more decimal places in the sum or difference of approximate numbers than there are in the term with the smallest number of decimal places (see also Arts. 2-42 to 2-44).

$$230.416 + 20.2178 + 3,040.2 = 3,290.8$$

2-48. Effect of Moving a Decimal Point. A shift of a decimal point is equivalent to multiplication or division by a power of 10.

Movement of the decimal point to the right one, two, three places, etc., multiplies the number by 10, 100, 1,000, etc.

Movement of the decimal point to the left one, two, three places, etc., divides the number by 10, 100, 1,000, etc. (multiplies by 0.1, 0.01, 0.001, etc.).

$$3.0361 \times 1,000 = 3,036.1$$
$$3.0361 \times 0.01 = 0.030361$$

2-49. Full Multiplication of Decimals. Ignore the decimals in multiplying. Then point off in the product, beginning at the right, as many decimal places as there were decimal places in both factors combined. Begin multiplication with the right-hand digit of the multiplier. Continue with the next digit, shifting the new product one decimal place to the left of the previous product. Repeat for the rest of the digits in the multiplier. Add the products.

• Multiply 36.84 by 1.235.

First multiply 36.84 by 5, then by 3, then by 2, then by 1. Shift each successive product one place to the left.

$$
\begin{array}{r}
3\ 6.8\ 4 = \text{multiplicand} \\
\times 1.2\ 3\ 5 = \text{multiplier} \\
\hline
1\ 8\ 4\ 2\ 0 \\
1\ 1\ 0\ 5\ 2 \\
7\ 3\ 6\ 8 \\
3\ 6\ 8\ 4 \\
\hline
4\ 5.4\ 9\ 7\ 4\ 0 = \text{product}
\end{array}
$$

Since there are two decimal places in the multiplicand and three in the multiplier, $2 + 3 = 5$ places are pointed off from the right of the product.

To check, interchange multiplier and multiplicand and multiply again. Or, add the digits in the multiplicand. If the result has more than one digit, add these. Continue until the sum consists of only one digit. Do the same for the multiplier. Find the product of the two one-digit sums. It should equal the one-digit sum of the product digits.

• In the preceding example,

$$
\begin{array}{llll}
3\ 6.8\ 4 & 3+6+8+4 = 21 & & 2+1 = 3 \\
\times 1.2\ 3\ 5 & 1+2+3+5 = 11 & & 1+1 = 2 \\
\hline
4\ 5.4\ 9\ 7\ 4\ 0 & 4+5+4+9+7+4+0 = 33 & & 3+3 = 6
\end{array}
$$

Since $3 \times 2 = 6$, the one-digit sum of the product, the multiplication checks. The decimal point can be checked by inspection.

2-50. Multiplication and Division of Approximate Numbers. (See also Arts. 2-42, 2-49, 2-51 to 2-53.) Keep no more digits in the product or quotient of approximate numbers than there are digits in the factor or divisor with the smallest number of digits. In the multiplication $320.2 \times 2.16 = 692$, for example, only three digits are given in the product. This is the same as the number of digits in 2.16, the factor with fewest digits.

2-51. Shortened Multiplication of Decimals. In accordance with Art. 2-50, carry no more digits than are required to maintain the accuracy of the least accurate factor. Begin the multiplication with the left-hand digit of the multiplier, instead of the right-hand digit as in full multiplication (Art. 2-49). Continue with the next multiplier digit, but shift its product one decimal place to the right. Repeat for the rest of the digits in the multiplier. Determine the position of the decimal point by inspection.

• Multiply 36.8 by 1.235.

Since 36.8 has three digits—fewer than 1.235—only three digits need be retained in the product. Multiply 1.235 first by 3, then by 6, then by 8. Shift each successive product one place to the right.

```
    1 .2 3 5
   ×3 6 .8
    3 7 .0      multiplication by 30
      7 .4      multiplication by 6
      1 .0      multiplication by 0.8
    4 5 .4      product with three digits
```

2-52. Full Division of Decimals. Shift the decimal point in the divisor to the right to eliminate the decimals. Shift the decimal point in the dividend to the right an equal number of decimal places; add zeros if the dividend has fewer decimal places than the divisor. Carry out the division, aligning the decimal point of the quotient with that of the dividend.

• Divide 45.4974 by 1.235. Give the answer to two decimal places.

Shift the decimal point to the right three places in dividend and divisor to rid the latter of decimals. Add a zero to the dividend so that the quotient can be obtained to two decimal places.

```
                        3  6 .8  4  = quotient
Divisor = 1 2 3 5|4 5 4 9 7 .4 0  = dividend
               −3 7 0 5            (3 × 1235)
                 8 4 4 7
                −7 4 1 0           (6 × 1235)
                 1 0 3 7 4
                −9 8 8 0           (8 × 1235)
                   4 9 4 0
                  −4 9 4 0         (4 × 1235)
```

• Divide 19.2 by 0.1875.

Shift the decimal point to the right four places in dividend and divisor to rid the latter of decimals; add three zeros to the dividend.

$$
\begin{array}{r}
1\ 0\ 2.4\ =\text{quotient}\\
\text{Divisor} = 1\ 8\ 7\ 5\overline{)1\ 9\ 2\ 0\ 0\ 0.0}\ =\text{dividend}\\
\underline{1\ 8\ 7\ 5}\\
4\ 5\ 0\ 0\\
\underline{3\ 7\ 5\ 0}\\
7\ 5\ 0\ 0\\
\underline{7\ 5\ 0\ 0}
\end{array}
$$

To check a division, multiply the quotient by the divisor and add the remainder, if there is one. The result should be equal to the dividend. Another check is to add the digits in the divisor; if the sum has more than one digit, add these and continue until the sum consists of only one digit. Do the same for dividend, remainder, and quotient.   Multiply the one-digit sums of divisor and quotient and add the remainder sum.   The result should be equal to the dividend one-digit sum.

• As an example, 422,352 divided by 1,868 yields 226 with a remainder of 184.

$$
\begin{array}{lllll}
1{,}868 & 1+8+6+8 = & 23 & 2+3 = & 5\\
\times 226 & 2+2+6 = & 10 & 1+0 = & \times 1\\
& & & & \overline{5}\\
+184 & 1+8+4 = & 13 & 1+3 = & +4\\
& & & & \overline{9}\\
422{,}352 & 4+2+2+3+5+2 = 18 & & 1+8 = & 9 \quad\text{check}
\end{array}
$$

2-53. Shortened Division of Decimals.   In accordance with Art. 2-50, carry no more digits than are required to maintain the accuracy of the least accurate of the given numbers.   The number of digits in the divisor can be reduced by one in each successive step until it has as many digits as the dividend.

• Divide 45.5 by 1.23498.

Since 45.5 has three digits—fewer than 1.23498—the quotient need be computed to no more than three digits.   To begin, shift the decimal point five places in both numbers to rid the divisor of decimals.

$$
\begin{array}{r}
3\ 6.8\ =\text{quotient}\\
\text{Divisor} = 1\ 2\ 3\ 4\ 9\ 8\overline{)4\ 5\ 5\ 0\ 0\ 0\ 0.0}\ =\text{dividend}\\
\underline{3\ 7\ 0\ 4\ 9\ 4}\\
\text{Divisor} = 1\ 2\ 3\ 5\ 0\ 0\overline{)\ 8\ 4\ 5\ 0\ 6\ 0}\\
\underline{7\ 4\ 1\ 0\ 0\ 0}\\
\text{Divisor} = 1\ 2\ 3\ 5\ 0\ 0\ \overline{|1\ 0\ 4\ 0\ 6\ 0\ 0}\\
9\ 8\ 8\ 0\ 0\ 0
\end{array}
$$

• Divide 45.4974 by 1.24.

Since 1.24 has only three digits—fewer than has 45.4974—the quotient need have no more than three digits. Shift the decimal point two places to the right in both numbers to rid the divisor of decimals.

$$
\begin{array}{r}
3\ \ 6.6\ \ 9 = 3\ \ 6.7 = \text{quotient}\\
\text{Divisor} = 1\ 2\ 4\overline{|4\ 5\ 4\ 9.7\ 4} = \text{dividend}\\
3\ 7\ 2\\
\hline
8\ 2\ 9\\
7\ 4\ 4\\
\hline
8\ 5\ 7\\
7\ 4\ 4\\
\hline
1\ 1\ 3\ 4
\end{array}
$$

2-54. Decimal Point in Slide-rule Calculations.   Ignore decimal points in reading slide-rule scales.   Determine the decimal point in the product or quotient by inspection or by treating the numbers as powers of 10.

• Evaluate $\dfrac{21,600 \times 0.0459}{31.60}$.

The slide-rule product is approximately 314 with decimal point position undetermined.   Since $21,600 \approx 2 \times 10^4$, $0.0459 \approx 5 \times 10^{-2}$, and $31.60 \approx 3 \times 10$, the answer should be approximately

$$2 \times 10^4 \times 5 \times 10^{-2} \div 3 \times 10 = 1\tfrac{2}{3} \times 10^{4-2-1} \approx 3 \times 10$$

Hence $\dfrac{21,600 \times 0.0459}{31.60} = 31.4$

2-55. Multiplication by Slide Rule.   Ignore decimal points until the final product is obtained, then locate the decimal in the answer as indicated in Art. 2-54.

1. With C and D scales.   Set either index (numeral 1) of the C scale over either factor on the D scale.   Move the hairline along the C scale to the next factor.   The reading under the hairline on the D scale is the product.   (If the setting is off the scale, use the opposite index.)   To continue multiplying, set an index under the last position of the hairline, then shift the indicator along the C scale to the next factor.

• Using C and D scales, multiply $2 \times 4 \times 3$.

Set the C-scale left index over 2 on the D scale and slide the indicator until the hairline is over 4 on the C scale.   Place the right index of the C scale under the hairline; then move the indicator to 3 on the C scale.   Read 24 under the hairline on the D scale.

2. With C, D, CF, and DF scales.   Proceed as for C and D scales alone.   If a setting would fall off the scale, instead of resetting the slide, move the indicator until the hairline is over the factor on the CF scale. Read the product under the hairline on the DF scale.   (The readings on

the folded scales, CF and DF, are π times the readings aligned with them on the C and D scales, respectively.)

• Multiply $2 \times 6 \times 9$, using C, D, CF, and DF scales.

Set the C-scale left index over 2 on the D scale. Both 6 and 9 on the slide are now beyond the D scale, so move the indicator over 6 on the CF scale. Next, shift the CF index under the hairline; then move the indicator until it is over 9 on the CF scale (9 on the C scale still is beyond the D scale). Read 108 on the DF scale.

3. With inverted scales CI and CIF. This is equivalent to division by a reciprocal. Set the hairline over one factor on the D (or DF) scale. Move the slide until the other factor, on the CI (or CIF when using DF) scale, also is under the hairline. The index on the CI (or CIF) scale will be aligned with the product on the D (or DF) scale.

• With CI or CIF scales, multiply $2 \times 6 \times 9$.

Set the hairline over 2 on the D scale. Move the slide until 6 on the CI scale also is under the hairline. Then shift the indicator to 9 on the CF scale (9 on the C scale is off the D scale). Read 108 on the DF scale.

2-56. Division by Slide Rule. Ignore decimal points until the quotient is obtained, then locate the decimal point as indicated in Art. 2-54.

1. With C and D scales. Set the hairline over the dividend on the D scale. Move the slide until the divisor on the C scale also is under the hairline. The index (numeral 1) on the C scale will be aligned with the quotient on the D scale.

• Divide 54 by 9.

Set the hairline over 54 on the D scale. Move the slide until 9 on the C scale is under the hairline. The C-scale right index is then over the quotient, 6, on the D scale.

2. With C, D, CF, and DF scales. Proceed as for C and D scales, either with those scales or with the CF and DF scales. (Readings on these folded scales are π times the readings aligned with them on the C and D scales, respectively.) When an index of the C scale is aligned with the answer on the D scale, the index on the CF scale also is aligned with the same answer on the DF scale.

3. With inverted scales, CI and CIF. This is equivalent to multiplication by a reciprocal. Set an index of the CI scale over the dividend on the D scale (or an index of the CIF scale over the dividend on the DF scale). Move the indicator to the divisor on the CI (or CIF) scale and read the quotient on the D (or DF) scale.

• Divide 540 by 9×15.

Set 9 on the C scale over 540 on the D scale. Move the indicator to 15 on the CI scale. Read 4 under the hairline on the D scale.

2-57. Powers and Roots by Slide Rule. Ignore the decimal point until the final product is obtained (Art. 2-54). When extracting square roots, mark the digits off in pairs on both sides of the decimal point and note whether there are one or two digits remaining at the extreme left. (In decimals, after the pairs are marked off, do not count a zero to the left of the first digit as a digit.) When extracting cube roots, mark the digits off in triplets on both sides of the decimal point and note whether there are one, two, or three digits remaining at the extreme left (not counting as digits the one or two zeros to the left of the first digit in decimals).

1. Squares. Set the hairline over the given number on the D scale. Read the square under the hairline on the A scale.

• Square 3,700.

$3,700 \approx 4 \times 10^3$; so $(3,700)^2 \approx 16 \times 10^6$. Set the hairline over 37 on the D scale; read 1369 under it on the A scale. Hence, the answer is 13.69×10^6, or 13,690,000.

• Square 0.201.

By inspection, the answer will be about 0.04. Set the hairline over 201 on the D scale; read 404 under it on the A scale. The answer is 0.0404.

2. Square Roots. Set the hairline over the given number on the A scale. If there is only one digit remaining at the left when the number is marked off in pairs, use the left half of the A scale; if there are two digits, use the right half of the A scale. Read the square root under the hairline on the D scale.

• Extract the square root of 4.

There is only one digit, so use the left half of the A scale. Set the hairline over 4 and read 2 under it on the D scale.

• Extract the square root of 3,600.

$3,600 = 36'00$; there are two digits at the extreme left, so use the right half of the A scale. With the hairline over 36 on the A scale, read 6 on the D scale. By inspection, the answer is 60.

• Extract the square root of 0.090.

$0.090 = 0.09'0$; there is only one digit at the extreme left, so use the left half of the A scale. Read 3 on the D scale. By inspection, the answer is 0.3.

• Extract the square root of 0.518.

$0.518 = 0.51'8$; there are two digits at the extreme left, so use the right half of the A scale. Read 72 on the D scale. By inspection, the answer is 0.72.

3. Cubes. Set the hairline over the given number on the D scale. Read its cube under the hairline on the K scale.

• Cube 8,100.

$8,100 \approx 8 \times 10^3$; so $(8,100)^3 \approx 8^3 \times 10^9 \approx 500 \times 10^9 \approx 5 \times 10^{11}$. Set the hairline over 81 on the D scale; read 532 on the K scale. The answer is 5.32×10^{11}.

• Cube 0.33.

$0.33 \approx 3 \times 10^{-1}$; hence, $(0.33)^3 \approx 3^3 \times 10^{-3} \approx 30 \times 10^{-3} \approx 3 \times 10^{-2}$. Set the hairline over 33 on the D scale; read 360 on the K scale. The answer then is $3.60 \times 10^{-2} = 0.0360$.

4. **Cube Roots.** Set the hairline over the given number on the K scale. If there is only one digit at the left when the number is marked off in triplets, use the left third of the K scale; if there are two digits, use the middle third; if there are three digits, use the right third. Read the cube root under the hairline on the D scale.

• Find the cube root of 81.

Since there are only two digits in 81, use the middle third of the K scale. With the hairline over 81, read 433 on the D scale. By inspection, the answer is 4.33.

• Extract the cube root of 512,000.

$512,000 = 512'000$. Since there are three digits at the left, use the right third of the K scale; read 800 on the D scale. Since $512,000 = 512 \times 10^3$, the cube root equals $\sqrt[3]{512} \times \sqrt[3]{10^3} = 8.00 \times 10 = 80.0$.

• Compute the cube root of 0.003 375.

$0.003\ 375 = 0.003'375$. Since there is only one digit at the left, use the left third of the K scale; read 1500 on the D scale. By inspection, the answer is 0.1500.

5. **Powers and roots with log-log scales.** Raising to powers and extracting roots with these scales are equivalent to multiplication and division with C and D scales. For numbers greater than 1, use the C scale and the LL1, LL2, and LL3 scales; for decimals, use the B scale and the LL0 scale. If the given number or the answer should fall off the scales, multiply or divide the given number by a convenient power of 10 and carry out the desired operations. Then apply a correction factor to the slide-rule result. The correction factor is found by performing the same operations on the power of 10 as on the given number.

• Compute 3^5.

Set the left index of the C scale over 3 on the LL3 scale. Move the indicator to 5 on the C scale and read 243 on LL3.

• Raise 1.11 to the twentieth power.

Set the left index of the C scale over 1.11 on the LL2 scale. Move the indicator to 2 on the C scale. Now the square of 1.11 is under the hairline on the LL2 scale and the 0.2 power is under it on the LL1 scale, but the twentieth power, the answer, is on the LL3 scale. Read $(1.11)^{20} = 8.08$.

• Extract the fourth root of 256.

Set 4 on the C scale over 256 on the LL3 scale. Under the left index of the C scale, read the answer, 4, on the LL3 scale. (Directly under it on the LL2 scale is the 0.4 root; on the LL1 scale is the 0.04 root.)

• Compute $\sqrt[5]{250,000}$.

Since 250,000 is beyond the LL3 scale, divide it by 100,000 to bring it within the range of the slide rule ($100,000 = 10^5$ and $\sqrt[5]{10^5} = 10$). Consequently, the slide-rule result will have to be multiplied by 10 for the correct answer. Set 5 on the C scale over 2.5 on the LL2 scale, and under the left index of the C scale read 1.201 on the LL2 scale. Hence the answer is $1.201 \times 10 = 12.01$.

• Raise 0.75 to the fourth power.

Align the left index of the B scale with 0.75 on the LL0 scale. Move the indicator to 4 on the B scale and read 0.317 on the LL0 scale.

• Extract the sixth root of 0.20.

Set 6 on the left half of the B scale under 0.20 on the LL0 scale. Read 0.764 on the LL0 scale over the left index of the B scale.

• Compute the fourth root of 0.000 040 9.

Since the given number is too small to be read on the LL0 scale, multiply it by 10,000 ($10,000 = 10^4$ and $\sqrt[4]{10^4} = 10$). Consequently, the slide-rule result will have to be divided by 10. Set 4 on the B scale under 0.409 on the LL0 scale, and read 0.80 on the LL0 scale above the left index of the B scale. The answer is $0.80/10 = 0.080$.

2-58. Conversion of Decimal to Per Cent. Multiply a number in decimal form by 100 to convert it to per cent.

• Convert 0.13 to per cent.

$$0.13 \times 100 = 13 \text{ per cent}$$

2-59. Conversion of Fraction to Per Cent. Change the fraction to a decimal (Art. 2-40); then multiply by 100.

• Change ¾ to per cent.

$$¾ = 0.75 \qquad 0.75 \times 100 = 75 \text{ per cent}$$

2-60. Percentage Computations. The terms per cent and rate are often used interchangeably, as in rate of interest.

In percentage computations, the number of which a percentage is taken is called the base. Percentage equals base times rate. $p = br$.

• What is 25 per cent of 480?

Here, $b = 480$ and $r = 25$ per cent $= 0.25$. Hence, the percentage is

$$p = br = 480 \times 0.25 = 120$$

• What per cent of 36 is 12?

Here, $b = 36$ and $p = 12$.

$$r = p/b = {}^{12}\!/_{36} = \tfrac{1}{3} = 0.333 \qquad 0.333 \times 100 = 33.3 \text{ per cent}$$

• 40 per cent of what number equals 16?

Here, $r = 40$ per cent $= 0.4$ and $p = 16$.

$$b = p/r = 16/0.4 = 40$$

2-61. Profit and Loss. To find per cent profit or loss, divide the difference between the initial value (cost) and the final value (selling price) by the initial value and multiply by 100. If the initial value is less than the final value, the difference is profit; if greater, the difference is loss.

• A store buys an item for $10 and sells it at a 25 per cent profit. How much does it make on the sale? What is the selling price?
Profit = $10 $\times$ 0.25 = $2.50.
Selling price = $10 + $2.50 = $12.50.

• If a store sells an item for $25 and loses $5 on a sale, what is the per cent loss?
Cost price = $25 + $5 = $30.
Per cent loss = $\frac{5}{30}$ = $\frac{1}{6}$ = 0.1667 = 16.67 per cent.

• If a store makes a 40 per cent profit, equal to $16, on a sale, how much did the item cost the store? What was the selling price?
Cost price = $16/0.4 = $40.
Selling price = $40 + $16 = $56.

2-62. Discounts. Multiply the list price by the discount rate to determine the discount. Selling price is list price less discount.

• If an item is listed for $100, less a 20 per cent discount, what is the selling price?
Discount = $100 $\times$ 0.20 = $20.
Selling price = $100 − $20 = $80.

2-63. Interpolation. Estimation of a value lying between two values given in a table.

A common method of interpolating when the interval between given values is small is to assume that a linear relationship holds in the interval.

If (x_1, y_1) and (x_2, y_2) are value pairs given in a table and the value of y is to be found for a value of x between x_1 and x_2, then

$$y \approx y_1 + \frac{x - x_1}{x_2 - x_1}(y_2 - y_1) \approx y_2 - \frac{x_2 - x}{x_2 - x_1}(y_2 - y_1)$$

• If a table gives the square of 30 as 900 and that of 31 as 961, what is the square of 30.2?

$$30.2^2 \approx 900 + \frac{30.2 - 30}{31 - 30}(961 - 900) \approx 912$$

• What is the square of 30.8?

$$30.8^2 \approx 961 - \frac{31 - 30.8}{31 - 30}(961 - 900) \approx 949$$

2-64. Extrapolation. Estimation of a value lying outside the range of values given in a table.

A common method of extrapolation is to assume that a linear relationship holds in the interval between values given.

If (x_1, y_1) and (x_2, y_2) are value pairs given in a table and the value of y is to be found for a value of x not between x_1 and x_2, then if $x > x_2 > x_1$,

$$y \approx y_2 + \frac{x - x_2}{x_2 - x_1} (y_2 - y_1)$$

If $x < x_1 < x_2$, then

$$y \approx y_1 - \frac{x_1 - x}{x_2 - x_1} (y_2 - y_1)$$

• If a table gives the square of 99 as 9,801 and that of 100 as 10,000 what is the square of 102?

$$102^2 \approx 10,000 + \frac{102 - 100}{100 - 99} (10,000 - 9,801) \approx 10,400$$

• What is the square of 97?

$$97^2 \approx 9,801 - \frac{99 - 97}{100 - 99} (10,000 - 9,801) \approx 9,400$$

BIBLIOGRAPHY

Griffin, Harriet: "Elementary Theory of Numbers," McGraw-Hill Book Company, Inc., New York, 1954.

Palmer, C. I., and S. F. Bibb: "Practical Mathematics," 4th ed., McGraw-Hill Book Company, Inc., New York, 1952.

Rice, H. S., and R. M. Knight: "Technical Mathematics with Calculus," McGraw-Hill Book Company, Inc., New York, 1957.

Wagner, R. W.: "Introductory College Mathematics," McGraw-Hill Book Company, Inc., New York, 1957.

Chapter 3

ALGEBRA

3-1. Existence Law for Addition.　Addition of any two numbers a and b always gives a single number c.　$a + b = c$.

3-2. Commutative Law.　Algebraic numbers can be added or multiplied in any order.　$a + b \equiv b + a; \quad ab \equiv ba$.

3-3. Associative Law.　The sum or product of three or more algebraic terms does not depend on the grouping of the terms.

$$a + b + c \equiv a + (b + c) \equiv (a + b) + c$$
$$abc \equiv a(bc) \equiv (ab)c \equiv (ac)b$$

3-4. Distributive Law.　$a(b + c) = ab + ac$.

3-5. Operations with Zero and Negative Numbers.　A number or letter without a sign in front of it is assumed to be positive.

$$a + 0 \equiv a \qquad\qquad 0 - a \equiv -a$$
$$a + (-a) \equiv 0 \qquad\qquad -(-a) \equiv a$$
$$a \cdot 0 \equiv 0 \qquad\qquad a(-b) \equiv -ab$$
$$0/a \equiv 0, \text{ if } a \neq 0 \qquad (-a)(-b) \equiv ab$$

If $ab = 0$, then either $a = 0$ or $b = 0$ or $a = b = 0$.　$a/0$ is undefined (Art. 7-2).

3-6. Order Relation.　If a and b are any real numbers, then either $a < b$ or $a = b$ or $a > b$.　If c is a third real number, and if $a < b$ and $b < c$, then $a < c$.

3-7. Axioms

1. If equals are added to equals, the sums are equal.
2. If equals are subtracted from equals, the differences are equal.
3. If equals are multiplied by equals, the products are equal.
4. If equals are divided by equals (except zero), the quotients are equal.
5. Like powers or like roots of equals are equal.
6. Numbers or terms equal to the same number or equal numbers are equal.
7. The whole equals the sum of its parts.

3-8. Signs of Aggregation.　Parentheses ( ), brackets [ ], braces { }, and the vinculum ‾ are used to indicate that the expressions included are to be treated as a whole.

25

A sign of aggregation preceded by a plus sign may be removed or inserted without changing the sign of any term within it. But when a sign of aggregation preceded by a minus sign is removed or inserted, the sign of every enclosed term is changed.

$$a + (b - c + d) \equiv a + b - c + d$$
$$a - (b - c + d) \equiv a - b + c - d$$

3-9. Variable. A symbol representing any value of a given set of elements. (In $A = \pi r^2$, where A is the area of a circle with radius r, and $\pi = 3.14159 \cdots$, A and r are variables. If the change in A is being studied as r is varied, then r is the independent variable and A is the dependent variable. A is called a function of r.)

3-10. Constant. A variable consisting of one element only.

An absolute or numerical constant always has the same value. An arbitrary constant or parameter has one value under certain conditions and different values under other conditions.

• A symbol representing the number 11 or $\sqrt{2}$ or $3.14159 \cdots$ is an absolute constant. In the expression $e = kP$, where e is the elongation of a bar due to a load P, k is a constant for a specific material but varies with different materials; hence, k is a parameter.

3-11. Proof by Mathematical Induction. Assume the proposition to be true for some number and show that it then must be true for the next higher one. Then prove that the proposition is true for a particular number.

• To prove that

$$1^2 + 2^2 + 3^2 + \cdots + n^2 = \tfrac{1}{6}n(n + 1)(2n + 1) \tag{3-1}$$

assume that it holds for some number k. Then, for $k + 1$, by Eq. (3-1),

$$\begin{aligned}
1^2 + 2^2 + 3^2 + \cdots + (k + 1)^2 &= \tfrac{1}{6}k(k + 1)(2k + 1) + (k + 1)^2 \\
&= (k + 1)[\tfrac{1}{6}k(2k + 1) + k + 1] \\
&= \tfrac{1}{6}(k + 1)(2k^2 + k + 6k + 6) \\
&= \tfrac{1}{6}(k + 1)(2k^2 + 7k + 6) \\
&= \tfrac{1}{6}(k + 1)(k + 2)(2k + 3) \\
&= \tfrac{1}{6}(k + 1)[(k + 1) + 1][2(k + 1) + 1]
\end{aligned}$$

So if the proposition holds for k, it holds for $k + 1$. But it is true for $k = 1$. So it also must be true for $k = 2$, $k = 3$, etc.

3-12. Monomial. A single term: a, ab, or x^2y^3.

3-13. Binomial. Two terms that are added or subtracted: $a + b$, $x^2 + y^3$, or $ab + x^4y^2$.

3-14. Polynomial. Two or more terms added or subtracted: $a + b + c$ or $x^2 + 3x - 5xy + y^2$.

3-15. Rational Term. One not containing the root of a variable: x^2, $1/y$, or u.

3-16. Irrational Term. One containing a root of a variable: $a^{1/6}$ or $\sqrt{x}$ (see also Art. 3-41).

3-17. Integral Term. One in which the variable does not occur in the denominator: x/y is integral with respect to x but not with respect to y.

3-18. Addition (Algebraic Sum). If two numbers have the same sign, add their absolute values. If they have opposite signs, subtract their absolute values. (The absolute value of a number is the number without its sign.)

$$5 + 6 = 11 \qquad 5 + (-6) = -1$$

Monomials that are alike, or similar, can be added or subtracted by adding or subtracting coefficients.

$$3ab + 6ab - 2ab \equiv 7ab$$

If the monomials are not alike the operations can only be indicated.

$$3ab + 6bc - ab \equiv 2ab + 6bc$$

To add polynomials containing similar terms, write them so that like terms are in the same column. Then add the terms in each column.

$$
\begin{array}{l}
3ab + 4ac - 3bc \\
2ab - 3ac + 2bc \\
-ab + 2ac + 6bc \\
\hline
4ab + 3ac + 5bc
\end{array}
$$

3-19. Subtraction (Algebraic Difference). Change the sign of the subtrahend (number to be subtracted) and add algebraically (Art. 3-18).

$$
\begin{array}{l}
15 - (-6) = 15 + 6 = 21 \\
15 - (+6) = 15 - 6 = 9 \\
a - (b - c) \equiv a - b + c
\end{array}
$$

3-20. Multiplication. A product is the result obtained by multiplying together two or more quantities. Each of the quantities is a factor of the product.

If there is no sign between two letters or a number and a letter, multiplication is understood.

Any factor is the coefficient of the product of the other factors.

The product of two numbers with like signs is positive; the product of two numbers with different signs is negative.

$$(+6)(+5) = +30 \qquad (-6)(-5) = +30 \qquad (+6)(-5) = -30$$

The product of an odd number of negative factors is negative; the product of an even number of negative factors is positive.

The product of a monomial and a polynomial is obtained by multiplying each term of the polynomial by the monomial.

$$a(b + c - d) \equiv ab + ac - ad$$

The product of two polynomials is obtained by multiplying each term of one by each term of the other and adding. The work can be carried out conveniently in a manner similar to multiplication in arithmetic:

$$
\begin{array}{r}
3\ x\ \ + 2\ y + 1 \\
x + y \\
\hline
3\ x\ y + 2\ y^2 + y \\
3\ x^2 + 2\ x\ y\ \ \qquad\qquad + x \\
\hline
3\ x^2 + 5\ x\ y + 2\ y^2 + y + x
\end{array}
$$

When the literal part of the product of two polynomials can be obtained by inspection, the multiplication can be carried out with the coefficients alone.

• Multiply $x^3 + 2x - 5$ by $x^2 - 5x + 1$. Denote the coefficients of missing powers by 0. Multiply the coefficients of the largest powers first.

$$
\begin{array}{r}
1 + 0 + 2 - \ \ 5 \\
1 - 5 + \ \ 1 \\
\hline
1 + 0 + 2 - \ \ 5 \\
-5 + 0 - 1\ 0 + 2\ 5 \\
1 + \ \ 0 + \ \ 2 - 5 \\
\hline
1 - 5 + 3 - 1\ 5 + 2\ 7 - 5
\end{array}
$$

The answer is $x^5 - 5x^4 + 3x^3 - 15x^2 + 27x - 5$.

3-21. Common Algebraic Products.   (See also Arts. 3-22 and 3-37.)

$$
\begin{aligned}
a(b + c + d) &\equiv ab + ac + ad \\
(a + b)(c + d) &\equiv ac + ad + bc + bd \\
(x + a)(x + b) &\equiv x^2 + (a + b)x + ab \\
(a + b)^2 &\equiv a^2 + 2ab + b^2 \\
(a - b)^2 &\equiv a^2 - 2ab + b^2 \\
(a + b)(a - b) &\equiv a^2 - b^2 \\
(ax + by)(cx + dy) &\equiv acx^2 + (bc + ad)xy + bdy^2 \\
(a + b + c)^2 &\equiv a^2 + b^2 + c^2 + 2ab + 2ac + 2bc \\
(a + b)^3 &\equiv a^3 + 3a^2b + 3ab^2 + b^3 \\
(a - b)^3 &\equiv a^3 - 3a^2b + 3ab^2 - b^3
\end{aligned}
$$

3-22. Factors.   Terms that are multiplied together to yield a given expression (see also Arts. 3-33 and 2-16 to 2-19).

Prime factors are integers that have no integer factors other than themselves and unity. For example, the prime factors of $12a^2b^3$ are 2, 3, a, and b, with 2 present twice, a twice, and b three times. Note that 12 also can be factored into $\frac{1}{2}$ and 24, or $\frac{2}{3}$ and 18, or $\sqrt{18}$ and $\sqrt{8}$.

$$kx + ky + kz \equiv k(x + y + z)$$
$$x^2 + px + q \equiv x^2 + (a + b)x + ab \equiv (x + a)(x + b)$$
$$nx^2 + px + q \equiv abx^2 + (ad + bc)x + cd \equiv (ax + c)(bx + d)$$
$$x^2 - y^2 \equiv (x - y)(x + y)$$
$$x^3 - y^3 \equiv (x - y)(x^2 + xy + y^2)$$
$$x^3 + y^3 \equiv (x + y)(x^2 - xy + y^2)$$
$$x^4 - y^4 \equiv (x^2 + y^2)(x - y)(x + y)$$
$$x^5 - y^5 \equiv (x - y)(x^4 + x^3y + x^2y^2 + xy^3 + y^4)$$
$$x^6 - y^6 \equiv (x^3 + y^3)(x^3 - y^3) \equiv (x^2 - y^2)(x^4 + x^2y^2 + y^4)$$
$$x^n - y^n \equiv (x - y)(x^{n-1} + x^{n-2}y + x^{n-3}y^2 + \cdots + y^{n-1})$$

If n is odd,

$$x^n + y^n \equiv (x + y)(x^{n-1} - x^{n-2}y + x^{n-3}y^2 - \cdots + y^{n-1})$$

3-23. Division. (See also Arts. 3-24 and 3-32.) The process of finding a factor if a product and another factor are known. The divisor is the given factor and the quotient is the desired factor. The dividend is the given product.

Division is equivalent to multiplication by the reciprocal of the divisor.

When dividend and divisor have like signs, the quotient is positive; if they have unlike signs, the quotient is negative.

To divide two polynomials:

1. Arrange terms in descending powers of a common letter.

2. Divide the first term of the dividend by the first term of the divisor and write the result as the first term of the quotient.

3. Multiply the divisor by this term and subtract the product from the dividend.

4. Treat the difference as a new dividend and proceed as in the first three steps.

5. Continue until the difference, or remainder, is zero, or the highest power of the letter according to which the dividend was arranged is less than the power of the same letter in the divisor.

• Divide $36x^3 + 24x^2 + 10x + 4$ by $3x + 1$.

$$
\begin{array}{l}
\quad\ \text{Divisor} \qquad\qquad \text{Dividend} \qquad\qquad \text{Quotient} \\
3x + 1\ |\ +36x^3 + 24x^2 + 10x + 4\ |\ 12x^2 \\
\qquad\ \ +36x^3 + 12x^2 \\
\qquad\qquad\quad\ \ +12x^2 + 10x + 4\ |\ 4x \\
\qquad\qquad\quad\ \ +12x^2 +\ \ 4x \\
\qquad\qquad\qquad\qquad\ \ +\ 6x + 4\ |2 \\
\qquad\qquad\qquad\qquad\ \ +\ 6x + 2 \\
\qquad\qquad\qquad\qquad\qquad\quad +\ 2
\end{array}
$$

Answer: $12x^2 + 4x + 2 + \dfrac{2}{3x + 1}$

3-24. Synthetic Division.   The division of a polynomial by a binomial of the form $x - r$ can be simplified by the following procedure:

1. Write the coefficients of the variable x in a horizontal line, with the coefficient of the highest power of the variable at the left and the coefficients of successively lower powers in order to the right.   Represent coefficients of missing powers by 0.

2. Write the first coefficient as the first term of the quotient.

3. Multiply it by the nonvariable term $(-r)$ of the binomial with sign changed, and add the product to the second coefficient.   This sum is the coefficient of the second term of the quotient.

4. Multiply this sum by the nonvariable term of the binomial with sign changed and add the product to the next coefficient.   Continue until the last coefficient has been added.

The last sum is the remainder, and the preceding sums in order are the coefficients of the quotient.

• Divide $2x^4 - 3x^3 - 5x^2 - 2x + 7$ by $x - 3$.

$$\begin{array}{l} \text{Dividend} = 2 - 3 - 5 - 2 + 7\underline{|3} = \text{divisor } (x - 3) \\ \phantom{\text{Dividend} = 2} + 6 + 9 + 12 + 30 \\ \hline \phantom{\text{Dividend} =} 2 + 3 + 4 + 10; + 37 \end{array}$$

The quotient is $2x^3 + 3x^2 + 4x + 10$; the remainder $+37$.

• Divide $x^6 + 3x^2 - 9$ by $x + 2$.

$$\begin{array}{l} 1 + 0 + 0 + 0 + 3 + 0 - 9\underline{|-2} \\ - 2 + 4 - 8 + 16 - 38 + 76 \\ \hline 1 - 2 + 4 - 8 + 19 - 38; + 67 \end{array}$$

The quotient is $x^5 - 2x^4 + 4x^3 - 8x^2 + 19x - 38$; the remainder is $+67$.

3-25. Algebraic Fractions.   To add, convert the fractions to a common denominator and add the resulting numerators.

$$\frac{a}{b} + \frac{c}{d} \equiv \frac{ad}{bd} + \frac{bc}{bd} \equiv \frac{ad + bc}{bd}$$

$$x + \frac{3}{y} \equiv \frac{xy}{y} + \frac{3}{y} \equiv \frac{xy + 3}{y}$$

To multiply, find the separate products of numerator and denominator. Cancel out factors common to numerator and denominator.

$$\frac{ax}{b} \frac{c}{dx} \equiv \frac{acx}{bdx} \equiv \frac{ac}{bd}$$

To divide, invert the divisor and multiply by the dividend.

$$\frac{a}{b} \div \frac{c}{d} \equiv \frac{a}{b} \frac{d}{c}$$

3-26. Partial Fractions. The several fractions whose sum equals a given fraction.

To separate a fraction into partial fractions when its denominator can be factored into different terms of the first degree, assume an unknown numerator for each factor. Solve for the unknowns by clearing of fractions and equating coefficients of like powers of the variable.

• Separate into partial fractions $(11x - 1)/(x^2 - 1)$.

The denominator has as factors $x + 1$ and $x - 1$. Assume these as the denominators of the desired partial fractions and assume unknown numerators A and B, respectively.

$$\frac{11x - 1}{x^2 - 1} \equiv \frac{A}{x + 1} + \frac{B}{x - 1}$$

Clear of fractions by multiplying both sides of the identity by $x^2 - 1$.

$$11x - 1 = A(x - 1) + B(x + 1) = (A + B)x - (A - B)$$

Equate coefficients of like powers.

$$A + B = 11$$
$$-(A - B) = -1$$

These equations yield $A = 6$, $B = 5$. Therefore

$$\frac{11x - 1}{x^2 - 1} \equiv \frac{6}{x + 1} + \frac{5}{x - 1}$$

When the factors of the denominator are of the first degree but some are repeated, assume unknown numerators for each factor. But if a term is present twice, make the corresponding denominators the term and its second power. If a term is present three times, make the corresponding denominators the term and its second and third powers, etc. Solve for the unknowns by clearing of fractions and equating coefficients of like powers of the variable.

For quadratic factors such as $ax^2 + bx + c$ in the denominator, take as the numerator a term of the form $Ax + B$.

• Separate into partial fractions $(x^2 + 3x + 4)/(x + 1)^3$.

Since the denominator has $x + 1$ as a factor three times, the denominators of the desired fractions will be $x + 1$, $(x + 1)^2$, and $(x + 1)^3$. Assume the unknown numerators to be A, B, and C, respectively.

$$\frac{x^2 + 3x + 4}{(x + 1)^3} \equiv \frac{A}{x + 1} + \frac{B}{(x + 1)^2} + \frac{C}{(x + 1)^3}$$

$$x^2 + 3x + 4 = A(x + 1)^2 + B(x + 1) + C = Ax^2 + (2A + B)x + (A + B + C)$$

Equate coefficients of like powers.

$$A = 1$$
$$2A + B = 3 \qquad B = 1$$
$$A + B + C = 4 \qquad C = 2$$

The solution is then

$$\frac{x^2 + 3x + 4}{(x + 1)^3} \equiv \frac{1}{x + 1} + \frac{1}{(x + 1)^2} + \frac{2}{(x + 1)^3}$$

3-27. Continuous Fractions. These may have an infinite number of terms in numerator or denominator, as, for example,

$$\cfrac{1}{3 + \cfrac{2}{4 + \cfrac{3}{5 + \cdots}}}$$

An irrational number can be approximated by a continuous fraction.

• $\sqrt{2}$ may be expressed as a continuous fraction.

$$\sqrt{2} = 1 + \cfrac{1}{2 + \cfrac{1}{2 + \cfrac{1}{2 + \cdots}}}$$

This can be verified as follows: since the fraction has an infinite number of similar terms, any last portion of it must be equal to the fraction. Let this value be n. Hence,

$$n = \frac{1}{2 + n}$$

One solution of this is $n = \sqrt{2} - 1$. So $\sqrt{2} = 1 + n$.

If $b = N - a^2$, where a is the largest integer whose square is less than N, then

$$\sqrt{N} = a + \cfrac{b}{2a + \cfrac{b}{2a + \cfrac{b}{2a + \cdots}}}$$

• Express $\sqrt{8}$ as a continuous fraction.

Here, $N = 8$, $a = 2$ (since $3^2 = 9$), and $b = 8 - 2^2 = 4$. Then

$$\sqrt{8} = 2 + \cfrac{4}{4 + \cfrac{4}{4 + \cfrac{4}{4 + \cdots}}}$$

3-28. Proportion. Equality of two ratios. For $a:b = c:d$, read a is to b as c is to d. This is equivalent to $a/b = c/d$.

The first and last terms are called extremes, the other terms are called means. The last term also is known as the fourth proportional.

If quantities are directly proportional, then one quantity equals some constant k times the other ($x = ky$). If quantities are inversely proportional, that is, $x_1:x_2 = y_2:y_1$, then one quantity equals some constant c times the reciprocal of the other ($x = c/y$).

The product of the means equals the product of the extremes. If $a:b = c:d$, then $bc = ad$.

The mean proportional between two numbers equals the square root of their product. If $a:b = b:c$, then $b = \sqrt{ac}$. When

$$a:b = c:d$$

then

$$b:a = d:c \qquad \text{inversion}$$
$$a:c = b:d \qquad \text{alternation}$$
$$a + b:b = c + d:d \qquad \text{composition}$$
$$a - b:b = c - d:d \qquad \text{division}$$
$$a + b:a - b = c + d:c - d \qquad \text{composition and division}$$
$$a^n:b^n = c^n:d^n$$

If $a:b = c:d$ and $x:y = u:v$, then $ax:by = cu:dv$.

3-29. Direct Variation. A variable varies directly as another variable if their ratio is a constant. $x \propto y$ if $x/y = k$, where k is a constant.

• If $x \propto y$ and $x = 2$ when $y = 3$, what is the value of x when $y = 12$?

$$k = \frac{x}{y} = \frac{2}{3}$$
$$x = \frac{2y}{3} = \frac{2}{3} \times 12 = 8$$

3-30. Inverse Variation. A variable varies inversely as another variable if it varies directly as the reciprocal of the other. x varies inversely as y if $x \propto 1/y$. Hence, $xy = k$, where k is a constant.

• If x varies inversely as y and $x = 2$ when $y = 3$, what is the value of x when $y = 12$?

$$k = xy = 2 \times 3 = 6$$
$$x = \frac{6}{y} = \frac{6}{12} = \frac{1}{2}$$

3-31. Joint Variation. If x varies directly as y when all other factors are constant, and x varies directly as z when all other factors are constant, then x varies directly as yz. $x \propto yz$ if $x = kyz$, where k is a constant.

• The area A of a triangle varies directly as the base b and the altitude h. When the base is 10 and the altitude is 12, the area is 60. What is the area when the base is 5 and the altitude is 8?

$A \propto bh$; so

$$A_1 = kb_1h_1 = 10 \times 12k = 60$$
$$A_2 = kb_2h_2 = 5 \times 8k$$
$$\frac{A_2}{A_1} = \frac{A_2}{60} = \frac{5 \times 8k}{10 \times 12k} = \frac{40k}{120k}$$
$$A_2 = 40 \times {}^{60}\!/_{120} = 20$$

3-32. Remainder Theorem. If an integral rational polynomial in x is divided by $x - m$, the remainder can be obtained by substituting m for x in the polynomial.

$(a^3 - 9a^2 + 27a - 15)/(a - 3)$ has the remainder
$$3^3 - 9(3)^2 + 27(3) - 15 = 12$$
$(4x^2 - 9)/(x + 1)$ has the remainder $4(-1)^2 - 9 = -5$.

Conversely, the value of a polynomial in x for $x = m$ can be obtained by dividing the polynomial by $x - m$ (Art. 3-24).

3-33. Factor Theorem. (See also Art. 3-22). If an integral rational polynomial in x becomes zero when m is substituted for x, then $x - m$ is a factor of the polynomial.

• If $a = 3$, then $a^3 - 9a^2 + 27a - 27 = 0$. Hence, $a - 3$ is a factor of the polynomial. Note also that if $a - 3$ is a factor, then 27 must be divisible by 3.

To find binomial factors of a polynomial in any variable, find the factors of the terms not involving that variable and see which of these factors make the polynomial zero when they are substituted for the variable.

• Factor $p^3 - 7p^2 + 7p + 15$.

The factors of 15 are ± 1, ± 3, ± 5, and ± 15. If $p = +1$, the polynomial is not zero, but for $p = -1$, $(-1)^3 - 7(-1)^2 + 7(-1) + 15 = 0$. Therefore, $p + 1$ is a factor.

Division of the polynomial by $p + 1$ yields $p^2 - 8p + 15$. This can be factored into $(p - 5)(p - 3)$. Hence

$$p^3 - 7p^2 + 7p + 15 = (p + 1)(p - 5)(p - 3)$$

3-34. Factorials. The product $n(n - 1)(n - 2) \cdots 1$ is called factorial n. It is usually written $\lfloor n$ or $n!$.

$$5! = 1 \cdot 2 \cdot 3 \cdot 4 \cdot 5 = 120$$
$$1! = 1$$
$$0! = 1$$

The Stirling formula gives approximate values of $n!$ for large n.

$$n! \approx n^n e^{-n} \sqrt{2\pi n}$$

(See also Art. 10-5.)

3-35. Powers. Products of two or more equal factors.

$$xx = x^2 \qquad xxx = x^3 \qquad xxxx \cdots x = x^n$$

The base of a power is the number that is repeated as a factor. The base of x^4 is x, that of a^n is a.

An exponent is a number that indicates how many times the base appears in the multiplication. The exponent of x^4 is 4, that of a^n is n.

All powers of a positive quantity are positive. Even powers of a negative quantity are positive; odd powers are negative.

$$(+1)^n = +1 \qquad (-3)^2 = +9 \qquad (-3)^3 = -81$$

$$a^m a^n = a^{m+n} \qquad\qquad \left(\frac{a^n}{b^m}\right)^p = \frac{a^{np}}{b^{mp}}$$

$$\frac{a^m}{a^n} = a^{m-n} \qquad\qquad a^{1/q} = \sqrt[q]{a}$$

$$a^0 = b^0 = c^0 = \cdots = z^0 = 1 \qquad a^{p/q} = \sqrt[q]{a^p} = (\sqrt[q]{a})^p$$

$$(a^n)^m = a^{nm} \qquad\qquad a^{-p} = \frac{1}{a^p}$$

$$(a^n b^m)^p = a^{np} b^{mp} \qquad\qquad \frac{1}{a^{-p}} = a^p$$

3-36. Squaring a Number. When a calculating device or table of squares is not available, a rapid method of squaring a number is to employ the identity

$$a^2 \equiv (a + b)(a - b) + b^2$$

• Square 97.

Take $b = 3$.   Then $(97)^2 = (97 + 3)(97 - 3) + (3)^2 = 100 \times 94 + 9 = 9{,}409.$

• Square $15\frac{1}{4}$.

Take $b = \frac{3}{4}$.   Then

$$(15\tfrac{1}{4})^2 = (15\tfrac{1}{4} + \tfrac{3}{4})(15\tfrac{1}{4} - \tfrac{3}{4}) + (\tfrac{3}{4})^2 = 16 \times 14\tfrac{1}{2} + \tfrac{9}{16}$$
$$= 16(14 + \tfrac{1}{2}) + \tfrac{9}{16} = 224 + 8 + \tfrac{9}{16} = 232\tfrac{9}{16}$$

To square a number ending in 5, multiply the number less 5 by the number plus 5 and add 25.

$$(75)^2 = 70 \times 80 + 25 = 5{,}625$$
$$(4.5)^2 = 4 \times 5 + 0.25 = 20.25$$

It is also useful to remember that $n^2 = 1 + 3 + 5 + \cdots + (2n - 1)$; that is, the square of any integer n is equal to the sum of the first n odd numbers.

Also, the difference between the squares of two successive integers is equal to the sum of the two integers.

• Square 26.

$(25)^2 = 625$, so $(26)^2 = 625 + 25 + 26 = 676.$

• Square 99.

$(100)^2 = 10{,}000$, so $(99)^2 = 10{,}000 - 100 - 99 = 9{,}801.$

3-37. Binomial Theorem

$$(a + b)^n \equiv a^n + na^{n-1}b + \frac{n(n - 1)}{2!} a^{n-2}b^2 + \cdots$$
$$+ \frac{n(n - 1) \cdots (n - r + 1)}{r!} a^{n-r}b^r + \cdots + b^n$$
$$\equiv a^n + C(n,1)a^{n-1}b + C(n,2)a^{n-2}b^2 + \cdots + C(n,r)a^{n-r}b^r + \cdots + b^n$$

where $C(n,r) = n(n-1) \cdots (n-r+1)/r!$ is the combination of n things taken r at a time (Art. 16-2).

$$(a+b)^4 \equiv a^4 + 4a^3b + \frac{4 \cdot 3}{2!} a^2b^2 + \frac{4 \cdot 3 \cdot 2}{3!} ab^3 + b^4$$

$$\equiv a^4 + 4a^3b + 6a^2b^2 + 4ab^3 + b^4$$

$$(a-3)^3 \equiv a^3 + 3(-3)a^2 + \frac{3 \cdot 2}{2!} (-3)^2a + (-3)^3$$

$$\equiv a^3 - 9a^2 + 27a - 27$$

For $-1 \leqq x \leqq +1$,

$$(1+x)^{\frac{1}{2}} \equiv 1 + \tfrac{1}{2}x + \frac{\frac{1}{2}(-\frac{1}{2})}{2!} x^2 + \frac{\frac{1}{2}(-\frac{1}{2})(-\frac{3}{2})}{3!} x^3 + \cdots$$

$$\equiv 1 + \tfrac{1}{2}x - \tfrac{1}{8}x^2 + \tfrac{1}{16}x^3 - \cdots$$

To expand $(a+b+c)^n$ let $a+b = k$ and first expand $(k+c)^n$ by the binomial theorem. Then substitute $a+b$ for k in the expansion and use the theorem to expand the various powers of $a+b$.

3-38. Pascal's Triangle. An arrangement of the coefficients of the binomial theorem for integral exponents, beginning with zero.

$$
\begin{array}{l}
(a+b)^0 \\
(a+b)^1 \\
(a+b)^2 \\
(a+b)^3
\end{array}
$$

```
(a + b)⁰                        1
(a + b)¹                     1     1
(a + b)²                  1     2     1
(a + b)³               1     3     3     1
                    1     4     6     4     1
                 1     5    10    10     5     1
              1     6    15    20    15     6     1
           1     7    21    35    35    21     7     1
        1     8    28    56    70    56    28     8     1
```

In this triangle, any number except the 1s can be obtained by adding the two numbers to the left and right on the line above it.

3-39. Square of a Polynomial

$$(a+b+c+d+\cdots)^2 = a^2 + b^2 + c^2 + d^2 + \cdots$$
$$+ 2a(b+c+d+\cdots) + 2b(c+d+\cdots) + 2c(d+\cdots) + \cdots$$

In brief, add the sum of the squares of each term to twice the product of each term and the sum of the following terms.

3-40. Roots. This term is used in two senses:

1. The values of the variables that satisfy an equation. (The root of $x^2 - 6x + 9 = 0$ is $+3$.) (See Arts. 3-63 to 3-91.)

2. The equal factors of a power, as indicated by a radical sign ($\sqrt{}$). $\sqrt{9} =$ the square root of $9 = 3$, since $3^2 = 9$; $\sqrt[3]{64} = 4$, since $4^3 = 64$; $\sqrt[n]{a^n} = a$.

The number in front of the radical sign is called the index. It indicates what root is to be taken. If no index is given, 2 (or square root) is assumed.

$$\sqrt[n]{a} = a^{1/n}$$

The number of roots, real and imaginary, is equal to the index.

An even root of a positive quantity may be positive or negative ($\sqrt{4} = \pm 2$; $\sqrt[4]{16} = \pm 2$). Odd roots have the same sign as the quantity ($\sqrt[3]{-125} = -5$; $\sqrt[5]{-243} = -3$). It is customary, however, to restrict the sign of a root to the prefixed sign; that is, $\sqrt{2} = \pm 1.41 \cdots$, but $5\sqrt{2} + 2\sqrt{2} = 7\sqrt{2}$.

If a quantity is negative and the index is an even number, its roots are imaginary. ($\sqrt{-4} = 2\sqrt{-1} = \pm 2i$, where $i = \sqrt{-1}$.)

$$\sqrt[n]{a^m} = (a^m)^{1/n} = a^{m/n}$$
$$\sqrt[p]{a^m b^n} = (a^m b^n)^{1/p} = a^{m/p} b^{n/p}$$
$$\sqrt[p]{a^m/b^n} = (a^m/b^n)^{1/p} = a^{m/p} b^{-n/p}$$
$$\sqrt[n]{a} \cdot \sqrt[n]{a} = \sqrt[n]{a^2}$$
$$\sqrt[n]{\sqrt[m]{a}} = \sqrt[mn]{a}$$

3-41. Surds. Irrational roots, radicals for which the roots cannot be obtained exactly.

An irrational number cannot be expressed as a quotient of positive integers or positive rational fractions, but may be represented approximately by rational fractions (see Arts. 3-27, 3-48, 3-50, and 3-52).

3-42. Conjugate Surds. A pair of binomial quadratic surds that differ only in sign ($\sqrt{x} + \sqrt{y}$ and $\sqrt{x} - \sqrt{y}$).

The product of a pair of conjugate surds is rational.

$$(\sqrt{x} + \sqrt{y})(\sqrt{x} - \sqrt{y}) = x - y$$

3-43. Addition and Subtraction of Radicals (Surds). Transform until surds are similar, then add or subtract the coefficients.

$$\sqrt{\tfrac{1}{3}} + 2\sqrt{48} - \sqrt{12} = \tfrac{1}{3}\sqrt{3} + 8\sqrt{3} - 2\sqrt{3} = 6\tfrac{1}{3}\sqrt{3}$$

3-44. Multiplication of Radicals (Surds). Transform all surds to the same order, then multiply. Use fractional exponents instead of radicals when convenient.

$$5\sqrt{y^5}\sqrt[3]{x^4} = 5\sqrt[6]{y^{15}}\sqrt[6]{x^8} = 5\sqrt[6]{x^8 y^{15}} = 5xy^2\sqrt[6]{x^2 y^3}$$
$$\text{or } 5\sqrt{y^5}\sqrt[3]{x^4} = 5x^{4/6}y^{5/2} = 5xy^2(x^{1/6}y^{1/2}) = 5xy^2(x^{2/6}y^{3/6}) = 5xy^2\sqrt[6]{x^2 y^3}$$

3-45. Division of Radicals (Surds). Rationalize the denominator and convert all surds to the same order.

$$\frac{2\sqrt[3]{x^4}}{5\sqrt{y^5}} = \frac{2\sqrt[3]{x^4}}{5\sqrt{y^5}}\frac{\sqrt{y}}{\sqrt{y}} = \frac{2\sqrt[3]{x^4}\sqrt{y}}{5y^3} = \frac{2x\sqrt[6]{x^2}\sqrt[6]{y^3}}{5y^3} = \frac{2x}{5y^3}\sqrt[6]{x^2y^3}$$

$$\frac{\sqrt{a}+\sqrt{b}}{\sqrt{a}-\sqrt{b}} = \frac{\sqrt{a}+\sqrt{b}}{\sqrt{a}-\sqrt{b}}\frac{\sqrt{a}+\sqrt{b}}{\sqrt{a}+\sqrt{b}} = \frac{a+b+2\sqrt{ab}}{a-b}$$

3-46. Square Root of a Polynomial. Calculation is based on the fact that $a + b \equiv \sqrt{a^2 + 2ab + b^2} \equiv \sqrt{a^2 + b(2a + b)}$. Also,

$$a + b + c \equiv \sqrt{a^2 + b(2a + b) + c[2(a + b) + c]}$$

Step 1:

$$a^2 + 2ab + b^2\underline{|a} = \text{first term of root}$$

Subtract: $\qquad a^2 = a^2$
$$\overline{0 + 2ab + b^2}$$

Step 2:

Trial divisor: $\qquad 2a + \,?\,|0 + 2ab + b^2\,\underline{|a + \,?\,} = \text{next term of root}$
Step 3:

Divisor: $\qquad 2a + b\,|0 + 2ab + b^2\,\underline{|a + b} = \text{square root}$
Subtract: $\quad b(2a + b) = \underline{+ 2ab + b^2}$
$$0 + 0 = \text{remainder}$$

• Find the square root of $9x^2 + 12x - 24xy - 16y + 16y^2 + 4$.

$$\begin{array}{r} a \quad\; b \quad\;\; c \\ 9x^2 + 12x - 24xy - 16y + 16y^2 + 4\,\overline{|3x + 2 - 4y} \end{array}$$

Subtract: $\qquad a^2 = (3x)^2 = 9x^2$
Divisor: $\quad 2a + b = 6x + 2|\;\; + 12x - 24xy - 16y + 16y^2 + 4$
Subtract: $\qquad\quad b(2a + b) = + 12x + 4$
Divisor: $2(a + b) + c = 6x + 4 - 4y|\; - 24xy - 16y + 16y^2$
Subtract: $\qquad\quad c[2(a + b) + c] = -24xy - 16y + 16y^2$
$$0\;\; + \;\;0 \;\;+ \;\;0$$

3-47. Square Root of a Number. Calculation is carried out as for a polynomial (Art. 3-46), with each pair of digits treated as a term.

• Find the square root of 5,241.76.
Mark off the pairs of digits starting at the decimal point and proceeding both to left and right.
Step 1:

$$52'41.76\,\overline{|7} = \text{first digit of root}$$

Subtract: $\qquad a^2 = 7^2 = \underline{49}$
$$3$$

Step 2:

$$\overset{\text{ab}}{}$$

Divisor: $2a + b = 2 \times 70 + 2 = 142 | 3'41 | 72$ = two digits of roots
Subtract· $b(2a + b) = 2 \times 142 = \underline{2\ 84}$
$$57$$

Step 3:

$$\overset{\text{abc}}{}$$

Divisor: $2(a + b) + c = 2 \times 720 + 4 = 1{,}444 | 57'76 | 72.4$ = root
Subtract: $c[2(a + b) + c] = 4 \times 1{,}444 = \underline{57\ 76}$
$$0$$

• Find the square root of 1.5625.

$$1.56'25 | 1.25 = \text{root}$$
$$\underline{1}$$
$$2 \times 10 + 2 = 22| \ \overline{56}$$
$$2 \times 22 = \underline{44}$$
$$2 \times 120 + 5 = 245| \ \overline{12\ 25}$$
$$5 \times 245 = \underline{12\ 25}$$
$$0$$

3-48. Approximate Square Root of a Number.   Estimate the root and substitute in one of the following formulas.   Accuracy can be improved by using the computed root as a new estimate.

$$x_1 = \frac{1}{2}\left(\frac{N}{x_0} + x_0\right)$$

$$x_1 = x_0 + \frac{N - x_0{}^2}{2x_0}$$

where N = given number
  x_0 = estimated root
  x_1 = first approximation

• Compute the square root of 5.
With $x_0 = 2$, $x_1 = \frac{1}{2}(N/x_0 + x_0) = \frac{1}{2}(\frac{5}{2} + 2) = 2.25$
With $x_1 = 2.25$, $x_2 = \frac{1}{2}[(5/2.25) + 2.25] = \frac{1}{2}(2.2222 + 2.25) = 2.2361$

• Compute the square root of 5,241.
With $x_0 = 70$, $x_1 = x_0 + (N - x_0{}^2)/2x_0 = 70 + (5{,}241 - 4{,}900)/140 = 72.4$.

3-49. Cube Root of a Number.   Calculation is based on the fact that $a + b \equiv \sqrt[3]{a^3 + 3a^2b + 3ab^2 + b^3} \equiv \sqrt[3]{a^3 + b[3a(a + b) + b^2]}$. Each triplet of digits is treated as a term.

• Find the cube root of 1,953.125.
Mark off the digit triplets starting at the decimal point and proceeding both to the left and the right.
Step 1:

$$\overset{a}{}$$
$$1'953.125 | 1 = \text{first digit of root}$$
Subtract: $a^3 = \underline{1}$
$$953$$

Step 2:

Trial divisor: $3a(a + b) + b^2 = \underline{3 \times 10(10 + ?) + ?}\overset{ab}{|953|1?}$

Step 3:

Divisor: $\underline{30(10 + 2) + 2^2}\overset{ab}{|953|12}$ = first two digits of root

Subtract: $b[3a(a + b) + b^2] = 2 \times 364 = \dfrac{728}{225}$

Step 4:

Trial divisor: $\underline{3 \times 120(120 + ?) + ?}\overset{abc}{|225'125|12?}$

Step 5:

Divisor: $360(120 + 5) + 5^2 = \underline{45,025}\overset{ab\ c}{|225'125|12.5}$ = cube root

Subtract: $5 \times 45,025 = \dfrac{225\ 125}{0}$

• Find the cube root of 16,194.277.

$$16'194.277|25.3 = \text{cube root}$$
$$2^3 = \ 8$$

Divisor: $60(20 + 5) + 5^2 = 1,525|\ 8\ 194$

Subtract: $5 \times 1,525 = \ 7\ 625$

$$750(250 + 3) + 3^2 = 189,759|569\ 277$$
$$3 \times 189,759 = \dfrac{569\ 277}{0}$$

If one cube root of a number is r, the two others are $r\omega$ and $r\omega^2$, where

$$\omega = -\tfrac{1}{2}(1 + \sqrt{-3})$$
$$\omega^2 = -\tfrac{1}{2}(1 - \sqrt{-3})$$

3-50. Approximate Cube Root of a Number. Estimate the root and substitute in the following formula:

$$x_1 = \frac{2N + x_0{}^3}{N + 2x_0{}^3}\, x_0$$

where N = given number

     x_0 = estimated root

     x_1 = first approximation

For more accurate results, repeat, substituting the calculated value for the estimated root in the formula.

• Compute the cube root of 1,953.

Since $12^3 = 1,728$ and $13^3 = 2,197$, let $x_0 = 12$. (13 would be as good a choice in this example.)

$$x_1 = \frac{2 \times 1,953 + 1,728}{1,953 + 2 \times 1,728} \times 12 = \frac{5,634}{5,409} \times 12 = 12.5$$

3-51. nth Root of a Number. Obtain the logarithm of the given number. Divide by n and determine the antilogarithm of the quotient.

• Extract the fifth root of 188.69.

$$\text{Log } 188.69 = 2.27575$$
$$2.27575/5 = 0.45515$$
$$\text{Antilog } 0.45515 = 2.852 = \sqrt[5]{188.69}$$

• Compute the tenth root of 0.002345.

$$\text{Log } 0.002345 = 7.37014 - 10 = 97.37014 - 100$$
$$0.1 \log 0.002345 = 9.73701 - 10$$
$$\text{Antilog } 9.73701 - 10 = 0.5458 = \sqrt[10]{0.002345}$$

3-52. Approximate nth Root of a Number. Estimate the root and substitute in the following formula:

$$x_1 = \frac{(n+1)N + (n-1)x_0{}^n}{(n-1)N + (n+1)x_0{}^n} x_0$$

where N = given number

n = index of the radical

x_0 = estimated root

x_1 = first approximation

For greater accuracy, repeat, substituting the calculated value for the estimated root in the formula.

• Compute the fifth root of 188.69.
Since $2^5 = 32$ and $3^5 = 243$, let $x_0 = 3$ and $n = 5$.

$$x_1 = \frac{6 \times 188.69 + 4 \times 243}{4 \times 188.69 + 6 \times 243} \times 3 = \frac{2,104.14}{2,212.76} \times 3 = 2.853$$

3-53. Roots of a Fraction. Rationalize the denominator, then extract the root of the numerator.

$$\bullet \frac{1}{\sqrt{3}} = \frac{1}{\sqrt{3}} \frac{\sqrt{3}}{\sqrt{3}} = \frac{\sqrt{3}}{3} = \pm \frac{1.7321}{3} = \pm 0.5774$$

$$\frac{1}{4 - \sqrt{5}} = \frac{1}{4 - \sqrt{5}} \frac{4 + \sqrt{5}}{4 + \sqrt{5}} = \frac{4 + \sqrt{5}}{16 - 5} = \frac{4 + 2.236}{11} = \frac{6.236}{11} = 0.567$$

3-54. Function. Relationship between two or more variables and constants. ($3x^3 + 6$ is a function of x; $3z + 4y$ is a function of y and z.)

$f(x)$ represents function of x. This notation indicates "function of" and not multiplication. $P(x)$ is P function of x.

$f(a)$ is formed by substituting a for x in $f(x)$. If $f(x) = x^2 - 2x + 1$, $f(2) = 2^2 - 2(2) + 1 = 1$.

If $f(x)$ is a rational integral function, the remainder when $f(x)$ is divided by $x - a$ is $f(a)$. (See Art. 3-32.)

• If, for example, $f(x) = x^3 - 9x^2 + 27x - 27$, then $f(5) = 8$.

$$
\begin{array}{r}
1 - 9 + 27 - 27 \underline{|5} \\
+ 5 - 20 + 35 \\
\hline
1 - 4 + 7 \underline{|+8} = \text{remainder} = f(5)
\end{array}
$$

(For synthetic division, as used here, see Art. 3-24.)

A function is symmetrical with respect to two variables if it is unchanged by interchanging the variables.

A single-valued function is one that has a single value for any value of the independent variable. $y = 2x$ is a single-valued function.

A multivalued function has more than one value for any given value of the independent variable. $y = \pm x^{1/2}$ is a multivalued function.

An implicit function is one that must be solved to express one variable in terms only of the other. $x^2 + 3y - y^2 = 9$ defines y as an implicit function of x.

An explicit function is one that gives a variable directly in terms of the other. $y = 12 - x^2$ gives y as an explicit function of x.

3-55. Continuous Function. A function $f(x)$ is continuous if for any value of x an infinitesimal change in x produces an infinitesimal change in $f(x)$. That is, if $h \to 0$,

$$f(x + h) - f(x) \to 0$$

The function ax^m is continuous if m is an integer.

The function $a_0 x^n + a_1 x^{n-1} + \cdots + a_n$ is continuous if n is an integer.

3-56. Graph of a Function. A curve showing how the magnitude of a function of one variable varies with the value of the variable.

To plot the graph of a function, lay off to a selected scale along a horizontal axis (abscissa or x axis) the values of the variable. At each value, plot vertically (parallel to the y axis) the corresponding value of the function (ordinate). Connect the points thus determined.

The intersection of the x and y axes is called the origin. The values of the abscissas and ordinates are called the coordinates of the points plotted.

• Plot the graph of $y = f(x) = 2x^2 - 4$.

By synthetic division or direct substitution compute $f(0), f(1), f(2), f(3), \ldots,$ $f(-1), f(-2), \ldots.$

x	y
0	−4
±1	−2
±2	+4
±3	+14
±4	+28

To a convenient scale, indicate values of x on a horizontal line or x axis, and indicate values of y along a line perpendicular to it (y axis). As shown in Fig. 3-1, for each value of x in the above table plot the corresponding value of y. Connect the points with a smooth curve. The result is a parabola.

3-57. Degree of a Term. The number of literal factors in a term.

The degree of a term also equals the sum of the exponents of the literal factors. (The degree of $4a^2b^3$ is $2 + 3 = 5$.)

3-58. Identity. A statement in symbolic form that holds for all values of the variables involved. For example,

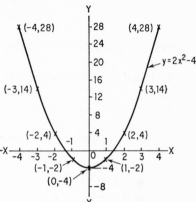

Fig. 3-1

$$(a + b)(a - b) \equiv a^2 - b^2$$

is true regardless of the numbers substituted for a and b.

3-59. Equation A relationship of equality that holds only for certain values of the variable involved. For example,

$$x + 4 = 11$$

holds only for $x = 7$.

An equation with one unknown is a statement in symbolic form that two functions of one variable have the same value for certain values of the variable.

3-60. Linear Equation. An equation containing only the first power of the unknowns.

$$3x + 4y = 70$$

3-61. Quadratic Equation. An equation in which the highest degree of any term is 2.

$$x^2 + 2xy + y^2 = 7$$

3-62. Cubic Equation. An equation in which the highest degree of any term is 3.

$$x^3 + 3x^2y + 6xy^2 - 3y^3 = 1$$

3-63. Roots (Solutions) of an Equation. A commensurable or rational root of an equation is a value of an unknown that satisfies the equation and is either an integer or a fraction. Any other solution is an incommensurable root and is an irrational, imaginary, or complex number.

If a rational integral function $f(x)$ is divisible by $x - a$, then a is a root of the equation $f(x) = 0$.

• If, for example, 3 is a solution of $x^3 - 9x^2 + 27x - 27 = 0$, then division of this function by $x - 3$ leaves no remainder (see Arts. 3-24 and 3-33).

$$
\begin{array}{r}
1 - 9 + 27 - 27\underline{|3} \\
+ 3 - 18 + 27 \\
\hline
1 - 6 + 9 + \underline{|0}
\end{array}
$$

Every rational integral equation of the nth degree has n roots. Roots occurring more than once are called multiple roots. Some of the roots may be imaginary or complex numbers.

If a is a root of $f(x) = 0$, division by $x - a$ yields a quotient that is a function of next-lower degree containing the remainder of the roots.

$$f(x) = (x - r_1)(x - r_2) \cdots (x - r_n) = 0$$

where $r_1, r_2, \ldots, r_n$ are all the roots of $f(x) = 0$.

3-64. Relations between Roots and Coefficients of an Equation. For equations written in the form

$$x^n + p_1 x^{n-1} + p_2 x^{n-2} + \cdots + p_{n-1} x + p_n = 0$$

the following statements hold:

The sum of the roots is equal to the coefficient of the second term p_1 with sign changed.

The sum of the products of the roots taken two at a time is equal to the coefficient of the third term p_2.

The sum of the products of the roots taken m at a time multiplied by $(-1)^m$ is equal to the coefficient of the $(m + 1)$th term p_m.

The product of all the roots and $(-1)^n$ is equal to the last term p_n.

If the second term is missing, the sum of the roots is zero.

If there is no constant term, at least one root is zero.

If all the coefficients are integers, the roots cannot be fractions. In addition, the integral roots are factors of the constant term.

If the coefficients are all real and a complex number $a + bi$ is a root, its conjugate $a - bi$ also is a root.

If the coefficients are all rational and $a + \sqrt{b}$ is a root, then its conjugate $a - \sqrt{b}$ also is a root.

If the signs of all the coefficients are positive, the equation has no positive roots.

If the signs of a complete equation (no terms missing) are alternately positive and negative, the equation has no negative roots.

3-65. Transformation of Equations. An equation can be transformed into another equation whose roots have a certain relation to the roots of the original equation by the substitution of a new variable for the original variable.

To transform

$$f(x) = x^n + p_1x^{n-1} + p_nx^{n-2} + \cdots + p_{n-1}x + p_n = 0 \quad (3\text{-}2)$$

into an equation whose roots are those of $f(x)$ with their signs changed, let $x = -y$; that is, change the signs of the coefficients of all odd powers of x.

• Transform $x^5 - 3x^4 + 4x^3 + 2x^2 - 6x - 1 = 0$ into an equation whose roots have signs opposite those of the original roots.

The required equation, obtained by changing the signs of coefficients of odd powers, is $-y^5 - 3y^4 - 4y^3 + 2y^2 + 6y - 1 = 0$.

To transform $f(x)$ into an equation whose roots are m times those of $f(x)$, let $x = y/m$. The required equation is obtained by multiplying the second term p_1 by m, the third term p_2 by m^2, . . . , the kth term p_{k-1} by m^{k-1}, and the constant term p_n by m^n.

• The equation whose roots are ten times those of $x^3 - 9x^2 + 27x - 27 = 0$ is $y^3 - 90y^2 + 2{,}700y - 27{,}000 = 0$.

To diminish the roots of an equation $f(x) = 0$ in the form of Eq. (3-2) by a number h, divide $f(x)$ by $x - h$ until the remainder does not involve x. This remainder will be the constant term of the required equation. Now, divide the rest of the quotient by $x - h$. The remainder will be the coefficient of the first power of the variable in the transformed equation. Continue dividing the quotients by $x - h$, each time finding the remainder as the coefficient of the next higher power.

• Transform $x^4 + 2x^2 - 17x + 12 = 0$ into an equation whose roots are each greater by 2.

Here $h = -2$. Use synthetic division (Art. 3-24).

$$
\begin{array}{r}
1 + 0 + 2 - 17 + 12\underline{|-2} \\
\underline{- 2 + 4 - 12 + 58} \\
1 - 2 + 6 - 29| + 70 \\
\underline{- 2 + 8 - 28|} \\
1 - 4 + 14| - 57 \\
\underline{- 2 + 12|} \\
1 - 6| + 26 \\
\underline{- 2|} \\
1 - 8
\end{array}
$$

The required equation is $y^4 - 8y^3 + 26y^2 - 57y + 70 = 0$.

3-66. Descartes's Rule of Signs. The number of positive roots in an equation

$$f(x) = x^n + p_1x^{n-1} + \cdots + p_n = 0$$

cannot exceed the number of variations (changes in sign between successive terms); the number of negative roots cannot exceed the number of permanences (successive terms with the same sign).   If a power of x is missing, represent the term by zero with the sign of the preceding term, but in testing for negative roots, transform the equation into one whose roots have signs opposite those of $f(x)$ before applying the rule.

• What is the least possible number of imaginary roots of

$$P(x) = x^6 - 3x^3 + 2x - 9 = 0?$$

With the missing terms filled in with zeros, $P(x)$ becomes $x^6 + 0x^5 + 0x^4 - 3x^3 - 0x^2 + 2x - 9$.   This has three variations, so there can be no more than three positive roots.

To estimate the negative roots, transform the equation by letting $x = -y$. $P(-y) = y^6 + 0y^5 + 0y^4 + 3y^3 + 0y^2 - 2y - 9$.   It has one variation, so it has no more than one positive root, indicating that the original equation $P(x)$ has at most one negative root.

Hence, $P(x)$ can have no more than $3 + 1$, or 4, real roots.   Since $P(x)$ is a sixth-degree equation, it has six roots.   Consequently, it must have at least $6 - 4$, or 2, imaginary roots.

3-67. Locating Roots of Equations with One Unknown.   The first step is to find out as much about the roots as possible by applying the relations between roots and coefficients given in Art. 3-64.   Then apply Descartes's rule of signs (Art. 3-66).

An upper limit for real roots can be obtained by grouping terms so that each group contains not more than one negative term.   Determine a value of the unknown that makes each group positive.

To obtain a lower limit, transform the equation into one with roots having signs opposite to those of the original equation (Art. 3-65) and proceed as for an upper limit.

• Find the upper and lower limits of the roots of

$$5x^5 + 2x^4 - 3x^3 + 2x^2 - x - 48 = 0$$

Application of Descartes's rule of signs indicates that there can be at most three positive roots.   Group the terms.

$$(5x^5 - 3x^3) + (2x^4 - x) + (2x^2 - 48)$$
$$= 5x^3(x^2 - 0.6) + 2x(x^3 - 0.5) + 2(x^2 - 24) = 0$$

Note that any value of x larger than about 5 makes every term positive.   This grouping then indicates that 5 is an upper limit.   A better grouping, however, would be:

$$(5x^5 - 48) + (2x^4 - 3x^3) + (2x^2 - x)$$
$$= 5(x^5 - 9.6) + 2x^3(x - 1.5) + 2x(x - 0.5) = 0$$

All terms will be positive if $x \geq (9.6)^{1/5} \geq 1.57$.   This is the upper limit.

The transformed equation is $5y^5 - 2y^4 - 3y^3 - 2y^2 - y + 48 = 0$. This has two variations, so there can be no more than two negative roots in the original equation. Group the terms.

$$(2y^5 - 2y^4) + (y^5 - 3y^3) + (y^5 - 2y^2) + (y^5 - y) + 48 = 0$$
$$2y^4(y - 1) + y^3(y^2 - 3) + y^2(y^3 - 2) + y(y^4 - 1) + 48 = 0$$

Note that all terms will be positive if $y = -x \geqq (3)^{\frac{1}{2}}$. Hence, the lower limit is $-\sqrt{3}$.

If a function $f(x)$ is continuous (Art. 3-55) and $f(a)$ and $f(b)$ have opposite signs, at least one root must lie between a and b.

Every equation $x^n + p_1 x^{n-1} + \cdots + p_n = 0$ of odd degree has at least one real root with sign opposite to that of the constant term p_n.

Every equation $x^n + p_1 x^{n-1} + \cdots + p_n = 0$ of even degree and with a negative constant term p_n has at least one positive and one negative root.

3-68. Transposition of Terms of an Equation. A term may be shifted from one side of an equation to another with change of sign.

• If $6x + 3 = 4x + 8$, then $6x - 4x + 3 = 8$ and $6x - 4x = 8 - 3$.

3-69. Solution of Linear Equation with One Unknown. Transpose the terms involving the unknown to the left side; transpose all other terms to the right. Add the terms on each side and divide both sides by the coefficient of the unknown.

• Solve $6x - 4 = 3x + 2$.
Transposition of terms involving x to the left and of constant terms to the right yields $6x - 3x = 2 + 4$; addition yields $3x = 6$; division by 3 yields $x = 2$.

3-70. Solution of Equations by Factoring. If an equation is in the form of a polynomial in one variable equated to zero, it may sometimes be solved by factoring the polynomial. (See also Art. 3-22.) Each of the factors may be set equal to zero to obtain a value of the variable.

• Solve $x^2 + x - 6 = 0$.
Factor the left side of the equation to obtain $(x + 3)(x - 2) = 0$. Since the polynomial will be zero when either $x + 3 = 0$ or $x - 2 = 0$, or both, $x = -3$ and $x = +2$ are solutions.

3-71. Solution of Quadratic Equations with One Unknown. If the first power is not present, collect the variable terms on the left side of the equation and sum. Collect the constant terms on the right, sum, and divide both sides by the coefficient of the unknown. Take the square root of both sides.

• Solve $3x^2 + 12 = 5x^2 - 6$.
Transposition of terms involving x to the left side and of constant terms to the right yields $5x^2 - 3x^2 = 12 + 6$ after division by -1. Hence, $2x^2 = 18$.

Division by 2 gives $x^2 = 9$. The square root operation gives the solution: $x = \pm 3$. The $\pm$ sign is necessary because either $(+3)^2$ or $(-3)^2$ equals 9.

If the first power is present, reduce the equation to the form

$$x^2 + px = q \qquad\qquad (3\text{-}3)$$

Make the left side of the equation a perfect square by adding to both sides $p^2/4$, the square of one-half the coefficient of x. Extract the square root of both sides and solve the resulting first-degree equations.

• Solve $x^2 + x - 6 = 0$.
First transpose the constant term to put this equation in the form of Eq. (3-3): $x^2 + x = 6$. (If x^2 had a coefficient other than 1, the first step would be to divide both sides of the equation by it.) To make the left side of the transposed equation a perfect square, add the square of one-half the coefficient of x to both sides.

$$x^2 + x + \tfrac{1}{4} = (x + \tfrac{1}{2})^2 = 6 + \tfrac{1}{4} = 6.25$$

Take the square roots of both sides to obtain $x + 0.5 = \pm 2.5$, from which $x = +2$ and -3.

3-72. Formula for Solution of Quadratic Equations. Quadratic equations with one unknown also can be solved by formula. Write the equation in the form

$$ax^2 + bx + c = 0 \qquad\qquad (3\text{-}4)$$

The solution is

$$x = \frac{-b \pm \sqrt{b^2 - 4ac}}{2a} \qquad\qquad (3\text{-}5)$$

• Solve $6x^2 + 5x - 21 = 0$.
Comparison with Eq. (3-5) indicates that $a = 6$, $b = 5$, and $c = -21$.

$$x = \frac{-5 \pm \sqrt{5^2 - 4 \times 6(-21)}}{2 \times 6} = \frac{-5 \pm \sqrt{529}}{12} = \frac{-5 \pm 23}{12}$$

Hence the solutions are $x = +{}^{18}\!/_{12} = +1.5$ and $x = -{}^{28}\!/_{12} = -{}^{7}\!/_{3}$.

If the equation, when written in the form of Eq. (3-4), can be factored by inspection, the solution can be readily obtained by setting each of the factors equal to zero (Art. 3-70).

3-73. Slide-rule Solution of Quadratic Equations. Write the equation in the form $x^2 + px = x(x + p) = q$. Set the index of the C scale over q on the D scale. Move the indicator until the reading on the D scale coincides with the sum of that number and p on the CI scale. The D-scale reading is one solution. The other solution can be obtained by dividing $-q$ by the first solution.

• Solve $x^2 + x = 6$.
This is equivalent to $x(x + 1) = 6$. Set the C-scale index over 6 on the D scale. Move the indicator and look for a reading on the CI scale that is greater

by 1 than the D-scale reading. This happens when the indicator is over 2 on the D scale, for then the indicator also is over $2 + 1 = 3$ on the CI scale. It also happens when the indicator is over 3 on the D scale if the readings are considered negative, since then $-3 + 1 = -2$, which is indicated on the CI scale. The second solution can be obtained by dividing -6 by 2, the first solution. Thus the solutions are $x = +2$ and -3.

3-74. Cardan's Solution of Cubic Equations. Reduce the equation to the form

$$x^3 + px + q = 0 \qquad (3\text{-}6)$$

To do this when given the general form $a_0x^3 + a_1x^2 + a_2x + a_3 = 0$, diminish the roots by $-a_1/3a_0$. (See Art. 3-65.) One solution of Eq. (3-6) is:

$$x = \sqrt[3]{-\frac{q}{2} + \sqrt{\frac{q^2}{4} + \frac{p^3}{27}}} + \sqrt[3]{-\frac{q}{2} - \sqrt{\frac{q^2}{4} + \frac{p^3}{27}}} = r + r'$$

The other two roots are $r\omega + r'\omega^2$ and $r\omega^2 + r'\omega$, where

$$\omega = -\tfrac{1}{2}(1 + \sqrt{-3})$$
$$\omega^2 = -\tfrac{1}{2}(1 - \sqrt{-3})$$

• Solve the equation $x^3 - 6x^2 - 12x + 112 = 0$.

The first step is to eliminate the coefficient -6 by diminishing the roots by $-(-6)/3 = 2$.

$$
\begin{array}{r}
1 - 6 - 12 + 112\underline{|2} \\
\underline{+ 2 - 8 - 40} \\
1 - 4 - 20|+ 72 \\
\underline{+ 2 - 4|} \\
1 - 2| - 24 \\
\underline{+ 2|} \\
1 + 0
\end{array}
$$

The transformed equation is $y^3 - 24y + 72 = 0$. Here $p = -24$ and $q = 72$.

$$y = \sqrt[3]{-\frac{72}{2} + \sqrt{\frac{72^2}{4} + \frac{-24^3}{27}}} + \sqrt[3]{-\frac{72}{2} - \sqrt{\frac{72^2}{4} + \frac{-24^3}{27}}}$$

$$= \sqrt[3]{-36 + \sqrt{784}} + \sqrt[3]{-36 - \sqrt{784}} = \sqrt[3]{-8} + \sqrt[3]{-64}$$

$$= -2 - 4 = -6$$

The other two roots are $-2\omega - 4\omega^2 = 3 - \sqrt{-3}$ and $-2\omega^2 - 4\omega = 3 + \sqrt{-3}$. Since $x = y + 2$, the solutions are $x = -4, 5 \pm \sqrt{-3}$.

3-75. Solution of Fractional Equations. Clear both sides of fractions by multiplying each term by the least common denominator. Then solve the resulting equation. Check the solutions to make sure that no extraneous roots have been introduced by multiplying by the variable in clearing fractions.

• Solve $2/(x + 1) + 1/(x - 1) = 1$.
Convert to a common denominator (Art. 3-25).

$$\frac{2(x - 1)}{(x - 1)(x + 1)} + \frac{x + 1}{(x - 1)(x + 1)} = \frac{(x - 1)(x + 1)}{(x - 1)(x + 1)}$$

Multiply both sides of the equation by the denominator $(x - 1)(x + 1)$.

$$2(x - 1) + x + 1 = x^2 - 1$$
$$x^2 - 3x = 0$$
$$x = 0, 3$$

Substitution of these values in the original fractional equation shows that they satisfy it; therefore they are the correct solutions.

3-76. Solution of Radical Equations. Transform the equation to a rational equation by raising both sides of the equation to equal powers. This operation may introduce new or extraneous roots, so each solution of the resulting rational equation must be checked by substitution in the original radical equation.

• Solve $\sqrt{a + 9} + \sqrt{2a + 17} = 12$.
To simplify the work, transpose one of the radicals to the right side of the equation.

$$\sqrt{a + 9} = 12 - \sqrt{2a + 17}$$

Square both sides.

$$a + 9 = 144 - 24\sqrt{2a + 17} + 2a + 17$$

Transpose the remaining radical to the left side of the equation; transpose all other terms to the right.

$$24\sqrt{2a + 17} = a + 152$$

Again square both sides.

$$576(2a + 17) = a^2 + 304a + 23{,}104$$
$$a^2 - 848a + 13{,}312 = 0$$
$$(a - 16)(a - 832) = 0$$
$$a = 16$$

$a = 832$ will not satisfy the original radical equation.

3-77. Solution of Equation with One Unknown by Graph. Transpose all terms to the left side of the equation, leaving zero on the right. Let y equal the function on the left side of the equation. Plot the graph of y. (See Art. 3-56.) The intersections of this curve, if there are any, with the x axis are the roots of the original equation.

If the approximate values of the roots can be estimated, it is a good idea to use an exaggerated scale and plot the curve only in the regions of the estimated values.

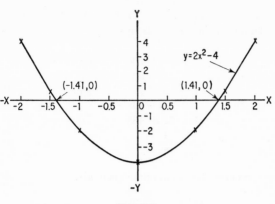

FIG. 3-2

• Solve $2x^2 = 4$ graphically.

Let $y = 2x^2 - 4$. Plot y as a function of x in the region 0 to ± 2. The curve intersects the x axis at $x = \pm 1.41$, so the solution of $2x^2 = 4$ is $x = \pm 1.41$.

3-78. Horner's Method of Solving Equations. First locate the roots approximately, either graphically (Art. 3-77) or by determining upper and lower limits (Art. 3-67). Continually narrow down the range in which the function changes signs by diminishing the roots of the given equation by each digit determined (Art. 3-65). To simplify the calculation, remove the decimals by multiplying the roots by 10, 100, 1,000, etc., as necessary.

• Find a root between 2 and 3 of $x^3 - 12x^2 + 45x - 53 = 0$.
Diminish the roots by 2.

$$
\begin{array}{r}
1 - 12 + 45 - 53\underline{|2} \\
+\ 2 - 20 + 50 \\
\hline
1 - 10 + 25| - \ 3 \\
+\ 2 - 16| \\
\hline
1 - \ 8|+ \ 9 \\
+\ 2| \\
\hline
1 - \ 6 \\
\end{array}
$$

The transformed equation is $y^3 - 6y^2 + 9y - 3 = 0$. Since the next digit to be obtained is a decimal, calculations would be simpler if the roots were multiplied by 10 to eliminate decimals. The new equation becomes

$$y^3 - 60y^2 + 900y - 3{,}000 = 0$$

A first approximation to its root can be obtained by neglecting powers of y higher than the first. $900y - 3{,}000 = 0$; $y = 3.3$.

Try $y = 4$. Division of the left side of the equation by $y - 4$ yields a remainder with a negative sign, the same as the constant term. On the other hand, division by $y - 5$ leaves a positive remainder.   Therefore the first decimal of the original equation is 4.   Diminish the new roots by 4.

$$
\begin{array}{l}
1 - 60 + 900 - 3{,}000 \underline{|4} \\
 + 4 - 224 + 2{,}704 \\
\hline
1 - 56 + 676| - 296 \\
 + 4 - 208| \\
\hline
1 - 52| + 468 \\
 + 4| \\
\hline
1 - 48
\end{array}
$$

Multiply the new roots by 10 to eliminate decimals.

$$z^3 - 480z^2 + 46{,}800z - 296{,}000 = 0$$

A first approximation to z is $296{,}000/46{,}800 = 6+$.   Since the function changes sign between 6 and 7, the second decimal is 6.   Diminish the roots by 6.

$$
\begin{array}{l}
1 - 480 + 46{,}800 - 296{,}000 \underline{|6} \\
 + 6 - 2{,}844 + 263{,}736 \\
\hline
1 - 474 + 43{,}956| - 32{,}264 \\
 + 6 - 2{,}808| \\
\hline
1 - 468| + 41{,}148 \\
 + 6| \\
\hline
1 - 462
\end{array}
$$

The next decimal is then $10 \times 32{,}264/41{,}148 = 7+$.   The required root is 2.467

3-79. Newton's Method of Solving Equations.   This method is applicable not only to high-degree algebraic equations, but also to other types, including trigonometric, exponential, and other transcendental, (non-algebraic) equations.

Write the given equation in the form $f(x) = 0$.   Next locate the roots approximately, either graphically (Art. 3-77) or by determining upper and lower limits (Art. 3-67).   Then assume a value of an unknown root and substitute in

$$x_1 = x_0 - \frac{f(x_0)}{f'(x_0)}$$

where x_0 = assumed value

$f(x_0)$ = value of given function for $x = x_0$

$f'(x_0)$ = value of first derivative of function for $x = x_0$ (Arts. 7-3 to 7-5)

x_1 = second approximation to the root

x_1 may be substituted for x_0 in this formula to obtain a third approximation x_2.

• Find a root between 2 and 3 of $x^3 - 12x^2 + 45x - 53 = 0$.

Let $x = 2$. Find $f(2)$ by dividing by $x - 2$ (Art. 3-32).

$$\begin{array}{r} 1 - 12 + 45 - 53\underline{|2} \\ +\ 2 - 20 + 50 \\ \hline 1 - 10 + 25\underline{|-\ 3} = f(2) \end{array}$$

$f'(x) = 3x^2 - 24x + 45$. Find $f'(2)$ by dividing by $x - 2$.

$$\begin{array}{r} 3 - 24 + 45\underline{|2} \\ +\ 6 - 36 \\ \hline 3 - 18\underline{|+\ 9} = f'(2) \end{array}$$

$$x_1 = 2 - \left(\frac{-3}{9}\right) = 2 + \tfrac{1}{3} = 2.3$$

As a second trial, let $x_0 = 2.3$. Divide $f(x)$ and $f'(x)$ by $x - 2.3$ to find $f(2.3) = -0.813$ and $f'(2.3) = 5.67$.

$$x_2 = 2.3 - (-0.813/5.67) = 2.3 + 0.14 = 2.44$$

As a third trial, let $x_0 = 2.44$, $f(2.44) = -0.116416$, $f'(2.44) = 4.3008$.

$$x_3 = 2.44 - (-0.116416/4.3008) = 2.44 + 0.027 = 2.467$$

3-80. Modified Newton's Method. Faster convergence than with Newton's method (Art. 3-79) can sometimes be obtained with the approximation

$$x_1 = x_0 - \frac{f'(x_0)}{f''(x_0)} \pm \sqrt{\left[\frac{f'(x_0)}{f''(x_0)}\right]^2 - \frac{2f(x_0)}{f''(x_0)}}$$

where $x_0 =$ trial value of unknown root

$f(x_0) =$ value of function for $x = x_0$

$f'(x_0) =$ value of first derivative of function for $x = x_0$ (Arts. 7-3 to 7-5)

$f''(x_0) =$ value of second derivative of function for $x = x_0$ (Art. 7-8)

$x_1 =$ second approximation to root

The sign to use before the radical is the one opposite that of the term $-f'(x_0)/f''(x_0)$. [If $f''(x_0)$ is small relative to $f(x_0)$ and has the same sign, use Newton's method.]

• Find a root between 2 and 3 of $x^3 - 12x^2 + 45x - 53 = 0$.

Let $x = 2$. Find $f(2)$ by dividing by $x - 2$ (Art. 3-32).

$$\begin{array}{r} 1 - 12 + 45 - 53\underline{|2} \\ +\ 2 - 20 + 50 \\ \hline 1 - 10 + 25\underline{|-\ 3} = f(2) \end{array}$$

$f'(x) = 3x^2 - 24x + 45;\ f'(2) = 9$.

$f''(x) = 6x - 24;\ f''(2) = -12$.

$f'(2)/f''(2) = 9/(-12) = -0.75;\ 2f(2)/f''(2) = 2(-3)/(-12) = 0.5$.

$$x_1 = 2 + 0.75 - \sqrt{0.75^2 - 0.5} = 2.75 - 0.25 = 2.50$$

As a second trial, let $x_0 = 2.5$. $f(2.5) = 0.125; f'(2.5) = 3.75; f''(2.5) = -9$. $f'(2.5)/f''(2.5) = 3.75/(-9) = -0.417;\ 2f(2.5)/f''(2.5) = 2 \times 0.125/(-9) = -0.027778$.

$$x_2 = 2.5 + 0.417 - \sqrt{0.417^2 + 0.027778} = 2.917 - 0.449 = 2.468$$

3-81. Steinman's Method of Solving Equations. This has the advantage over Newton's method (Arts. 3-79 and 3-80) of not requiring calculus for the solution of algebraic equations.

First locate the roots approximately, either graphically (Art. 3-77) or by determining upper and lower limits (Art. 3-67). Then write the equation in the form

$$a_0 x^n = a_1 x^{n-1} + a_2 x^{n-2} + \cdots + a_n$$

Assume a value of an unknown root and substitute in

$$x_1 = \frac{a_1 + (2a_2/x_0) + (3a_3/x_0^2) + (4a_4/x_0^3) + \cdots}{a_0 + (a_2/x_0^2) + (2a_3/x_0^3) + (3a_4/x_0^4) + \cdots}$$

where x_0 = assumed value
x_1 = second approximation

• Find a root between 2 and 3 of $x^3 - 12x^2 + 45x - 53 = 0$.
The equation is equivalent to

$$x^3 = 12x^2 - 45x + 53$$

$a_0 = 1$, $a_1 = 12$, $a_2 = -45$, and $a_3 = 53$. Let $x_0 = 2$.

$$x_1 = \frac{12 + 2(-45)/2 + 3 \times 53/2^2}{1 - 45/2^2 + 2 \times 53/2^3} = \frac{12 - 45 + 39.75}{1 - 11.25 + 13.25} = \frac{6.75}{3.00} = 2.25$$

As a second trial, let $x_0 = 2.25$. Substitution in the formula gives $x_2 = 2.45$. A third trial with this value yields $x_3 = 2.467$.

3-82. Simultaneous Equations. A group of equations with several unknowns that can be satisfied by the same values of the unknowns. (See also Art. 13-2.)

Independent equations represent different relations between the unknowns. If they are linear and also consistent, they have only one solution. If they have no solution, the equations are inconsistent.

If the equations are satisfied by an unlimited number of values of the unknowns, the equations are dependent. This occurs when there are more equations than unknowns or when the determinant (see Arts. 13-7 to 13-13) formed by the coefficients of the unknowns is zero.

Consider, for example, the system of equations

$$2x - 3y + 4z = 8 \qquad (3\text{-}7)$$
$$x + 2y - 3z = -4 \qquad (3\text{-}8)$$
$$4x + y - 2z = 1 \qquad (3\text{-}9)$$

The determinant of the coefficients is

$$\begin{vmatrix} 2 & -3 & 4 \\ 1 & 2 & -3 \\ 4 & 1 & -2 \end{vmatrix} = -8 + 4 + 36 - 32 + 6 - 6 = 0$$

Hence the equations are dependent.

A system of equations is consistent if both the matrix of the coefficients (see Arts. 13-1 and 13-16) and the matrix of the coefficients and constant terms (augmented matrix) have the same rank (the largest determinants not zero formed from these matrices have equal numbers of terms).

• Consider, for example, the system of equations

$$2x + 3y = 1 \qquad\qquad\qquad (3\text{-}10)$$
$$x - 2y = 4 \qquad\qquad\qquad (3\text{-}11)$$
$$4x - y = 6 \qquad\qquad\qquad (3\text{-}12)$$

The matrix of the coefficients is

$$\begin{vmatrix} 2 & 3 \\ 1 & -2 \\ 4 & -1 \end{vmatrix}$$

Its rank is 2, since it has at least one second-order determinant that is not zero. The augmented matrix is

$$\begin{vmatrix} 2 & 3 & 1 \\ 1 & -2 & 4 \\ 4 & -1 & 6 \end{vmatrix}$$

Its rank is 3, since it forms a third-order determinent that is not zero. Since the ranks of the two matrices are not the same, the equations are inconsistent.

3-83. Solution of Simultaneous Linear Equations. A common method is to eliminate one unknown at a time, until only one remains. When the value of this unknown is determined, the others can be found successively. (See also Art. 13-15.)

Elimination of one unknown at a time can be carried out by combining the equations in pairs, by addition or subtraction, substitution or comparison. (In dealing with equations of the form $a/x + b/y + \cdots = c$, let $1/x = x'$, $1/y = y'$, etc., and solve for x' and y' first.)

• Solve the following system of equations:

$$3x + 2y - 2z = 4 \qquad\qquad\qquad (3\text{-}13)$$
$$2x - 3y + 3z = 7 \qquad\qquad\qquad (3\text{-}14)$$
$$4x + 5y - 4z = 7 \qquad\qquad\qquad (3\text{-}15)$$

Eliminate x by addition or subtraction.

$2 \times$ Eq. (3-13) $- 3 \times$ Eq. (3-14):

$$13y - 13z = -13 \tag{3-16}$$

Eq. (3-15) $- 2 \times$ Eq. (3-14):

$$11y - 10z = -7 \tag{3-17}$$

Eliminate y by substitution.

From Eq. (3-16):

$$y = \frac{13z - 13}{13} = z - 1 \tag{3-18}$$

Substitute y from Eq. (3-18) in Eq. (3-17):

$$11(z - 1) - 10z = 11z - 11 - 10z = -7$$
$$z = 4 \tag{3-19}$$

Eliminate z by comparison.

From Eq. (3-16):

$$z = \frac{13y + 13}{13} = y + 1 \tag{3-20}$$

From Eq. (3-17):

$$z = \frac{11y + 7}{10} \tag{3-21}$$

Equate Eqs. (3-20) and (3-21):

$$y + 1 = \frac{11y + 7}{10} \tag{3-22}$$
$$10y + 10 = 11y + 7$$
$$y = 3$$

Substitute $y = 3$ and $z = 4$ in Eq. (3-13).

$$3x + 2 \times 3 - 2 \times 4 = 4$$
$$x = 2$$

3-84. Solution of Simultaneous Equations by Determinants. n linear equations with n unknowns can be solved by evaluating the ratio of two determinants for each unknown. (See Arts. 13-7 to 13-13.) The denominator for each unknown is the determinant of the coefficients of the unknowns in the system of equations. For a specific unknown, the numerator differs from the denominator only in that the constant terms of the system of equations replace the coefficients of that unknown.

• Solve the following system of equations:

$$x - 2y + 3z = 2$$
$$2x \qquad - 3z = 3$$
$$x + y + z = 6$$

The denominator of the solution is

$$D = \begin{vmatrix} 1 & -2 & 3 \\ 2 & 0 & -3 \\ 1 & 1 & 1 \end{vmatrix} = 19$$

For the numerator of the solution of each unknown replace its column of coefficients in D by the constant terms 2, 3, 6.

$$x = \frac{\begin{vmatrix} 2 & -2 & 3 \\ 3 & 0 & -3 \\ 6 & 1 & 1 \end{vmatrix}}{19} = \frac{57}{19} = 3$$

$$y = \frac{\begin{vmatrix} 1 & 2 & 3 \\ 2 & 3 & -3 \\ 1 & 6 & 1 \end{vmatrix}}{19} = \frac{38}{19} = 2$$

$$z = \frac{\begin{vmatrix} 1 & -2 & 2 \\ 2 & 0 & 3 \\ 1 & 1 & 6 \end{vmatrix}}{19} = \frac{19}{19} = 1$$

3-85. Solution of One Linear, One Quadratic Equation. Solve for one of the unknowns in the linear equation and substitute in the quadratic.

• Solve the following system of equations:

$$3x - 2y = 5 \tag{3-23}$$
$$x^2 + 2xy + y^2 = 25 \tag{3-24}$$

From Eq. (3-23):
$$x = \frac{2y + 5}{3} \tag{3-25}$$

Substitute Eq. (3-25) in Eq. (3-24):

$$\left(\frac{2y + 5}{3}\right)^2 + 2\left(\frac{2y + 5}{3}\right)y + y^2 = 25$$
$$4y^2 + 20y + 25 + 12y^2 + 30y + 9y^2 = 9 \times 25 = 225$$
$$25y^2 + 50y - 200 = 0$$
$$y = 2, -4 \qquad \text{(see Arts. 3-71 to 3-73)}$$

From Eq. (3-25):
$$x = 3, -1$$

3-86. Solution of Simultaneous Quadratic Equations. If x and y are two unknowns and if two of the quantities $x + y$, $x - y$, and xy are given, the third can be obtained from $(x + y)^2 - 4xy \equiv (x - y)^2$, and the equations may be solved.

• Solve the following system of equations:

$x + y = 9$		(3-26)
$xy = 20$		(3-27)

Square Eq. (3-26): $\qquad x^2 + 2xy + y^2 = 81 \tag{3-28}$
$4 \times$ Eq. (3-27): $\qquad\qquad\qquad 4xy = 80 \tag{3-29}$
Eq. (3-28) $-$ Eq. (3-29): $\quad x^2 - 2xy + y^2 = 1 \tag{3-30}$
Square root of Eq. (3-30): $\qquad x - y = \pm 1 \tag{3-31}$
Add and subtract Eqs. (3-26) and (3-31):

$$x = 5 \qquad x = 4$$
or $$y = 4 \qquad y = 5$$

If all the terms are of the same degree in both equations, or if the quadratic terms can be eliminated by manipulation of the equations, the solution can be converted to that for one linear and one quadratic equation (Art. 3-85). If only one term is of different degree, manipulate the equations to eliminate it.

• Solve the following system of equations:

$$4x^2 - 9xy + 3y^2 - 2y = 0 \tag{3-32}$$
$$2x^2 - 2xy - y^2 - 6y = 0 \tag{3-33}$$

$3 \times$ Eq. (3-32): $12x^2 - 27xy + 9y^2 - 6y = 0 \tag{3-34}$

Eq. (3-34) − Eq. (3-33): $10x^2 - 25xy + 10y^2 = 0 \tag{3-35}$

Factor Eq. (3-35): $(10x - 5y)(x - 2y) = 0 \tag{3-36}$

Equate the factors to zero: $x = \dfrac{y}{2} \text{ or } 2y \tag{3-37}$

Substitute Eq. (3-37) in Eq. (3-33) and solve for y:

$$y = 0, 2$$
$$x = 0, 4$$

3-87. Graphical Solution of Two Equations with Two Unknowns.

Fig. 3-3

Plot both equations. (See Art. 3-56.) The coordinates of the intersections of the curves are the solutions.

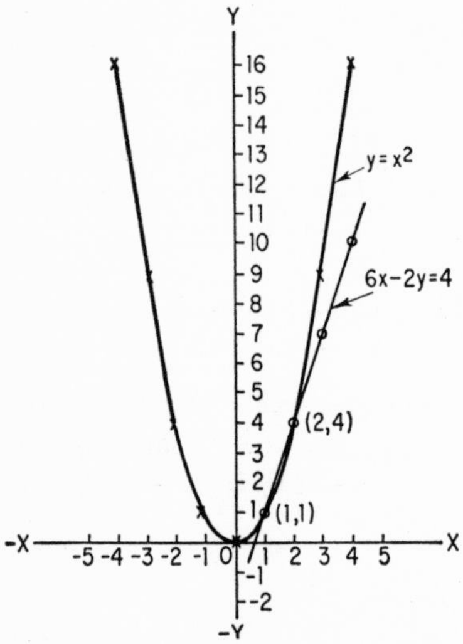

• Solve graphically $y = x^2$ and $6x - 2y = 4$.

Compute x and y for both equations and plot the values.

$y = x^2$		$6x - 2y = 4$	
x	y	x	y
0	0	1	1
± 1	1	2	4
± 2	4	3	7
± 3	9	4	10
± 4	16		

The equations are plotted in Fig. 3-3. The graphs intersect at (2,4) and (1,1). Hence, the solutions are $x = 2$, $y = 4$ and $x = 1$, $y = 1$.

3-88. Linear Simultaneous Equations with More Equations than Unknowns. Write the equations in the form

$$a_{11}x + a_{12}y + a_{13}z + \cdots + c_1 = 0$$
$$a_{21}x + a_{22}y + a_{23}z + \cdots + c_2 = 0$$
$$a_{31}x + a_{32}y + a_{33}z + \cdots + c_3 = 0$$

If a system of n of these equations in $n - 1$ unknowns is to have a single solution, the determinant of the coefficients of the unknowns and the constants must be zero.

$$\begin{vmatrix} a_{11} & a_{12} & a_{13} & \ldots & c_1 \\ a_{21} & a_{22} & a_{23} & \ldots & c_2 \\ a_{31} & a_{32} & a_{33} & \ldots & c_3 \end{vmatrix} = 0$$

3-89. Integer Solutions of Linear Equation with Two Unknowns (Diophantine Analysis). If $x = r_1$ and $y = r_2$ are a solution of

$$ax + by = c$$

then the equation will also be satisfied by

$$x = r_1 + kb \tag{3-38}$$
$$y = r_2 - ka \tag{3-39}$$

where k is any positive or negative number. If the solution must consist only of integers, find one solution and select integral values for k. If, in addition, the solution must contain only positive integers, k may be restricted to a finite number of values.

The equation has a solution only if c is divisible by the greatest common divisor of a and b, including 1.

A solution can usually be found in the following manner:

1. Divide both sides of the equation by a or b, whichever is smaller.

2. Equate the fractional part of the left side to the fractional part of the right side, or the fraction plus an unknown integer n.

3. Solve for the unknown or the unknown and n.

4. Compute the other unknown from the initial equation.

• Solve in positive integers $12x + 7y = 109$. $\tag{3-40}$

Eq. (3-40) ÷ 7: $\qquad x + \tfrac{5}{7}x + y = 15 + \tfrac{4}{7}$ $\tag{3-41}$

If the fractional parts of the two sides of Eq. (3-41) are equated, the solution yields a fractional x. Let

$$\tfrac{5}{7}x = \tfrac{4}{7} + n \tag{3-42}$$

where n is a positive integer. Clear Eq. (3-42) of fractions.

$$5x - 7n = 4 \tag{3-43}$$

Equation 3-43 must now be solved for positive integral values of x and n. Divide Eq. (3-43) by 5.

$$x - n - \tfrac{2}{5}n = \tfrac{4}{5} \tag{3-44}$$

Equate the fractional parts of Eq. (3-44) to obtain a solution for n.

$$\frac{-2n}{5} = \frac{4}{5} \qquad n = -2 \tag{3-45}$$

Since n must be positive, examine the general solution $n = -2 + 5k'$ [see Eq. (3-38)] for the lowest possible positive value. If $k' = 1$, then $n = 3$, the least positive solution. Substitute this in Eq. (3-43) and solve for x. The result is $x = 5$, and from Eq. (3-40), $y = 7$. The general solution of Eq. (3-40) then is:

$$x = 5 + 7k \qquad (3\text{-}46)$$
$$y = 7 - 12k \qquad (3\text{-}47)$$

If y is to be positive, $k \leq 0$, according to Eq. (3-47). If x is to be positive, $k \geq 0$. The only possible value, therefore, is $k = 0$, and the only positive solution in integers of Eq. (3-40) is $x = 5$, $y = 7$.

3-90. Integer Solutions of Linear Equation with Three Unknowns. If the equation

$$ax + by + cz = k$$

is to have an integer solution, k must be divisible by the greatest common divisor of a, b, and c, which may be 1. If this divisor is other than 1, divide both sides of the equation by it to reduce the equation to a form in which 1 is the greatest common divisor of a, b, and c. If d is the greatest common divisor of a and b,

$$cz + dt = k \qquad (3\text{-}48)$$

where t is an integer. Solutions of this equation are

$$z = z_0 - dw \qquad (3\text{-}49)$$
$$t = t_0 + cw \qquad (3\text{-}50)$$

where z_0, t_0 are one solution of Eq. (3-48) and w is an integer. Then x and y can be found from

$$x = \frac{d - bm}{a}(t_0 + cw) - \frac{b}{d}v \qquad (3\text{-}51)$$

$$y = m(t_0 + cw) + \frac{a}{d}v \qquad (3\text{-}52)$$

where v is an arbitrary integer and m is an integer that makes the expression $(d - bm)/a$ an integer. (Evaluate by first dividing numerator and denominator by d and then solve for m as in Art. 3-89.)

• Solve $28x + 21y + 3z = 183$, where x, y, and z are integers.

Here $a = 28$, $b = 21$, and $c = 3$. The greatest common divisor of these numbers and 183 is 1, so the equation can have an integer solution.

The greatest common divisor of a and b is, by inspection, $d = 7$. Substitution in Eq. (3-48) yields

$$3z + 7t = 183 \qquad (3\text{-}53)$$

To solve this for integral roots, divide both sides of the equation by 3.

$$z + 2t + \frac{t}{3} = 61 \qquad (3\text{-}54)$$

Since the right side of Eq. (3-54) is an integer, an obvious solution is

$$t_0 = 0 \qquad (3\text{-}55)$$
$$z_0 = 61 \qquad (3\text{-}56)$$

Substitution of Eqs. (3-55) and (3-56) in Eqs. (3-49) and (3-50) gives the general solution for z and t:

$$z = 61 - 7w \qquad (3\text{-}57)$$
$$t = 0 + 3w \qquad (3\text{-}58)$$

where w is an arbitrary integer.

Next, evaluate $(d - bm)/a$ for Eq. (3-51) by determining an integral value of m that makes this expression an integer.

$$\frac{d - bm}{a} = \frac{7 - 21m}{28} = \frac{1 - 3m}{4} = \frac{1}{4} - \frac{3m}{4} \qquad (3\text{-}59)$$

Let $3m/4 = \frac{1}{4} + n$, where n is an integer. Solve for m.

$$m = \frac{4n}{3} + \frac{1}{3} = n + \frac{n}{3} + \frac{1}{3}$$

Let $n/3 = -\frac{1}{3}$. This gives $n = -1$; hence, $m = -1 - \frac{1}{3} + \frac{1}{3} = -1$. Then

$$\frac{d - bm}{a} = \frac{1}{4} - \frac{3(-1)}{4} = 1 \qquad (3\text{-}60)$$

Substitution of $m = -1$, Eq. (3-58), and Eq. (3-60) in Eqs. (3-51) and (3-52) yields the solutions for x and y; z was given in Eq. (3-57).

$$x = 1(0 + 3w) - 21v/7 = 3w - 3v \qquad (3\text{-}61)$$
$$y = -1(0 + 3w) + 28v/7 = -3w + 4v \qquad (3\text{-}62)$$
$$z = 61 - 7w \qquad (3\text{-}63)$$

where w and v are arbitrary integers.

3-91. Integer Solutions of $x^2 + y^2 = z^2$. If k is an arbitrary integer and m and n are integers that have no common divisor other than 1 and are not both odd numbers $(m > n)$, the general solution is

$$x = 2kmn$$
$$y = k(m^2 - n^2)$$
$$z = k(m^2 + n^2)$$

• For some typical solutions, assume $k = 1$

m	n	x	y	z
2	1	4	3	5
3	2	12	5	13
4	1	8	15	17
4	3	24	7	25

3-92. Inequalities. The sense of an inequality is not changed if both members are increased or diminished by the same number. If a is greater than b $(a > b)$, then $a + m > b + m$.

A term may be transposed from one side of an inequality to the other with change of sign. If c is less than $a + b$ $(c < a + b)$, then $c - a < b$.

The sense of an inequality is not changed if both sides are multiplied or divided by the same positive number. If $a > b$, $ma > mb$ and $a/m > b/m$.

The sense of an inequality is reversed if both sides are multiplied or divided by the same negative number. If $a - b < c - d$, then multiplying both sides by -1 results in $b - a > d - c$.

If $a > b$ and $c > d$, then $a + c > b + d$. Similarly, if $a < b$ and $c < d$, then $a + c < b + d$.

If a, b, c, and d are positive and $a > b$ and $c > d$, then $ac > bd$.

If $a > b$, then $a^n > b^n$, if the signs are not changed.

If $a > b$ and $c = d$, then $a \pm c > b \pm d$, but $c - a < d - b$.

If $a > b$ and $b > c$, then $a > c$.

3-93. Solution of a Conditiona¹ Inequality. So've a conditional inequality in the same manner as an equation, but change the sense of the inequality if the signs of both sides are changed by multiplication or division.

• Solve $6x - 12 > x^2 - 6$.

Transpose: $-x^2 + 6x - 6 > 0$
Multiply by -1: $x^2 - 6x + 6 < 0$

From the quadratic equation solution (Art. 3-72),

$$\left(x - \frac{6 - \sqrt{36 - 4 \times 6}}{2}\right)\left(x - \frac{6 + \sqrt{36 - 24}}{2}\right) < 0$$

If this product is to be less than zero, one of the factors on the left must be negative, and dividing through by it changes the sense of the inequality. Take the second factor as negative because in that factor a larger negative number is subtracted from x. Then

$$x - \frac{6 - \sqrt{12}}{2} > 0 \quad \text{from which} \quad x > \left(\frac{6 - 2\sqrt{3}}{2} = 3 - \sqrt{3}\right)$$

Divide through by the first factor.

$$x - \frac{6 + \sqrt{12}}{2} < 0 \quad \text{from which} \quad x < \left(\frac{6 + 2\sqrt{3}}{2} = 3 + \sqrt{3}\right)$$

Hence the solution is $3 + \sqrt{3} > x > 3 - \sqrt{3}$.

3-94. *e.* An irrational number defined by the infinite series

$$e = 1 + \frac{1}{1!} + \frac{1}{2!} + \frac{1}{3!} + \frac{1}{4!} + \cdots \approx 2.71828$$

See also Art. 1-7.

$$e = \lim [(1 + x)^{1/x}]_{x \to 0}$$

3-95. Logarithms. If $b^x = n$, then x is the logarithm of n to the base b.

$$x = \log_b n$$

Logarithms to the base 10 are called common or Briggs logarithms.
Logarithms to the base e (Art. 3-94) are called natural logarithms.

$$\log_e n = \frac{\log_{10} n}{\log_{10} e} = \frac{\log_{10} n}{0.434294} = 2.302585 \log_{10} n$$

See also Art. 1-7.

The logarithm of 1 to any base is zero ($\log_b 1 = 0$).

The logarithm of the base is 1 ($\log_b b = 1$).

The logarithm of a product equals the sum of the logarithms of the factors ($\log ab = \log a + \log b$).

The logarithm of a quotient equals the logarithm of the dividend less the logarithm of the divisor ($\log a/b = \log a - \log b$).

The logarithm of a power equals the exponent multiplied by the logarithm of the number ($\log n^p = p \log n$).

The logarithm of the root of a number equals the logarithm of the number divided by the index [$\log \sqrt[p]{n} = \log n^{1/p} = (1/p) \log n$].

If the logarithm of a number is written as a decimal, the characteristic of the logarithm is the part on the left of the decimal point and the mantissa is the part on the right.

The characteristic of a common logarithm can be written by inspection. For numbers greater than 1:

Range	Characteristic
1–9.999 $\cdots$	0
10–99.999 $\cdots$	1
100–999.999 $\cdots$	2
1,000–9,999.999 $\cdots$	3

The characteristic is one less than the number of digits on the left of the decimal point.

For fractions, the characteristics are negative, but it is more convenient to write them as positive.

Range	Characteristic	
0.1–0.999 $\cdots$	9.	−10
0.01–0.0999 $\cdots$	8.	−10
0.001–0.00999 $\cdots$	7.	−10

To obtain the characteristic, subtract from 9 the number of zeros between the decimal point and the first digit and add -10.

If two numbers are identical except for location of the decimal point, their logarithms have identical mantissas but different characteristics (log 25 = 1.39794; log 2.5 = 0.39794; log 0.25 = 9.39794 $-$ 10; log 0.025 = 8.39794 $-$ 10).

Tables of common logarithms consist only of mantissas. To find the logarithm of a number, obtain the mantissa from a table, determine the characteristic by inspection, and add the mantissa as a decimal after the characteristic. To find an antilogarithm (a number whose logarithm is given), find the mantissa in a table of logarithms, determine the corresponding number, and point off the decimal point as indicated by the characteristic.

A cologarithm of a number is zero less the logarithm of a number. Addition of a colog is equivalent to subtraction of a logarithm and therefore to division by the number.

Obtain numbers beyond the range of a log table by interpolation.

• Evaluate $(3.163^4 \times 0.2061 \sqrt[3]{9,563})/305.1^2$.

$$
\begin{array}{llll}
\log 3.163^4 & = 4 \log 3.163 = 4 \times 0.5001 & = & 2.0004 \\
\log 0.2061 & = & = & 9.3141 - 10 \\
\log \sqrt[3]{9,563} & = \tfrac{1}{3} \log 9,563 = \tfrac{1}{3} \times 3.9806 & = & \underline{1.3269} \\
& & & 12.6414 - 10 \\
-\log 305.1^2 & = -2 \log 305.1 = -2 \times 2.4844 & = & \underline{-4.9688} \\
& & \log \text{answer} = & 7.6726 - 10 \\
& & \text{Answer} = & 0.004716
\end{array}
$$

• Solve $4^x = 18$.

$$x \log 4 = \log 18$$
$$x = \frac{\log 18}{\log 4} = \frac{1.2553}{0.6021} = 2.085$$

To change from common logarithms to logarithms to any other base, divide the common logarithm of the number by the common logarithm of the base.

3-96. Logarithmic Equations. If $x = B^y$ then y is a logarithmic function of x. $y = \log_B x$.

• Solve $43(1.04)^x = 85$.

Take logarithms of both sides.

$$\log 43 + x \log 1.04 = \log 85$$
$$x = \frac{\log 85 - \log 43}{\log 1.04} \approx \frac{1.9294 - 1.6335}{0.0170} \approx 17.41$$

An equation involving the logarithm of an expression that contains an unknown is called a logarithmic equation.

• Solve $\log_{10}(x^2 - 21) = 2$.

By the definition of logarithm,

$$x^2 - 21 = 10^2 = 100$$
$$x = \pm 11$$

3-97. Simple Interest. If I is the simple interest on a principal P, r is the interest rate, and t is the time (in periods), then

$$I = Prt$$

The simple amount at the end of a given number of interest periods is

$$A = P(1 + rt)$$

• How much interest will be paid in 2 years on \$1,000 invested at 4 per cent interest, paid annually? What will the simple amount be at the end of that time?

$$I = Prt = \$1,000 \times 0.04 \times 2 = \$80$$
$$A = P + Prt = \$1,000 + \$80 = \$1,080$$

3-98. Compound Interest. The amount A_n of a principal of P dollars invested at interest rate r compounded annually for n years is

$$A_n = P(1 + r)^n$$

If interest is compounded m times a year,

$$A_n = P\left(1 + \frac{r}{m}\right)^{mn}$$

At the end of mn periods the compound interest totals

$$I = P\left(1 + \frac{r}{m}\right)^{mn} - P = A_n - P$$

• How much interest will be paid in 20 years on \$4,000 invested at 8 per cent interest compounded semiannually?

The sum in 20 years will amount to

$$A_n = \$4,000\left(1 + \frac{0.08}{2}\right)^{2 \times 20} = \$4,000(1.04)^{40}$$
$$\log 1.04 = 0.0170$$
$$\underline{\times 40}$$
$$\log 1.04^{40} = \overline{0.6800}$$
$$+ \log 4,000 = 3.6021$$
$$\log A_n = \overline{4.2821}$$
$$A_n = \$19,150$$

The interest totals $\$19,150 - \$4,000 = \$15,150$.

3-99. Present Value of Sum Due in n Years. The sum P that, if invested now at a compound interest rate r, will amount to a given sum S

after n years, is

$$P = \frac{S}{(1 + r)^n} = S(1 + r)^{-n}$$

• What is the present value of $\$1,000$ due 8 years from now when an interest rate of 5 per cent compounded semiannually is obtainable?

$S = \$1,000; r = 0.05/2 = 0.025; n = 2 \times 8 = 16.$

$$P = \$1,000(1 + 0.025)^{-16} = \$1,000 \times 0.6736 = \$673.60$$

This sum, if invested at 5 per cent interest compounded semiannually for 8 years, would amount to $\$1,000$.

3-100. Annuity. A fixed sum of money payable at equal intervals of time. The amount, or final value A_n, of an annuity of S dollars accumulated for n years at a compound interest rate r is

$$A_n = \frac{S}{r}[(1 + r)^n - 1]$$

• What is the value of an annuity of $\$40$ accumulated for 10 years at 5 per cent compound interest?

$S = \$40; r = 0.05; n = 10.$

$$A_{10} = \frac{\$40}{0.05}[(1 + 0.05)^{10} - 1] = \$800(1.6289 - 1) = \$503.14$$

3-101. Present Value of an Annuity. The present value P of an annuity of S dollars for n years is the sum that, if invested now at a compound interest rate r, will equal the amount of the annuity in n years.

$$P = \frac{S}{r}\left[1 - \frac{1}{(1 + r)^n}\right]$$

• Find the present value of an annuity of $\$800$ compounded at 4 per cent for 10 years.

$S = \$800; r = 0.04; n = 10.$

$$P = \frac{\$800}{0.04}\left[1 - \frac{1}{(1.04)^{10}}\right] = \$20,000(1 - 0.675616) = \$6,487.68$$

BIBLIOGRAPHY

Allendoerfer, C. B., and C. O. Oakley: "Principles of College Algebra and Analysis," McGraw-Hill Book Company, Inc., New York, 1959.

Dull, R. W., and Richard Dull: "Mathematics for Engineers," 3d ed., McGraw-Hill Book Company, Inc., New York, 1951.

Harper, F. S.: "Mathematics of Finance," International Textbook Company, Scranton, Pa., 1946.

Palmer, C. I., and S. F. Bibb: "Practical Mathematics," 4th ed., McGraw-Hill Book Company, Inc., New York, 1952.

Rice, S., and R. M. Knight: "Technical Mathematics with Calculus," McGraw-Hill Book Company, Inc., New York, 1957.

Wagner, R. W.: "Introductory College Mathematics," McGraw-Hill Book Company, Inc., New York, 1957.

Chapter 4

PLANE AND SOLID GEOMETRY

4-1. Geometric Proofs. Most proofs in Euclidean geometry are developed by deductive reasoning from axioms (Arts. 3-7 and 4-3), undefined fundamental quantities and relations, defined quantities and relations, and previously proved theorems and corollaries. This reasoning usually takes the form:

If a statement X is true, then another statement Y is true.

X is true.

Therefore, Y is true.

A geometric proof of a proposition consists of a statement of given facts with a sketch of the figures involved, a note of what is to be proved, and a series of statements, each accompanied by the reason it is true, connecting the given facts with the relations to be proved.

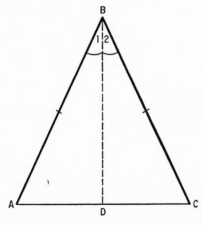

Fig. 4-1

• Prove that the base angles of an isosceles triangle are equal.

Given: Triangle ABC with $AB = BC$.

Prove: Angle A = angle C.

Proof: Draw BD bisecting angle B.

67

Statement	*Reason*
1. $AB = BC$.	1. Given.
2. Angle 1 = angle 2.	2. By construction.
3. $BD = BD$.	3. By identity.
4. Triangles ABD and CBD are congruent.	4. Two triangles are congruent if two sides and the included angle of one are equal respectively to two sides and the included angle of the other.
5. Therefore, angle A = angle C.	5. Corresponding parts of congruent triangles are equal.

Some geometric proofs are indirect. These are based on the assumption that either a statement or its denial is true (law of the excluded middle). This type of reasoning takes the form:

Either X is true, or *not* X is true.

Not X is false.

Therefore, X is true.

Generalization of this method of proof leads to proof by elimination: list all the possible alternatives and prove that all but one cannot be true.

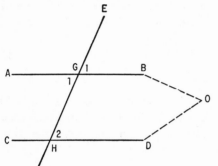

FIG. 4-2

• Prove that if two straight lines are cut by a transversal and if the alternate interior angles are equal, then the lines are parallel.

Given: Lines AB and CD cut by line EF at G and H, respectively; angle 1 = angle 2.

Prove: AB is parallel to CD.

Proof: Either AB does not meet CD or it intersects it at some point O; if the lines meet at O, they form triangle GHO.

Statement	*Reason*
1. Angle 1 > angle 2.	1. An exterior angle of a triangle is greater than either opposite interior angle.
2. But angle 1 = angle 2.	2. Given.
3. AB cannot meet CD.	3. The supposition that they meet leads to a contradiction of a given fact.
4. Therefore, AB is parallel to CD.	4. By definition.

4-2. Geometric Analysis. When a geometric proof or construction method is not obvious at the start, a detailed plan of solution is needed. To develop such a plan:

1. Make a sketch showing given figures and given parts (preferably emphasized by heavy lines) and parts to be proved or constructed.

2. List given facts and the relations that can be drawn from them by inspection. Draw auxiliary lines, if necessary, to relate given parts with those involved in a proof or construction.

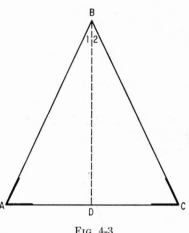

FIG. 4-3

3. List what is to be accomplished and the most likely ways of establishing desired relationships, taking into account the given and implied relationships.

4. If the solution is not visualized at this stage, assume that the problem has been solved. Then list the relationships that have to be proved to obtain the solution, until enough given or implied relationships are listed to make the proof apparent.

• Prove that if two angles of a triangle are equal, the sides opposite these angles are equal.

Analysis:

1. The sketch (Fig. 4-3) shows triangle ABC with equal angles A and C emphasized.
2. *Given:* Angle A = angle C.
3. *Prove:* $AB = BC$.
4. Do this by showing that AB and BC are corresponding sides of congruent triangles, one containing angle A, the other containing angle C. Two such triangles are formed when line BD is drawn, bisecting angle B or perpendicular to AC.

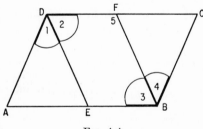

FIG. 4-4

• Prove that the bisectors of opposite angles of a parallelogram are parallel.

Analysis:

1. The sketch (Fig. 4-4) shows parallelogram $ABCD$, with DE bisecting angle D and BF bisecting angle B.
2. *a. Given:* $ABCD$ is a parallelogram.
   Implied: (1) AB is parallel to DC; AD is parallel to BC.
   (2) $AB = CD$; $AD = BC$
   (3) Angle A = angle C; angle B = angle D.

(4) Angle A is supplementary to angle B; angle C is supplementary to angle D.

  b. Given: Angle 1 = angle 2; angle 3 = angle 4.
  Implied: Angle 1 = angle 2 = angle 3 = angle 4.

3. *Prove: DE* is parallel to *BF.*   Do this by showing:

  a. A transversal cutting *DE* and *BF* forms equal alternate interior angles or equal corresponding angles or supplementary interior angles on the same side.
  b. DE and *BF* are opposite sides of a parallelogram.
  c. DE and *BF* are perpendicular to the same line or parallel lines.

If *DF* is considered a transversal to *DE* and *BF*, angles 2 and 5 are interior angles on the same side.   Are they supplementary angles?   Yes, because angle 5 is supplementary to angle 3, which is equal to angle 2.   Since interior angles on the same side of a transversal are supplementary, *DE* must be parallel to *BF*.

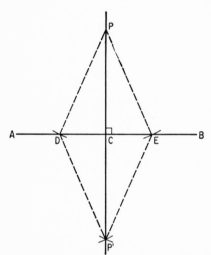

Fig. 4-5

• Construction: Draw a line through a given point perpendicular to a given straight line.

Analysis:

1. Fig. 4-5 shows given point *P* and given line *AB.* Line *PP'* is to be drawn perpendicular to *AB*, intersecting *AB* at *C.*
2. *Given:* Line *AB* and point *P* not on *AB.*
3. *Draw: PP'* perpendicular to *AB.*
4. Do this by making:
   a. P and *P'* equidistant from two points on *AB.*
   b. PC the altitude of an isosceles triangle.

If *P* is chosen as a center and arcs are swung with a convenient radius, intersecting *AB* at two points *D* and *E*, then *P* will be equidistant from *D* and *E*. If, now, arcs are swung with the same radius and with *D* and *E* as centers, their intersection will be a point *P'*, also equidistant from *D* and *E*.   Hence, *PP'* will be perpendicular to *AB.*

4-3. Axioms of Euclidean Geometry.   In addition to the axioms of algebra (Art. 3-7), plane and solid Euclidean geometry are based on the following axioms (after D. Hilbert):

1. Undefined quantities and relations
   Points, straight lines, planes.
   Incidence (lying on), being in, between, congruence, being parallel, continuous.

2. Axioms of incidence
   a. Given any two points, there exists a straight line lying on them.
   b. Given any two points, there exists at most one straight line lying on them.
   c. There are at least two points on a given straight line. There are at least three points that do not lie on a straight line.
   d. If three points do not lie on a straight line, there exists a plane on which these points lie.
   e. If three points do not lie on a straight line, there exists at most one plane on which these points lie.
   f. If two points on a straight line lie on a plane, then every point on the line lies on that plane.
   g. If two planes have a point in common, they have at least one more point in common.
   h. There are at least four points that do not lie on a plane.
3. Axioms of order
   a. If a point B is between points A and C, then A, B, and C lie on a straight line, and B also is between C and A.
   b. If A and C are two points on a straight line AC, there exists at least one point B on AC such that C is between A and B.
   c. If three points lie on a straight line, at most one of them lies between the other two.
   d. (Pasch's axiom.) Let A, B, and C be three points not on a straight line. In the plane determined by A, B, and C, let a straight line DE on which the points A, B, and C do not lie intersect line AB in a point between A and B. Then, either a point on DE is between A and C or it is between B and C.
4. Axioms of congruence
   a. If A and B are points on a straight line a and if A' is a point on a straight line a', then on either side of A' and on a' is a point B' such that the interval $A'B'$ is congruent to AB ($A'B' = AB$).
   b. If $A'B' = AB$ and $A''B'' = AB$, then $A'B' = A''B''$.
   c. Let B lie between A and C. Let B' lie between A' and C'. If $AB = A'B'$ and $BC = B'C'$, then $AC = A'C'$.
   d. Given an angle (h,k), where h and k are half rays (sides of the angle) on a plane α. Given also a straight line on a plane α' and a half ray h' on α' emanating from a point O'. Then, there exists on α' exactly one half ray k' emanating from O' such that the angle (h',k') is congruent to (h,k) and has its interior on a given side of the given straight line. Every angle is congruent to itself.
   e. If AB, AC, and angle BAC of triangle ABC are congruent, respectively, to $A'B'$, $A'C'$, and angle $B'A'C'$ of triangle $A'B'C'$, then angles ABC and ACB are congruent, respectively, to angles $A'B'C'$ and $A'C'B'$.

5. Axiom of parallels

If a straight line and a point not on the line lie on a given plane, there exists only one straight line on the plane and on the point that does not intersect the given line.

6. Axioms of continuity

a. (Archimedean axiom.) If AB and CD are intervals of a straight line, there are points A_1, A_2, $\ldots$, A_n on AB such that AA_1, A_1A_2, $\ldots$, $A_{n-1}A_n$ and CD are congruent to each other and B is between A and A_n.

b. (Axiom of completeness.) It is not possible to add to a system of points, lines, and planes other elements such that the system thus generalized forms a new geometry in which all the preceding axioms are valid.

c. (Axiom of neighborhood.) Given an interval AB of a straight line, then there exists a triangle in the interior of which is no interval congruent to AB.

4-4. Distance. A line segment or interval is the portion of a straight line between two of its points. These points are called the end points; the line segment is designated by the letters given these end points. For example, if the points are marked A and B, the line segment is denoted by AB or BA.

The distance between two points along one or more line segments connecting them is the sum of the lengths of the line segments.

A straight line is the shortest distance between two points. A corollary of this is: the sum of two sides of a triangle is greater than the third side.

The perpendicular is the shortest line that can be drawn from a given point to a line or plane. The common perpendicular between two lines not in the same plane is the shortest line between them.

4-5. Equidistance. Every point on the perpendicular bisector of a line is equidistant from the ends of the line.

Every point on the bisector of an angle is equidistant from the sides of the angle.

The bisectors of the angles of a triangle meet at a point equidistant from the sides (center of inscribed circle).

The perpendicular bisectors of the sides of a triangle meet at a point equidistant from the vertices of the triangle (center of circumscribed circle).

A circle is a plane closed curve with all points equidistant from a fixed point in the plane.

A sphere is a solid bounded by a surface with all points equidistant from a point within.

Oblique lines drawn from a point to a plane, meeting the plane at equal distances from the foot of the perpendicular from the point, are equal.

The locus of points equidistant from a circle is a line through the center perpendicular to the plane of the circle.

The locus of points equidistant from the vertices of a triangle is the line through the center of the circumscribed circle perpendicular to the plane of the triangle.

The locus of points equidistant from two given points is the plane perpendicular, at its midpoint, to the line joining them.

Parallel lines or planes are everywhere equidistant from each other.

The locus of points equidistant from two parallel lines or planes is a plane that perpendicularly bisects the line segment between and perpendicular to the given lines or planes.

4-6. Plane Angles. Straight lines all having one point in common are called rays or beams. The portion of a ray to one side of a given point on the ray is called a half ray. The half ray is said to emanate from the given point.

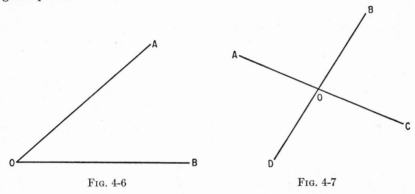

Fig. 4-6   Fig. 4-7

An angle is defined by a pair of half rays emanating from a given point. The half rays are called the sides of the angle and the given point is called the vertex. (In Fig. 4-6, half rays OA and OB are the sides of angle AOB; point O is the vertex.)

A vertical angle is a pair of rays (Fig. 4-7).

4-7. Equal Plane Angles. If two angles are equal, their sides and vertices can be made to coincide.

If two straight lines intersect, the vertical angles are equal. In Fig. 4-7, angle AOD = angle BOC and angle AOB = angle DOC.

Corresponding angles of congruent or similar polygons are equal.

The angles of a regular polygon are equal. Hence, the angles of an equilateral triangle are equal, as are the angles of a square.

Angles that have equal complements or supplements are equal.

Base angles of an isosceles triangle are equal.

Angles inscribed in the same segment of a circle are equal.

Tangents drawn to a circle from an external point make equal angles with the line joining the point with the center of the circle.

If two parallel lines are cut by a transversal, the alternate interior angles and the corresponding angles are equal. In Fig. 4-8, AB and CD

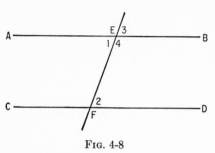

are parallel lines; angles 1 and 2 are equal alternate interior angles; angles 2 and 3 are equal corresponding angles.

If the sides of one angle are parallel to the sides of another angle, and if both pairs of parallels extend in the same direction or in opposite directions, the angles are equal.

FIG. 4-8

If two angles of one triangle equal two angles of another triangle, their third angles are equal.

The opposite angles of a parallelogram are equal.

In the same circle or equal circles, equal arcs define equal central angles. Inscribed angles intercepting equal arcs are equal.

4-8. Supplementary Angles. A straight angle is an angle the sides of which extend in opposite directions and lie on a straight line. Two angles are supplementary if their sum is a straight angle.

If two parallel lines are cut by a transversal, interior angles on the same side of the transversal are supplementary. In Fig. 4-8, angles 3 and 4 are supplementary adjacent angles; they have one side in common and their other sides lie on a straight line. Angles 2 and 4 are supplementary because they are interior angles on the same side of EF, the transversal to parallel lines AB and CD.

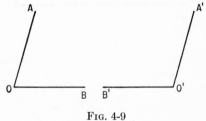

FIG. 4-9

If the sides of one angle are parallel to the sides of another angle, and if one pair of parallels extend in the same direction from the vertices and the other pair extend in opposite directions, the angles are supplementary (Fig. 4-9).

Consecutive angles of a parallelogram are supplementary.

4-9. Complementary Angles. A right angle is either of two equal supplementary angles, that is, half of a straight angle. Two angles are complementary if their sum is a right angle.

The acute angles of a right triangle are complementary.

4-10. Acute and Obtuse Angles. Angles smaller than a right angle are called acute. Those larger than a right angle, but less than a straight angle, are called obtuse.

4-11. Perpendiculars. Either side of a right angle is said to be perpendicular to the other side. Thus, the two sides of a right triangle that include the right angle are perpendicular to each other.

The diagonals of a square are perpendicular to each other, as are the sides. Adjacent sides of a rectangle are perpendicular.

The bisector of the vertex angle of an isosceles triangle is a perpendicular bisector of the base. It is also an altitude.

An altitude is a line from the vertex of a triangle perpendicular to the opposite side. A triangle has three altitudes; they meet at a point called the orthocenter.

If a line is perpendicular to one of two parallels, it is also perpendicular to the other.

Every point equidistant from the ends of a line lies on the perpendicular bisector of the line.

The perpendicular bisector of a chord of a circle passes through the center of the circle.

A tangent to a circle is perpendicular to the radius drawn to the point of tangency. Conversely, a perpendicular to a tangent at the point of tangency passes through the center of the circle.

An angle inscribed in a semicircle is a right angle.

A line is perpendicular to a plane if it is perpendicular to every line in the plane that passes through its foot (its intersection with the plane). The plane is said to be perpendicular to the line.

Intersecting planes that form right dihedral angles are perpendicular to each other.

If a line is perpendicular to each of two other lines at their point of intersection, it is perpendicular to their plane.

All the perpendiculars that can be drawn to a given line at a given point lie on a plane perpendicular to the given line at that point. Only one plane can be perpendicular to a given line at a given point. Only one plane can be passed through an external point perpendicular to a given line.

Only one line can be perpendicular to a plane at a given point on the plane. Only one line can be drawn through an external point perpendicular to a given plane.

If one of two parallel lines is perpendicular to a plane, the other is also perpendicular to the plane.

If two planes are perpendicular to each other, a line on one of them perpendicular to their intersection is perpendicular to the other plane. A perpendicular to one of the planes at any point of their intersection will lie on the other plane. A perpendicular to one plane from any point on the other will lie on the other plane.

If a line is perpendicular to a plane, every plane containing this line is perpendicular to the plane.

If two intersecting planes are each perpendicular to a third plane, their intersection also is perpendicular to that plane.

There is only one plane perpendicular to a given plane and containing a given line not perpendicular to the given plane.

Only one common perpendicular can be drawn between two lines not in the same plane.

A plane tangent to a sphere is perpendicular to the radius of the sphere at the point of tangency.

The line joining the center of a sphere and the center of a circle of the sphere is perpendicular to the plane of the circle.

The line joining the centers of two intersecting circles is the perpendicular bisector of their common chord. The line between the centers of two intersecting spheres is perpendicular to the plane of the circle formed by the intersecting surfaces.

All arcs of great circles drawn through the pole of a given great circle are perpendicular to the given circle.

4-12. Projections. The projection of a point on a line or plane is the foot of the perpendicular from the point to the line or plane. The projection of a straight line is the line segment between the projections of the extremities of the given line (Fig. 4-10).

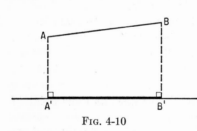

Fig. 4-10

The projection on a plane of a line oblique to the plane is a straight line. The projection of a perpendicular line is a point.

The acute angle a line makes with its projection on a plane is the least angle it makes with any line on the plane.

4-13. Parallels. (See also Art. 4-3.5.) Two lines in the same plane, two planes, or a line and a plane are parallel if they do not meet no matter how far extended.

If two straight lines are cut by a transversal, and if the alternate interior angles or the corresponding angles are equal, or if the interior angles on the same side of the transversal are supplementary, the lines are parallel. In Fig. 4-8, AB is parallel to CD if alternate interior angles 1 and 2 are equal or if corresponding angles 2 and 3 are equal. They are also parallel if interior angles 2 and 4 are supplementary.

Two straight lines perpendicular to the same line or plane are parallel. Two planes perpendicular to the same line or plane are parallel.

If two lines are parallel to a third line, they are parallel to each other. Also, two planes parallel to a third plane are parallel.

Opposite sides of a parallelogram are parallel. Thus, the opposite sides of a rhombus, square, or rectangle are parallel. So are the lateral edges of parallelepipeds and prisms and the elements of cylindric surfaces.

The line joining the midpoints of two sides of a triangle is parallel to the third side. More generally, if a line divides two sides of a triangle proportionally, it is parallel to the third side.

If two lines are parallel, every plane containing one of the lines, and only one, is parallel to the other line.

Through either of two lines not in the same plane, one plane, and only one, can be passed parallel to the other.

Through a given point, only one plane can be passed parallel to a given plane or to any two given lines in space.

The intersections of two parallel planes with a third plane are parallel lines.

If two intersecting lines are parallel to a plane, the plane of the lines is parallel to that plane.

If two angles not in the same plane have parallel sides lying on the same side of the straight line joining their vertices, the angles are equal and their planes are parallel.

The opposite faces of a parallelepiped are congruent and parallel.

4-14. Equal Lines. (See also Arts. 4-3.4 and 4-5.)

Corresponding sides of congruent triangles, polygons, and solids are equal.

Sides of regular polygons, such as equilateral triangles, squares, regular pentagons, and regular hexagons are equal. By definition, the sides of a rhombus are equal. So are the radii and diameters of a circle or sphere, the lateral edges of prisms, parallelepipeds, and regular pyramids, the edges of regular polyhedrons, and the elements of cylinders with parallel bases and right cones.

If two angles of a triangle are equal, the sides opposite these angles are equal. So are the bisectors of these angles and the altitudes and medians to the legs.

Opposite sides of a parallelogram are equal. The diagonals of a parallelogram bisect each other. The diagonals of a square are equal.

If three or more parallels cut off equal segments on one transversal, they cut off equal segments on every transversal. Thus, a line parallel to one side of a triangle bisecting a second side bisects the third side. The line joining the midpoints of two sides of a triangle is equal to one-

half the third side.   In Fig. 4-11, *DE* is drawn parallel to *AC* through
the midpoint of *AB*.   Consequently, *BE = EC* and

$$DE = AF = FC = \tfrac{1}{2} AC$$

In the same circle or in equal circles, equal arcs have equal chords.
Chords equidistant from the center are equal.

If a diameter is perpendicular to a chord, it bisects the chord and its
two arcs.

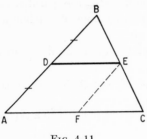

FIG. 4-11

Tangents drawn to a circle from an external point are equal.

If two circles intersect, the line of centers
is the perpendicular bisector of their common
chord.

Arcs of a circle intercepted by two parallel
lines are equal.

Parallel lines included between other parallel lines or parallel planes are equal.

Circles of a sphere equidistant from the center are equal.   All great
circles of a sphere are equal.   Every great circle bisects the spherical
surface.   Two great circles bisect each other.

The spherical distances from either pole of a circle to all points of the
circle are equal.   The straight lines from the pole to the circle also are
equal.

4-15. Unequal Lines.   (See also Art. 4-17.)   If one angle of a triangle
is greater than a second, the side opposite the larger angle is longer than
the side opposite the smaller angle.   In Fig. 4-11, angle *C* is larger than
angle *A*; hence, *BA*, the side opposite angle *C*, is larger than *BC*, the side
opposite angle *A*.

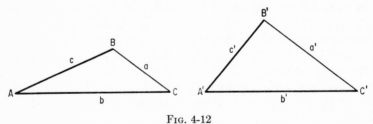

FIG. 4-12

If two sides of one triangle are equal respectively to two sides of
another triangle, but the included angles are unequal, the triangle with
the greater included angle has the longer third side.   In Fig. 4-12, sides
b and *c* of triangle *ABC* are equal respectively to sides *b'* and *c'* of triangle
A'B'C'.   But since angle *A'* is larger than angle *A*, side *a'* is longer
than *a*.

The sum of two sides of a triangle is greater than the third side.

In the same circle or in equal circles, the larger of two central angles has the longer arc. If both arcs are less than a semicircle, the longer arc has the longer chord.

The diameter of a circle is longer than any other chord. If two chords are at unequal distances from the center of a circle, the one closer to the center is longer.

If two oblique lines drawn from a point to a plane meet the plane at unequal distances from the foot of the perpendicular from the point, the more remote oblique line is the longer. Conversely, if two oblique lines from a point to a plane are unequal, the longer meets the plane at the greater distance from the foot of the perpendicular from the point.

If two planes cut a spherical surface at unequal distances from the center, the circle of intersection nearer the center is the larger.

Each side of a spherical triangle is less than the sum of the other two sides.

If two angles of a spherical triangle are unequal, the sides opposite are unequal; the longer side is opposite the larger angle.

The circumference of a circle is larger than the perimeter of any inscribed regular polygon but smaller than the perimeter of any circumscribed regular polygon.

4-16. Unequal Angles. An exterior angle of a triangle is greater than either opposite interior angle. In Fig. 4-13, BCD is an exterior angle of triangle ABC, formed by extending side AC. It is larger than angle A and angle B.

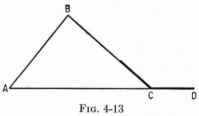

FIG. 4-13

If one side of a triangle is longer than a second side, the angle opposite the long side is greater than the angle opposite the short side. In Fig. 4-13, side BC is longer than side AB; hence, angle A, opposite BC, is larger than angle C, opposite AB.

If two sides of one triangle are equal, respectively, to two sides of another triangle, but the third sides are unequal, the triangle with the longer third side has the larger angle opposite that side. In Fig. 4-12, sides b and c of triangle ABC are equal respectively to sides b' and c' of triangle $A'B'C'$. But since side a' is longer than side a, angle A' is larger than angle A.

In the same circle or in equal circles, the longer arc defines the greater central angle.

4-17. Ratio and Proportion in Geometry. (See also Arts. 3-28, 4-18, and 4-19.) Two or more lines are divided proportionately if the ratio of the segments of one line equals the ratio of the segments of the other lines.

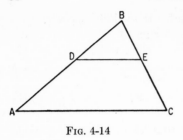

Fig. 4-14

Corresponding sides of similar polygons are proportional.

The medians of a triangle divide each other into segments in the ratio 1:2.

A line drawn through two sides of a triangle parallel to the third side divides the two sides proportionally. One side is to either of its intercepted segments as the other side is to its corresponding segment. In Fig. 4-14, DE is parallel to side AC of triangle ABC. Then

$$BD:DA = BE:EC; \qquad BA:BD = BC:BE; \qquad BA:DA = BC:EC$$

Corresponding altitudes, medians, and angle bisectors of similar triangles have the same ratio as any two corresponding sides.

Three or more parallel lines cut off proportional segments on any two transversals. If two lines are cut by parallel planes, the corresponding segments are proportional. In Fig. 4-15, three parallel planes cut line AD at B, J, and C. They cut line EH at F, K, and G. Then

$$BJ:JC = FK:KG$$

The bisector of an interior angle of a triangle divides the opposite side into segments proportional to

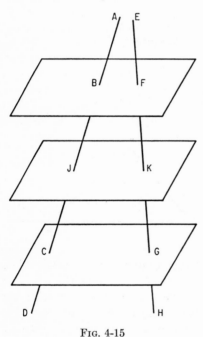

Fig. 4-15

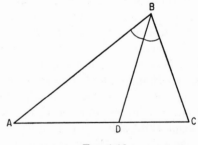

Fig. 4-16

the adjacent sides. In Fig. 4-16, BD bisects angle B of triangle ABC. Hence, $AB:AD = BC:CD$.

If two chords intersect inside a circle, the product of the segments of one equals the product of the segments of the other. In Fig. 4-17, chords AB and CD intersect at E. Hence, $AE:CE = ED:EB$ and $AE \times EB = CE \times ED$.

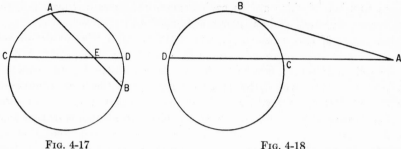

Fig. 4-17 Fig. 4-18

If a tangent and secant are drawn from a point outside a circle, the tangent is the mean proportional between the secant and its external segment. In Fig. 4-18, AB is a tangent and AD is a secant intersecting the circle at C and D. Then $AD:AB = AB:AC$.

The altitude on the hypotenuse of a right triangle is the mean proportional between the two segments of the hypotenuse; each leg of the triangle is the mean proportional between the hypotenuse and the segment adjacent to that leg. In Fig. 4-19, BD is perpendicular to the hypotenuse AC of right triangle ABC. Then $AD:BD = BD:DC$; $AC:AB = AB:AD$; $AC:BC = BC:CD$.

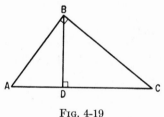

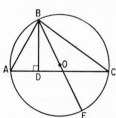

Fig. 4-19   Fig. 4-20

The perpendicular from any point on a circle to a diameter is the mean proportional between the segments of the diameter. In Fig. 4-20, CD is perpendicular to diameter AB. Hence, $AD:CD = CD:DB$.

In any triangle, the product of two sides equals the product of the diameter of the circumscribed circle and the altitude on the third side. In Fig. 4-21, BD is perpendicular to side AC of triangle ABC and BE is a diameter of the circumscribing circle. Then $BE:BC = AB:BD$; $BE \times BD = BC \times AB$.

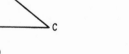

The areas of two similar polygons are to each other as the squares of any two corresponding sides or the squares of their perimeters.

The perimeters of two regular polygons of the

Fig. 4-21

same number of sides are to each other as their radii (distances from center to a vertex) and as their apothems (radii of inscribed circles).

The areas of two regular polygons of the same number of sides are to each other as the squares of their radii or apothems.

The circumferences of two circles are to each other as their radii or diameters. The ratio of the circumference of any circle to its diameter is a constant (represented by π, Greek pi). See Art. 1-7.

The areas of two circles are to each other as the squares of their radii or diameters.

4-18. Harmonic Elements. Two points A and B on a directed line determine a line segment the length AB of which is positive if the direction from A to B agrees with the positive direction on the line, and negative if the direction from A to B is opposite the positive direction of the line. Thus, $AB = -BA$.

A line segment is divided internally if the point of division lies between its extremities. It is divided externally if the point of division is on the extension of the segment. Thus, if a point P divides a directed line segment in the ratio r, where $r = AP/PB$, the ratio is positive if P is an internal point and negative if P is an external point.

Menelaus' Theorem. A necessary and sufficient condition that three points P, Q, and R on the respective sides or extensions of the sides BC, CA, and AB of triangle ABC be collinear is that

$$\frac{AR}{RB}\frac{BP}{PC}\frac{CQ}{QA} = -1$$

(see Fig. 4-22). In other words, if three points divide the three sides of a triangle into six segments such that the product of three nonconsecutive segments is equal to the product of the other three, the three points lie on a straight line.

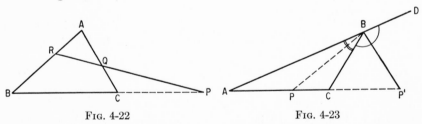

FIG. 4-22   FIG. 4-23

The bisector of an exterior angle of a triangle divides the opposite side externally into segments that are proportional to the other two sides. In Fig. 4-23, BP' bisects exterior angle CBD of triangle ABC and intersects side AC extended at P'. Then, $AB:CB = -AP':P'C$.

A line segment is divided harmonically if it is divided internally and externally in the same ratio.

In Fig. 4-23, the four collinear points A, P, C, and P' are harmonic if $AP:PC = -AP':P'C$. The pair of division points P and P' of the line segment AC are called harmonic conjugates of A and C. Points A and C may also be considered to be division points of line segment PP'.

A set of four concurrent lines that will divide any transversal harmonically is called a harmonic set.

The interior and exterior bisectors of an angle form, with the sides of the angle, a harmonic set of lines. In Fig. 4-23, take ABC as the angle, BP as its bisector, and BP' as the bisector of its adjacent supplementary angle. Then, if AP' is any line cutting the four lines BA, BP, BC, and BP', $AP:PC = -AP':P'C$.

4-19. Mean and Extreme Ratio (Golden Section). A line segment is divided internally in mean and extreme ratio if the greater segment is the mean proportional between the whole line and the smaller segment.

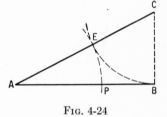

To divide a line AB in mean and extreme ratio, first draw at B a perpendicular CB equal to one-half AB (Fig. 4-24). Draw AC and mark off on it $CE = CB$. Finally, mark off $AP = AE$ on AB. Point P divides AB in mean and extreme ratio, since $AB:AP = AP:PB$.

FIG. 4-24

4-20. Concurrent Lines. Two or more lines that meet in a point are said to be concurrent.

In all triangles, the bisectors of the interior angles meet at a point. So do the perpendicular bisectors of the sides, the altitudes, and the medians to all three sides.

If three circles intersect in pairs, the three common chords pass through the same point.

Ceva's Theorem. A necessary and sufficient condition for the three lines joining points L, M, and N on the respective sides BC, CA, and AB of triangle ABC to the opposite vertices to be concurrent is that

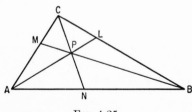

$$\frac{AM}{MC}\frac{CL}{LB}\frac{BN}{NA} = 1$$

FIG. 4-25

(see Fig. 4-25). In other words, if three points divide the three sides of a triangle into six segments such that the product of three nonconsecutive segments is equal to the product of the other three, the lines joining the three points to the opposite vertices of the triangle meet at a point.

The lines joining the vertices of a triangle to the points of tangency of the opposite sides with the inscribed circle of the triangle are concurrent.

4-21. Relations between Parts of a Triangle. (See also Arts. 4-22 to 4-27.) A triangle is a closed plane figure bounded by three straight lines. In Fig. 4-26, points A, B, and C are the vertices of triangle ABC; angles A, B, and C are called the angles of the triangle; the bottom side b is the base. Note that, for convenience, the side opposite angle A is marked a, the side opposite angle B is b, and the side opposite angle C is c.

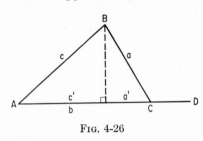

FIG. 4-26

A triangle with three unequal sides is scalene. A triangle with two equal sides is isosceles; a triangle with three equal sides is equilateral.

If one of the angles is 90°, the triangle is a right triangle. Triangles without a right angle are oblique.

A median is a line from a vertex of a triangle to the midpoint of the opposite side. A triangle has three medians. They meet at a point two-thirds of the distance from each vertex to the opposite side. The point of intersection of the medians is the center of gravity of the triangle.

An altitude is a line from a vertex of a triangle perpendicular to the opposite side. The three altitudes meet at a point called the orthocenter.

The bisectors of the angles of a triangle meet at a point equidistant from the sides (the center of inscribed circle).

The perpendicular bisectors of the sides of a triangle meet at a point equidistant from the vertices (the center of circumscribed circle).

The sum of all the angles of a triangle equals a straight angle.

An exterior angle is the angle between one side of a triangle and the adjacent side extended. An exterior angle equals the sum of the two opposite interior angles. In Fig. 4-26, exterior angle

$$BCD = \text{angle } A + \text{angle } B$$

There can be at most one right angle or one obtuse angle in a triangle.

In any triangle, the square of the side opposite an acute angle equals the sum of the squares of the other two sides minus twice the product of one of those sides and the projection of the other side on it. In Fig. 4-26, $a^2 = b^2 + c^2 - 2bc'$.

In any obtuse triangle, the square of the side opposite the obtuse angle equals the sum of the squares of the other two sides plus twice the product of one of those sides and the projection of the other side on it. In Fig. 4-27, $c^2 = a^2 + b^2 + 2a'b$.

In any triangle, the product of two sides equals the square of the bisector of the included angle plus the product of the segments of the third side. In Fig. 4-28, $ac = t^2 + mn$.

In any triangle, the product of two sides equals the product of the diameter of the circumscribed circle and the altitude on the third side (see Fig. 4-21, where $AB \times BC = BE \times BD$).

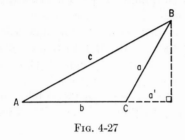

FIG. 4-27

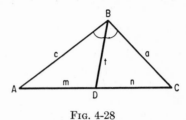

FIG. 4-28

The area of a triangle equals one-half the product of its base and altitude.

4-22. Isosceles Triangle. A triangle with two equal sides.

In the isosceles triangle in Fig. 4-29, side AB = side BC. These sides are called the legs of the triangle. The third side AC is the base. The angles opposite the legs, angles A and C, are the base angles.

The base angles of an isosceles triangle are equal. So are the bisectors of these angles and the altitudes and medians to the legs.

The exterior angle at the vertex is twice either base angle; its bisector is parallel to the base.

The altitude of an isosceles triangle is the perpendicular bisector of the base. It bisects the vertex angle.

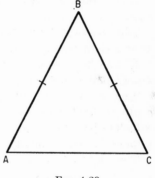

FIG. 4-29

4-23. Equilateral Triangle. (See also Art. 4-22.) A triangle with three equal sides.

An equilateral triangle is also equiangular. Each angle of the triangle is 60°; each exterior angle is 120°.

In an equilateral triangle, altitudes also are medians and angle bisectors. All altitudes, medians, and angle bisectors are equal.

4-24. Right Triangle. A triangle in which one angle is a right angle.

Right triangle ABC in Fig. 4-30 has a right angle at C. Sides a and b are called legs. c, the side opposite the right angle, is the hypotenuse.

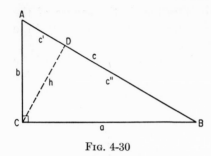

FIG. 4-30

Acute angles A and B are complementary. In an isosceles right triangle, each is 45°.

The hypotenuse of a right triangle is a diameter of the circumscribing circle. The midpoint of the hypotenuse is equidistant from the three vertices.

The altitude on the hypotenuse of a right triangle divides it into two triangles similar to each other and the given triangle. The altitude is the mean proportional between the two segments of the hypotenuse. (In Fig. 4-30, $c':h = h:c''$.) Each leg is the mean proportional between the hypotenuse and the segment adjacent to that leg. (In Fig. 4-30, $c:a = a:c''$ and $c:b = b:c'$.)

Pythagorean Theorem. The square of the hypotenuse of a right triangle equals the sum of the squares of the legs. (In Fig. 4-30, $c^2 = a^2 + b^2$.)

If one angle of a right triangle is 30°, the side opposite this angle is one-half the hypotenuse.

The area of a right triangle equals one-half the product of the legs.

Two right triangles are congruent if:

1. The hypotenuse and an acute angle of one equal the hypotenuse and an acute angle of the other.

2. A leg and an acute angle of one equal the corresponding leg and acute angle of the other.

3. The hypotenuse and a leg of one are equal respectively to the hypotenuse and a leg of the other.

Two right triangles are similar if:

1. An acute angle of one equals an acute angle of the other.

2. The legs are proportional.

4-25. Congruent Triangles. Corresponding parts of congruent triangles are equal.

Two triangles are congruent if:

1. Two angles and the included side of one are equal to the corresponding parts of the other.

2. Two sides and the included angle of one equal the corresponding parts of the other.

3. Three sides of one equal three sides of the other.

4. Two sides and the median or altitude to one of these sides in one triangle equal the corresponding parts of the other.

5. A side, an adjacent angle, and the bisector of that angle in one triangle equal the corresponding parts of the other.

6. Two angles and the side opposite one of these angles in one triangle equal the corresponding parts of the other.

(See Art. 4-24.)

4-26. Similar Triangles. Two triangles are similar if their angles are equal and their sides are proportional in accordance with the definition of similar polygons. But two triangles also may be similar if:

1. Corresponding angles are equal.
2. Two angles of one are equal respectively to two angles of the other.
3. Their sides are proportional.
4. Their sides are parallel.
5. Their sides are perpendicular to each other.
6. Two sides are proportional and the included angles are equal.

(See Art. 4-24.)

4-27. Equivalent Triangles. Two triangles are equivalent if their areas are equal.

The area of a triangle equals one-half the product of its base and altitude.

Two triangles are equivalent if:

1. They have equal bases and altitudes.
2. They have a common base and their vertices lie on a line parallel to the base.

4-28. Quadrilateral. A plane closed figure bounded by four straight lines (Fig. 4-31).

A trapezoid is a quadrilateral with two parallel sides (Fig. 4-32). The parallel sides are called bases; the nonparallel sides are called legs. The perpendicular between the bases is the altitude. The line joining the midpoints of the nonparallel sides is parallel to the bases. It is called the median and is equal to one-half the sum of the bases.

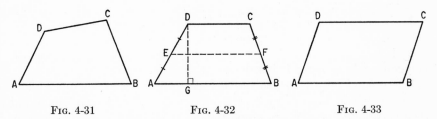

FIG. 4-31   FIG. 4-32   FIG. 4-33

The area of a trapezoid equals one-half the product of its altitude and the sum of the bases. It also equals the product of the altitude and median.

A parallelogram is a quadrilateral whose opposite sides are parallel in pairs (Fig. 4-33).

A rhombus is a parallelogram with equal sides.   A square is a rhombus the angles of which are right angles.

A rectangle is a parallelogram the sides of which are perpendicular to each other.   Thus a square also is a rectangle with equal sides.

The opposite sides of a parallelogram are equal, as are the opposite angles.   Consecutive angles are supplementary.

A diagonal divides a parallelogram into two congruent triangles.

If the opposite sides of a quadrilateral are equal in pairs, or if two sides are equal and parallel, the figure is a parallelogram.

The diagonals of a parallelogram bisect each other.   The diagonals of a rectangle are equal; those of a square, in addition, are perpendicular bisectors of each other.

The area of a parallelogram equals the product of its base and altitude. Any side may be considered as the base; the altitude is the perpendicular from the opposite side to the base.

The area of a rectangle equals the product of two adjacent sides.

The lines joining the midpoints of the consecutive sides of a quadrilateral form a parallelogram.

4-29. Polygons.   A polygon is a plane closed figure bounded by three or more straight lines.   Triangles (Arts. 4-21 to 4-27) and quadrilaterals (Art. 4-28) are polygons with three and four sides, respectively.

A polygon is convex if none of its angles is greater than 180°.   A five-sided polygon is called a pentagon; a six-sided polygon is called a hexagon; an eight-sided polygon is called an octagon; a ten-sided polygon is called a decagon.

The intersections of the sides are called the vertices and the angles formed by the sides are the angles of the polygon.   An exterior angle is the angle between one side of the polygon and the extension of an adjoining side.

A diagonal of a polygon is a straight line between any two nonconsecutive vertices.

An inscribed circle is a circle to which all the sides of a polygon are tangent.   The polygon is circumscribed about the circle.

A circumscribed circle is a circle passing through all the vertices of a polygon.   The polygon is inscribed in the circle.

If two polygons are congruent, their corresponding parts are equal.

Two polygons are similar if their corresponding angles are equal and their corresponding sides are proportional.   Similar figures have the same shape but not necessarily the same size.

Two polygons are equivalent if their areas are equal.

A regular polygon has equal sides and equal angles.   The centers of both the inscribed and circumscribed circles of a regular polygon coincide at a point called the center of the polygon.   The radius of a regular polygon is the radius of the circumscribed circle. The apothem is the radius of the inscribed circle.

The central angle of a regular polygon is the angle formed by two radii drawn to the extremities of any side. In a regular polygon of n sides, the central angle is $360/n$ degrees.

The sum of the angles of any convex polygon of n sides equals $(n - 2)$ straight angles.

Each angle of a regular polygon of n sides equals $(n - 2)/n$ straight angles.

The sum of the exterior angles of a convex polygon equals two straight angles.

If two polygons are similar, the perimeters are to each other as any two corresponding sides. The areas are to each other as the squares of any two corresponding sides or of the perimeters.

Two similar polygons may be divided into the same number of triangles by drawing diagonals from a vertex. These triangles will be similar to each other and similarly located.

If two polygons are comprised of the same number of similar triangles similarly placed, the polygons are similar.

Two regular polygons of the same number of sides are similar.

An equilateral polygon inscribed in a circle is a regular polygon.

If a circle is divided into three or more equal parts, the chords joining the successive points of division form an inscribed regular polygon. The tangents at these points form a circumscribed regular polygon.

If the midpoints of the arcs subtended by the sides of an inscribed regular polygon are joined to the adjacent vertices of the polygon, an inscribed regular polygon of double the number of sides is formed. If tangents are drawn at the midpoints of the arcs of adjacent points of tangency of the sides of a circumscribed regular polygon, a circumscribed regular polygon of double the number of sides is formed.

If the alternate vertices of an inscribed regular hexagon are joined, an inscribed equilateral triangle is formed.

The area of a regular polygon equals one-half the product of its apothem and perimeter.

4-30. Circles. (See also Arts. 4-31 and 4-42.) A circle is a plane closed curve with all points equidistant from a fixed point in the plane, called its center. The length of the circle is called its circumference.

A semicircle is one-half of a circle. A minor arc is an arc less than one-half of a circle; a major arc is greater than one-half of a circle.

A radius is a line from the center to any point on the circle. A diameter is a straight line through the center terminating in two points on the circle. A chord is any straight line with end points on the circle.

A central angle is an angle formed by two radii. An inscribed angle has its vertex on the circle and two chords as sides. In Fig. 4-34, angle AOB formed by radii OA and OB is a central angle. Angle CDE formed by chords DC and DE is an inscribed angle, as is angle BCD.

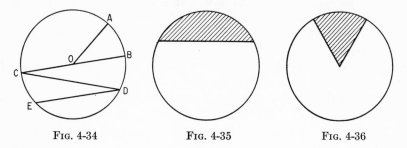

FIG. 4-34   FIG. 4-35   FIG. 4-36

A segment of a circle is an area bounded by an arc and its chord. The shaded area in Fig. 4-35 is a segment.

A sector of a circle is the area enclosed by two radii and the arc intercepted by them. The shaded area in Fig. 4-36 is a sector.

A secant is a straight line that intersects a circle in two points (a chord extended).

A tangent is a straight line that meets a circle at only one point. The length of a tangent is the distance from a given point on it to the point of tangency.

A straight line joining the centers of two circles is called the line of centers.

Two circles tangent to the same line at the same point are said to be tangent. The line is called the common tangent. The line of centers passes through the point of contact or tangency.

If two circles intersect, the line joining the points of intersection is the common chord. The line of centers is the perpendicular bisector of the common chord.

In the same circle or in equal circles, equal central angles define equal arcs. Equal chords subtend equal arcs. The larger of two central angles defines the larger arc. The longer of two chords subtends the longer minor arc.

If a diameter is perpendicular to a chord, it bisects the chord and the two arcs subtended by the chord.

In the same or equal circles, equal chords are equidistant from the center. The longer of two chords is closer to the center. The diameter, therefore, is longer than any other chord.

Only one circle can be drawn through three points not on a straight line.

A straight line perpendicular to the outer extremity of a radius is tangent to the circle. A line from the center of a circle perpendicular to a tangent passes through the point of tangency.

Tangents drawn to a circle from an external point are equal and make equal angles with the line joining the point with the center of the circle.

A circle may be circumscribed about any regular polygon. Also, a circle may be inscribed in any regular polygon.

The circumference of a circle equals πd or $2\pi r$, where d is a diameter and r a radius. $\pi = 3.14159 \cdots$. (See Art. 1-7.)

The area of a circle equals one-half the product of its radius and circumference (πr^2 or $\frac{1}{4}\pi d^2$).

The area of a sector of a circle (Fig. 4-36) equals one-half the product of the radius and its arc.

4-31. Measurement of Plane Angles. A degree is defined as $\frac{1}{90}$th of a right angle. A minute is $\frac{1}{60}$th of a degree and a second is $\frac{1}{60}$th of a minute.

A central angle is measured by its intercepted arc. An inscribed angle is measured by one-half its intercepted arc. In Fig. 4-34, central angle AOB is measured by arc AB. Inscribed angle CDE is measured by one-half arc CE.

An angle formed by two chords intersecting inside a circle is measured by one-half the sum of the two arcs it and its vertical angle intercept. In Fig. 4-37, angle APD is measured by $\frac{1}{2}$(arc AD + arc BC) and angle DPB is measured by $\frac{1}{2}$(arc BD + arc AC).

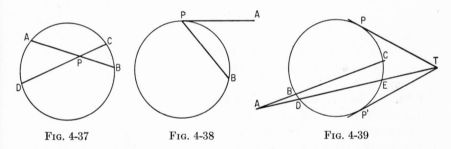

FIG. 4-37   FIG. 4-38   FIG. 4-39

An angle formed by a tangent and a chord drawn from the point of tangency is measured by one-half the intercepted arc. In Fig. 4-38, angle APB is measured by one-half arc PB.

An angle formed by two secants, or two tangents, or a secant and a tangent drawn from an external point to a circle is measured by one-half the difference of the intercepted arcs. In Fig. 4-39, angle CAE is measured by $\frac{1}{2}$(arc CE − arc BD). Angle PTP' is measured by $\frac{1}{2}$(arc $PBDP'$ − arc $PCEP'$). And angle PTD is measured by $\frac{1}{2}$(arc PBD − arc PCE).

4-32. Lines and Planes in Space. (See also Art. 4-3.) A straight line and a point not on the line determine a plane, as do two parallel lines. Only one plane can be passed through two intersecting straight lines.

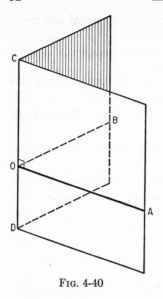

Fig. 4-40

If two planes meet, their intersection is a straight line.

4-33. Dihedral Angles. A pair of intersecting planes form a dihedral angle (Fig. 4-40). The intersection is called the edge and the planes are called faces.

Dihedral angles with a common edge and face are said to be adjacent.

If two planes meet, forming equal adjacent dihedral angles, each of the angles is a right dihedral angle.

The plane angle of a dihedral angle is the angle formed by two straight lines, one in each face of the dihedral angle, perpendicular to the edge at the same point. In Fig. 4-40, angle AOB, with sides OA and OB perpendicular to edge CD, is the plane angle of the dihedral angle.

The plane angle of a dihedral angle has the same magnitude no matter at what point on the edge of the dihedral angle it is drawn.

Two dihedral angles are equal if their plane angles are equal. They also have the same ratio as their plane angles. Thus, the plane angle may be taken as the measure of the dihedral angle.

The locus of points equidistant from the faces of a dihedral angle is the plane bisecting the angle.

4-34. Polyhedral Angles. A polyhedral angle is formed when three or more planes emanate from a point. In Fig. 4-41, the common point O is the vertex. The intersections of the planes OA, OB, OC, etc., are the edges. The planes are called the faces of the angle. Angles such as AOB, AOF, and BOC, formed by adjacent edges, are the face angles of the polyhedral angle. A polyhedral angle is convex if a section cut by a plane is a convex polygon.

A trihedral angle is a polyhedral angle with three faces. A tetrahedral angle has four faces.

Two polyhedral angles are equal if

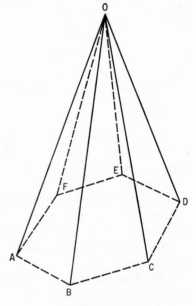

Fig. 4-41

they have equal parts correspondingly placed. Polyhedral angles with equal parts arranged in reverse order are said to be symmetric.

The sum of any two face angles of a trihedral angle is greater than the third face angle.

If three face angles of one trihedral angle are equal respectively to three face angles of another, the trihedral angles are equal or symmetric. Also, the dihedral angles of one are equal to the corresponding dihedral angles of the other.

The sum of the face angles of any convex polyhedral angle is less than four right angles.

4-35. Polyhedrons. A polyhedron is a solid bounded by planes (Fig. 4-42). The planes are the faces; the intersections of the planes are the edges; the intersections of the edges are the vertices. A polyhedron is convex if every section cut by a plane is a convex polygon.

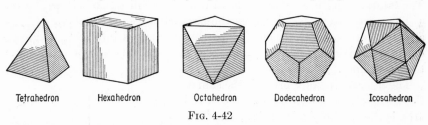

| Tetrahedron | Hexahedron | Octahedron | Dodecahedron | Icosahedron |

FIG. 4-42

A tetrahedron is a polyhedron with four faces. A hexahedron has six faces; an octahedron has eight faces; a dodecahedron has twelve faces; an icosahedron has twenty faces (Fig. 4-42).

The diagonal of a polyhedron is a line joining any two vertices not in the same face.

A regular polyhedron is a polyhedron with congruent regular polygons as faces and with equal polyhedral angles. Only five regular convex polyhedrons are possible: tetrahedron, hexahedron (cube), octahedron, dodecahedron, and icosahedron.

Two solids with equal volumes are equivalent. Two solids with equal parts similarly arranged are congruent. If the parts are equal, but arranged in reverse order, the solids are symmetric. Polyhedrons that have the same number of faces, respectively similar and similarly placed, and equal corresponding polyhedral angles, are similar.

In any polyhedron, the number of edges plus two equals the sum of the number of vertices and faces: $n + 2 = v + f$.

The sum of the face angles of any polyhedron with v vertices equals $(v - 2)$ times 360°.

4-36. Prismatoids. A prismatoid is a polyhedron that has for bases two polygons in parallel planes, and for each lateral face a triangle or trapezoid with one side common with one base and the opposite vertex or side common with the other base (Fig. 4-43). Its altitude is the distance between the parallel planes. Its midsection is a plane parallel to the bases and bisecting the altitude.

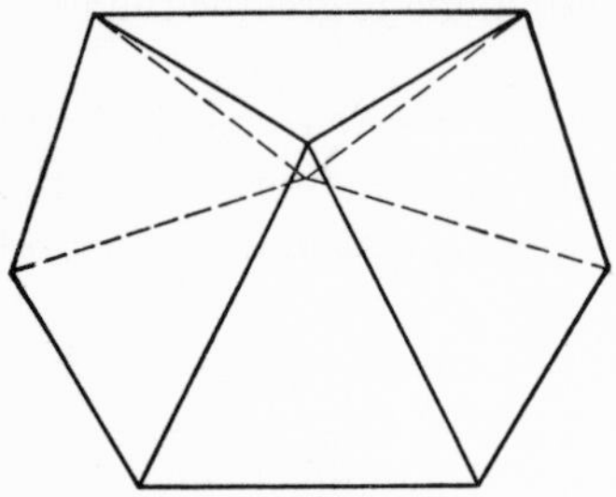

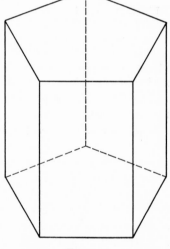

The volume V of a prismatoid equals $\frac{1}{6}h(b + b' + 4m)$, where b and b' are the areas of the two bases, m is the area of the midsection, and h is the altitude.

4-37. Prisms. (See also Arts. 4-35 and 4-38.) A prism is a polyhedron that has for bases two congruent polygons in parallel planes, and for each lateral face a parallelogram with two opposite sides in common with sides of the base polygons

Fig. 4-43

(Fig. 4-44). The intersections of the lateral faces are called lateral edges. The lateral edges are equal. The altitude is the distance between bases.

The prismatic surface between the parallel polygons is generated by a straight line (generatrix) moving so that it always intersects a side of one of the polygons (directrix) and always is parallel to a fixed line oblique to their planes.

The lateral area of a prism is the sum of the areas of the lateral faces. The total area is the sum of the lateral area and the areas of the two bases.

A right prism is a prism with lateral edges perpendicular to the bases. An oblique prism has lateral edges oblique to the bases.

A triangular prism is a prism with triangular bases. A quadrangular prism is a prism with quadrilateral bases.

A truncated prism is the part of a prism included between a base and a section made by an oblique plane. A section made by a plane perpendicular to the lateral edges is called a right section.

The sections of a prism made by parallel planes cutting all edges are congruent

Fig. 4-44

polygons. Every section made by a plane parallel to the bases is congruent to them. All right sections of a prism are congruent.

The lateral area of a prism equals the product of a lateral edge and the perimeter of a right section. The lateral area of a right prism equals the product of its altitude and the perimeter of its base.

Two prisms, or two truncated prisms, are congruent if three faces comprising a trihedral angle of one are respectively congruent to three faces comprising a trihedral angle of the other, similarly placed. Two right prisms are congruent if they have equal altitudes and congruent bases.

The volume of a prism equals the product of its base and altitude. The volume of a truncated triangular prism equals the product of its right section and one-third the sum of its lateral edges.

4-38. Parallelepipeds. (See also Arts. 4-35 and 4-37.) A parallelepiped is a prism with parallelograms as bases (Fig. 4-45). It is called a right parallelepiped if the lateral edges are perpendicular to the bases. It is a rectangular parallelepiped if it is a right prism with rectangles as bases. If all the faces are squares, the parallelepiped is a cube.

The opposite faces of a parallelepiped are congruent and parallel.

A plane passed through two diagonally opposite edges of a parallelepiped divides it into two equivalent triangular prisms.

The volume of a parallelepiped equals the product of its base and altitude. The volume of a rectangular parallelepiped is equal to the product of three mutually perpendicular sides.

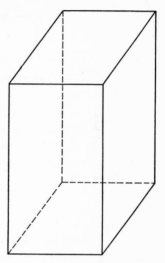

Fig. 4-45

4-39. Pyramids. (See also Art. 4-35.) A pyramid is a polyhedron with a polygon as base and triangles with a common vertex for the other faces (Fig. 4-46). The triangles are called lateral faces; their intersections are lateral edges. The altitude is the perpendicular distance from the vertex to the base.

A triangular pyramid (tetrahedron) is a pyramid with a triangular base. A quadrangular pyramid has a quadrilateral base.

A regular, or right, pyramid is a pyramid with a regular polygon for its base and an altitude that meets the base at its center. The slant height of a regular pyramid is the altitude of one of the lateral faces.

A frustum of a pyramid is the portion of a pyramid included between the base and a section parallel to the base (Fig. 4-47). The section also is called a base. The altitude is the perpendicular distance between bases.

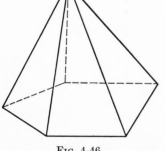

Fig. 4-46

A truncated pyramid is the portion of a pyramid included between the base and a plane cutting all the lateral edges.

In a regular pyramid, the lateral edges are equal and the lateral faces

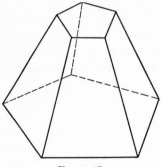

are congruent isosceles triangles; the slant height is the same for all lateral faces.

The lateral area of a pyramid is the sum of the areas of the lateral faces. The lateral area of a regular pyramid equals one-half the product of its slant height and the perimeter of its base. The lateral area of a frustum of a regular pyramid equals one-half the sum of the perimeters of the two bases multiplied by the altitude of a trapezoidal lateral face.

FIG. 4-47

If a pyramid is cut by a plane parallel to the base, the edges and altitude are divided proportionately. The section is a polygon similar to the base.

If two pyramids have equal altitudes and equivalent bases, sections made by planes parallel to the bases and at equal distances from the vertices are equivalent.

The volume of a pyramid equals one-third the product of its base and altitude.

The volume V of a frustum of a pyramid equals $\frac{1}{3}h(b + b' + \sqrt{bb'})$, where h is the altitude and b and b' are the areas of the two bases.

4-40. Cylinders. A cylindric surface is generated by a straight line (generatrix) moving so that it intersects a given curve (directrix) and is always parallel to a given line not in the plane of the curve. The generatrix in any position is called an element of the surface.

A cylinder is a solid bounded by a cylindric surface and two parallel planes (Fig. 4-48). All of its elements are equal. A right cylinder has all elements perpendicular to the bases.

A circular cylinder is a cylinder with a circular base. A cylinder of revolution is a right circular cylinder generated by one side of a rectangle when it is rotated about its other side.

FIG. 4-48

The line joining the centers of the bases of a circular cylinder is called its axis. It passes through the center of all sections parallel to the bases.

The altitude of a cylinder is the perpendicular distance between bases. Bases of a cylinder are congruent.

Every section of a cylinder made by a plane passing through an element is a parallelogram.   In a right cylinder, the section is a rectangle.

Any two parallel sections cutting all elements of a cylinder are congruent.

The lateral area of a circular cylinder equals the product of an element and the circumference of a right section of the cylinder.   The lateral area of a cylinder of revolution equals the product of the altitude and the circumference of the base: $S = \pi dh = 2\pi rh$, where h is the altitude and d and r are the diameter and radius of the base.

The volume of a cylinder equals the product of its base and altitude. The volume V of a circular cylinder equals $\pi r^2 h$.

4-41. Cones.   A conical surface is generated by a straight line (generatrix) moving so that it intersects a given curve (directrix) and passes through a given point (vertex).   The two parts of the conical surface on either side of the vertex are called nappes (see Fig. 4-49).   The generatrix in any position is called an element.

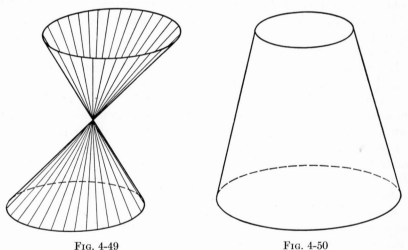

FIG. 4-49   FIG. 4-50

A cone is a solid bounded by one nappe of a conical surface and a plane cutting all elements.   Its altitude is the perpendicular from the vertex to the base.   The lateral area is the area of the conical surface between base and vertex.

A circular cone is a cone with a circular base.   Its axis is the line joining the vertex with the center of the base.   If the axis is perpendicular to the circular base, the cone is a right circular cone.   This is also known as a cone of revolution, since it is generated by the hypotenuse of a right triangle rotated about one leg.   Any element of such a cone is the slant height.

The frustum of a cone is the part included between the base and a plane parallel to the base (Fig. 4-50).   The section also is called a base.

The altitude is the perpendicular distance between bases. The slant height is the portion of an element of the cone included between the two bases.

Every section of a cone made by a plane passing through its vertex is a triangle.

A section made by a plane parallel to the base of a circular cone is a circle. The axis of a circular cone passes through the center of this circle (see also Art. 6-49).

The lateral area of a cone of revolution equals one-half the product of the slant height and the circumference of the base: $S = \pi r l$, where r is the radius of the base and l is the slant height.

The volume V of a circular cone equals $\frac{1}{3}\pi r^2 h$, where r is the radius of the base and h is the altitude of the cone.

The lateral area of a frustum of a cone of revolution equals one-half the sum of the circumferences of its bases multiplied by the slant height l: $S = \pi l(r + r')$, where r and r' are the radii of the bases.

The volume V of the frustum of a cone of revolution equals $\frac{1}{3}\pi h(r^2 + r'^2 + rr')$, where h is the altitude of the frustum and r and r' are the radii of the bases.

4-42. Spheres. A sphere is a solid bounded by a surface with all points equidistant from a point within (center). A hemisphere is one-half of a sphere.

A great circle is the intersection of a spherical surface and a plane through the center of a sphere. Only one great circle can pass through two given points on a sphere if the points are not the extremities of a diameter. The length of the minor arc of the great circle passing through the two points is the shortest distance along the spherical surface between those points. It is called the spherical distance. One-fourth of a great circle is a quadrant.

A small circle is the intersection of a sphere and a plane not through the center of the sphere. The poles of the circle are the extremities of a diameter of the sphere perpendicular to the plane of the circle. The spherical distance from the nearer pole of the circle to any point on it is called the polar distance of the circle.

Only one circle can be drawn through any three points on a sphere.

An inscribed sphere is a sphere tangent to all the faces of a polyhedron. A circumscribed sphere is a sphere on which lie all the vertices of a polyhedron. A sphere may be inscribed in or circumscribed about any tetrahedron or cube.

Only one spherical surface can be passed through four points not in the same plane.

If a plane intersects a sphere, the intersection is a circle. Circles of a sphere equidistant from the center are equal. If two circles are made by planes at unequal distances from the center, the circle nearer the center is larger.

Parallel circles have the same poles.

All great circles of a sphere are equal. Every great circle bisects the spherical surface. Two great circles bisect each other.

If the planes of two great circles are perpendicular, each circle passes through the poles of the other.

The polar distance of a great circle is a quadrant. A point on a sphere at the distance of a quadrant from each of two other points, not the extremities of a diameter, is a pole of the great circle through those points.

A plane perpendicular to a radius at its extremity is tangent to the sphere.

The intersection of two spherical surfaces is a circle with its plane perpendicular to the line joining the centers of the spheres; the center of the circle lies on that line.

The area S of the surface of a sphere equals $\pi d^2 = 4\pi r^2$, where d is a diameter and r is a radius of the sphere.

A zone is a portion of a spherical surface included between two parallel planes (Fig. 4-51). It is generated by an arc of a great circle rotating

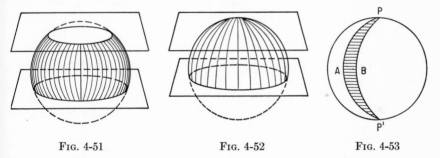

FIG. 4-51   FIG. 4-52   FIG. 4-53

about its diameter. If one plane is tangent to the sphere, a zone of one base is formed (Fig. 4-52). The altitude of a zone is the perpendicular distance between bases.

The area of a zone equals the product of its altitude and the circumference of the great circle of the sphere: $S = 2\pi rh$, where r is a radius of the sphere and h is the altitude of the zone.

A lune is the portion of a spherical surface bounded by two half great circles (Fig. 4-53). The angle of a lune is the angle between the bounding semicircles.

The area of a lune is to the area of the spherical surface as the angle of the lune is to 360°. Hence, the area S of a lune equals $\pi r^2\theta/90$, where r is a radius of the sphere and θ is the angle of the lune.

The volume of a sphere equals one-third the product of its area and radius. $V = \frac{1}{3}Sr = \frac{4}{3}\pi r^3 = \frac{1}{6}\pi d^3$, where S is the area of the sphere, r is the radius, and d is the diameter.

A spherical segment is a solid bounded by a zone of a sphere and its two parallel bases (Fig. 4-51) or a zone and one base (Fig. 4-52). The volume of a spherical segment equals one-half the sum of its bases multiplied by its altitude, plus the volume of a sphere with that altitude as diameter: $V = \frac{1}{6}\pi h(3r^2 + 3r'^2 + h^2)$, where h is the altitude and r and r' are the radii of the bases; if the segment has only one base, $r' = 0$.

A spherical sector is the portion of a sphere generated by the revolution of a circular sector about any diameter of the circle of which the

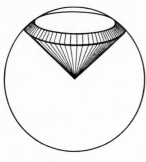

FIG. 4-54

sector is a part (Fig. 4-54). The zone generated by the arc is the base of the spherical sector. The volume of a spherical sector equals one-third the product of the area of the zone that forms its base and the radius of the sphere: $V = \frac{2}{3}\pi r^2 h$, where r is the radius of the sphere and h is the altitude of the zone.

4-43. Spherical Angles. A spherical angle is formed by two arcs of great circles emanating from a point. In Fig. 4-53, two great circles of a sphere, PAP' and PBP', intersect at P to form spherical angle APB and at P' to form spherical angle $AP'B$. As with plane angles, the point of intersection P or P' is called the vertex and the arcs are called the sides of the spherical angle.

A spherical angle is measured by the plane angle formed by the tangents to the arcs at the vertex. Thus, it has the same measure as the dihedral angle formed by the planes of its sides. A spherical angle also is measured by the arc of the great circle that has the vertex of the angle as pole and is included between the sides of the angle, extended if necessary.

4-44. Spherical Polygons. A spherical polygon is the portion of a spherical surface bounded by three or more arcs of great circles. If a convex polyhedral angle is formed at the center of the sphere by the planes of the sides of the spherical polygon, the polygon is convex. The points of intersection of the sides are the vertices of the polygon; the angles formed by the sides are the angles of the polygon.

The spherical excess of a polygon of n sides is the number of degrees the sum of the angles of the polygon exceeds $(n - 2)$ times 180°.

The sides of a spherical polygon, when measured in degrees, equal the corresponding face angles of the polyhedral angle formed by the planes of the sides. The angles of the polygon, in degrees, equal the corresponding dihedral angles of the polyhedral angle.

A spherical triangle is a spherical polygon with three sides. A polar triangle is a spherical triangle formed by arcs of great circles having as poles the vertices of another spherical triangle. The corresponding vertices lie on the same side of corresponding sides. In Fig. 4-55, spherical triangle ABC is the polar triangle of DEF. F is the pole of arc AB, D the pole of arc AC, and E is the pole of arc BC.

If one spherical triangle is the polar triangle of another, the second is the polar triangle of the first. Thus, in Fig. 4-55, DEF is the polar triangle of ABC and each angle of one is the supplement of the opposite side of the other. For example, in Fig. 4-55, angle A is the supplement of arc DF; angle F is the supplement of arc AB.

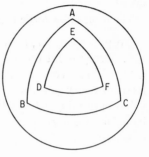

FIG. 4-55

A birectangular triangle is a spherical triangle with two right angles. A trirectangular triangle has three right angles.

Two spherical polygons on the same sphere or on equal spheres are symmetric if their corresponding parts are equal but arranged in reverse order. A symmetric polygon can be formed by drawing diameters to the vertices and connecting the other ends with arcs of great circles. In general, two symmetric polygons are not congruent; one is the reflection of the other.

Each side of a spherical triangle is less than the sum of the other two sides.

The sum of the sides of a spherical polygon is less than 360°.

The sum of the angles of a spherical triangle is greater than 180° and less than 540°. A spherical triangle may have one, two, or three right or obtuse angles. The sum of two angles of a spherical triangle must be less than 180° plus the third angle.

Two symmetric spherical triangles are equivalent in area.

Two triangles on the same sphere or on equal spheres are either congruent or symmetric if:

1. Two sides and the included angle of one are equal to two sides and the included angle of the other.

2. Two angles and the included side of one are equal to two angles and the included side of the other.

3. The sides of one equal the sides of the other. (Hence, the angles of one equal the angles of the other.)

4. The angles of one equal the angles of the other. (Hence, the corresponding sides are equal.)

The angles opposite the equal sides of an isosceles spherical triangle are equal. Conversely, the sides opposite equal angles of a spherical triangle are equal. If two sides are unequal, the angles opposite are unequal; the larger angle is opposite the longer side.

The area of a spherical polygon in spherical degrees equals its spherical excess.   In square units, the area $S = E\pi r^2/180$, where E is the spherical excess and r is the radius of the sphere.

4-45. Loci.   A locus consists of one or more straight or curved lines, planes, or surfaces that contain all points satisfying given conditions.

• What is the locus of all points equidistant from two given points?
The locus is the perpendicular bisector of the line segment joining the two points.

• What is the locus of all points inside a given angle and equidistant from its sides?
The locus is the angle bisector.

To prove a locus, show that:
1. Every point on the locus satisfies the given conditions.
2. Either every point that satisfies the given conditions lies on the locus or every point not on the locus does not satisfy the conditions.

To determine a locus that satisfies more than one condition, draw the locus for each condition.   The common intersection of the loci is the required locus.

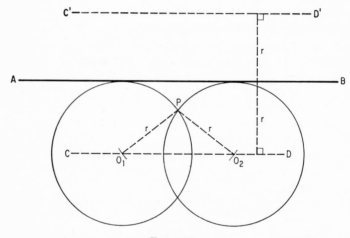

FIG. 4-56

• Draw a circle of given radius r passing through a given point P and tangent to a given line AB.

The center of the required circle can be found as the intersection of loci satisfying the two conditions:
1. Tangent to AB
2. Passing through P

The loci of the centers of all circles of radius r tangent to AB are CD and $C'D'$ in Fig. 4-56.   These are parallel to AB and at a distance r from it.   The locus of the centers of all circles of radius r passing through P is a circle of radius r with center at P. This circle intersects CD at O_1 and O_2; it does not meet $C'D'$. Since O_1 and O_2 are the intersections of the two loci, they are the centers of circles of radius r passing through P and tangent to AB.

4-46. Geometric Constructions. Only two instruments may be used: a straightedge and a compass.

To carry out a desired construction:

1. Assume the solution and draw on it in heavy lines the given components.

2. Find one or more parts of the figure (using construction lines if necessary) that are fully determined by the given components. Make these the basis of the construction.

3. Locate all other points on the figure.

4. Carry out the construction.

5. Prove that all requirements have been met.

4-47. How to Bisect an Angle. To divide an angle AOB into two equal parts, swing an arc CD with the vertex O as center (Fig. 4-57). (If the vertex is not accessible, draw inside the angle two lines parallel to the sides of the angle and at the same distance from them; then bisect the angle formed by these lines.) With C and D as centers and with a radius greater than one-half the distance CD swing two arcs intersecting at E. The straight line OE bisects the angle.

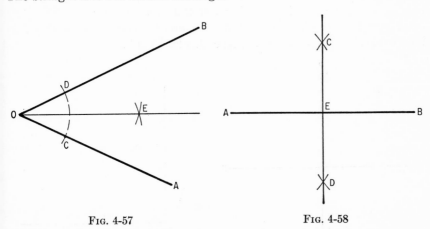

FIG. 4-57   FIG. 4-58

4-48. How to Bisect a Line. To divide a line AB into two equal parts, swing intersecting arcs with A and B as centers and with radius greater than one-half AB (Fig. 4-58). A line joining C and D, the intersections of the arcs, bisects AB.

4-49. How to Draw an Angle Equal to Another. On the given angle AOB (Fig. 4-59) swing an arc CD with O as center. Next, swing arc $C'E'$ with the same radius and with the center on a given point O', the vertex of the equal angle. Then, with C' as center and radius equal to the distance CD, swing an arc, intersecting arc $C'E'$ at D'. Draw $O'A'$ through D'. Angle $A'O'B'$ = angle AOB.

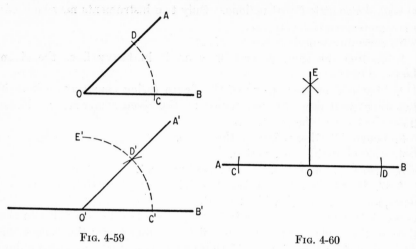

FIG. 4-59   FIG. 4-60

4-50. How to Draw a Perpendicular

1. At a point O on a line AB (Fig. 4-60). With O as center and any radius swing two arcs intersecting AB at C and D. With C and D as centers and any radius larger than the preceding one, swing two arcs intersecting at E. Line OE is perpendicular to AB.

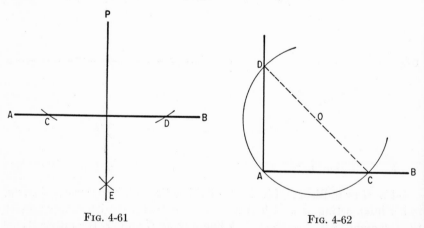

FIG. 4-61   FIG. 4-62

2. From a point P outside given line AB (Fig. 4-61). With P as center swing an arc intersecting AB at C and D. With C and D as centers and radius larger than one-half DC swing two arcs intersecting at E. Line PE is perpendicular to AB.

3. At an end point A on AB without prolonging the line (Fig. 4-62). Swing an arc with center at any point O (not on AB) and radius OA intersecting AB at C. Draw CO and extend it until it meets the arc at D. AD is perpendicular to AB. Or, construct a right triangle at A with sides in the ratio $3:4:5$.

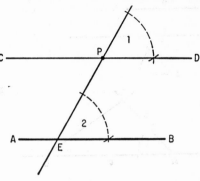

4-51. How to Draw a Parallel Line. Through a given point P (Fig. 4-63) draw a line intersecting at E given line AB to which a parallel is to be drawn. At P construct angle 1 = angle 2 (Art. 4-49). Side PD of angle 1 is parallel to AB.

FIG. 4-63

4-52. How to Construct a Triangle

1. Given three sides a, b, and c. (The sum of any two sides must be greater than the third.) Mark off $AC = b$ on a straight line (Fig. 4-64). With A as center and c as radius, swing an arc. With C as center and a as radius, swing another arc intersecting the first at B. Draw AB and BC to complete the triangle.

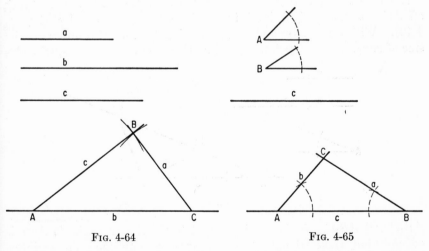

FIG. 4-64   FIG. 4-65

2. Given angles A and B and the included side c. Mark off $AB = c$ on a straight line (Fig. 4-65). Construct angle A at A and angle B at B (Art. 4-49). Prolong the sides of the angles until they meet at C, forming triangle ABC.

3. Given two sides b and c and the included angle A. Mark off $AB = c$ on a straight line (Fig. 4-66). Construct angle A at A (Art. 4-49).

Mark off $AC = b$ on the side of angle A. Draw BC to complete the triangle.

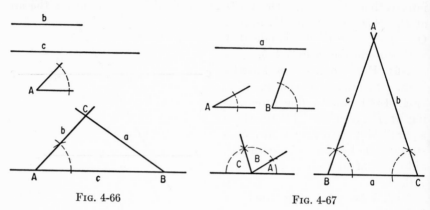

FIG. 4-66   FIG. 4-67

4. Given two angles A and B and an opposite side a. Add angles A and B and subtract from a straight angle to find angle C (Fig. 4-67). Mark off $BC = a$ on a straight line. Construct angle B at B and angle C at C (Art. 4-49). Prolong the sides of these angles until they meet at A, completing the triangle.

5. Given two sides a and b and angle A opposite one of them. Mark off $AC = b$ on a straight line (Fig. 4-68). Construct angle A at A (Art. 4-49). With C as center and radius a, swing an arc. If it intersects the side of angle A at two points B_1 and B_2 there are two solutions, triangles

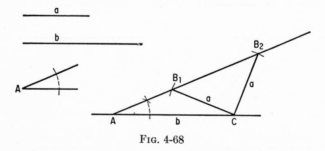

FIG. 4-68

AB_1C and AB_2C. If the arc is tangent to the side of angle A, there is one solution, right triangle ABC. If there is only one point of intersection, there is only one solution. If there is no intersection, there is no solution.

4-53. How to Divide a Line into Equal Parts. Draw AN' at an angle to the given line AN (Fig. 4-69). Start at A and use a convenient unit to mark off as many equal intervals on AN' as the number of parts into which AN is to be divided. Connect N', the end of the last interval, with N. At the end of each of the intervals draw a line parallel to NN' (Art. 4-51). These parallels mark off equal intervals on AN.

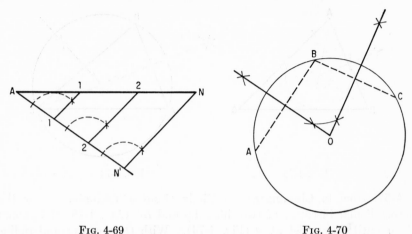

FIG. 4-69   FIG. 4-70

4-54. How to Locate the Center of a Circle. Select any three points A, B, and C on the circle (Fig. 4-70). Construct the perpendicular bisectors of the chords between them (Art. 4-48). The intersection of these bisectors is the center of the circle O.

4-55. How to Draw a Tangent to a Circle

1. At a point on the circle. Draw the radius to the given point T (Fig. 4-71). Erect a perpendicular to the radius at T (Art. 4-50.3).

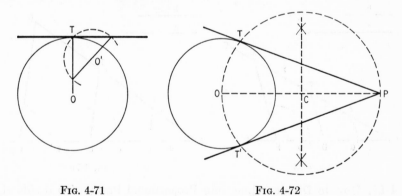

FIG. 4-71   FIG. 4-72

2. From a point outside the circle. Bisect the line from the given point P (Fig. 4-72) to the center O of the circle (Art. 4-48). With the midpoint C as center and with radius $OC = CP$, swing an arc intersecting the circle at T and T'. Lines PT and PT' are tangent to the circle.

4-56. How to Inscribe a Circle in a Triangle. Bisect any two angles A and B of the triangle (Art. 4-47). From the point of intersection O

(Fig. 4-73) draw a perpendicular OT to any side (Art. 4-50.2). With OT as radius and O as center, draw the inscribed circle.

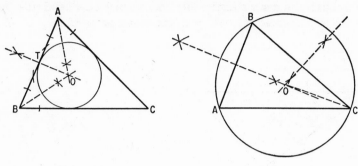

FIG. 4-73   FIG. 4-74

4-57. How to Circumscribe a Circle about a Triangle. Draw the perpendicular bisectors of two sides AB and BC (Art. 4-48) and extend them until they meet at O (Fig. 4-74). With O as center and radius equal to the distance from O to any vertex, draw the circumscribing circle.

4-58. How to Construct a Fourth Proportional. To find a line x such that $a:b = c:x$, lay off $AD = a$ and $DF = b$ on any line AB (Fig. 4-75). Draw AC at any acute angle to AB. Lay off $AE = c$. Draw DE; then draw a line parallel to it through F (Art. 4-51), intersecting AC at G. Then $EG = x$, the fourth proportional.

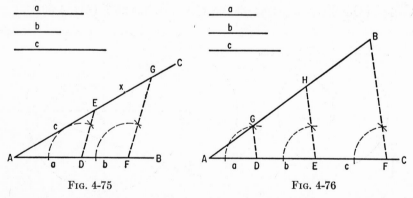

FIG. 4-75   FIG. 4-76

4-59. How to Divide a Line into Proportional Parts. To divide AB (Fig. 4-76) into parts proportional to lines a, b, c, . . . , draw line AC making any acute angle with AB. On AC, mark off $AD = a$, $DE = b$, $EF = c$, . . . ; connect the end of the last interval with B and then draw lines parallel to the connecting line through D,E,F, . . . , intersecting AB at G,H, . . . (Art. 4-51). AG, GH, and HB are the required parts, since

$$AG:GH:HB = a:b:c$$

4-60. How to Construct a Mean Proportional. To find a line x such that $a:x = x:b$, lay off $AC = a$ and $CD = b$ on any line AB (Fig. 4-77). On AD as diameter draw a semicircle. At C erect a perpendicular intersecting the semicircle at E. $EC = x$, the mean proportional.

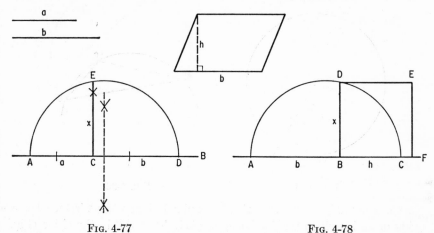

FIG. 4-77   FIG. 4-78

4-61. How to Draw a Square Equivalent to a Parallelogram. Construct the mean proportional x between the base b and altitude h of the parallelogram (Art. 4-60). Draw a square $BDEF$ with x as side (Fig. 4-78).

4-62. How to Inscribe a Square in a Circle. Draw any two diameters AC and BD perpendicular to each other (Fig. 4-79). The lines joining their extremities form an inscribed square $ABCD$.

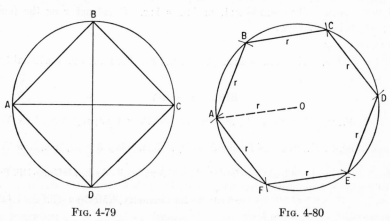

FIG. 4-79   FIG. 4-80

4-63. How to Inscribe a Regular Hexagon in a Circle. Start with a point A on the circle and mark off successive chords equal in length to the radius (Fig. 4-80). These chords form hexagon $ABCDEF$.

4-64. How to Inscribe a Regular Decagon in a Circle. Draw a radius to any point A on the circle (Fig. 4-81). Divide the radius in mean and extreme ratio (Art. 4-19). Start at A and mark off successive chords equal in length to the larger segment of the radius.

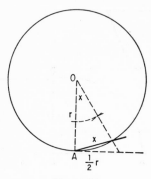

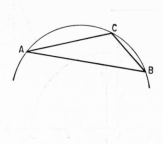

Fig. 4-81   Fig. 4-82

4-65. How to Inscribe a 15-sided Regular Polygon in a Circle. Draw a side of an inscribed regular hexagon (Art. 4-63) AB in the circle (Fig. 4-82). Next, from A, draw a side AC of the inscribed regular decagon (Art. 4-64). Then, BC is a side of the 15-sided polygon.

4-66. How to Multiply or Divide. Convert the product to a proportion, the division to a ratio.

• Multiply a by b.
Let $x = ab$. Then $1:b = a:x$. Construct x as the fourth proportional with the first term as unity (Art. 4-58).

• Divide a by b.
Let $x = a/b$. Then $a:b = x:1$, or $b:a = 1:x$. Construct x as the fourth proportional (Art. 4-58).

• Evaluate ab/c.
Let $x = ab/c$. Then $c:a = b:x$. Construct x as the fourth proportional (Art. 4-58).

BIBLIOGRAPHY

Adler, C. F.: "Modern Geometry," McGraw-Hill Book Company, Inc., New York, 1958.
Hemmerling, E. M.: "College Plane Geometry," John Wiley & Sons, Inc., New York, 1958.
Schnell, L. H., and Mildred Crawford: "Plane Geometry," 3d ed., McGraw-Hill Book Company, Inc., New York, 1953.
Schnell, L. H., and Mildred Crawford: "Solid Geometry," McGraw-Hill Book Company, Inc., New York, 1953.

Chapter 5

PLANE AND SPHERICAL TRIGONOMETRY

5-1. Angular Units. With 1° defined as $\frac{1}{90}$th of a right angle or $\frac{1}{180}$th of a straight angle (Art. 4-31), there are 360 degrees to a circle (sexagesimal system), 60 minutes to a degree, and 60 seconds to a minute.

In the centesimal system (French), a right angle is divided into 100 parts called grades. A minute is 0.01 grade. A circle contains 400 grades or 40,000 minutes.

In the circular or natural system, a circle is divided into 2π parts, called radians. Hence, the central angle of a circle measures one radian when the arc it intercepts is equal to the radius of the circle. Any central angle in radians equals $\theta = s/r$, where s is the intercepted arc and r is the radius.

$$1 \text{ radian} = \frac{180}{\pi} \text{ degrees (see Art. 1-7)}$$

$$1° = \frac{\pi}{180} \text{ radian}$$

A mil (artillery unit) is 1/6,400th of a circle (nearly 0.001 radian).

5-2. Size of Angles. Any angle may be considered as being formed by a fixed line (the base or axis) and another line (generatrix or radius vector) rotating counterclockwise about a fixed point (the vertex) on the axis (Fig. 5-1).

At the start, when the generatrix coincides with the axis, the angle is zero (degrees or radians). When the generatrix becomes perpendicular to the axis, the angle is a right angle, 90° or $\pi/2$ radians. When the generatrix becomes a continuation of the axis, the angle is a straight angle, 180° or π radians. When the generatrix again is perpendicular to the axis, the angle is 270° or $3\pi/2$ radians. When the generatrix returns to the axis, the angle is 360° or 2π radians. Every point on the rotating line, therefore, has generated a complete circle (Fig. 5-1e).

5-3. Quadrants. The circle generated by a point on a rotating line is divided into four quadrants for convenience in dealing with angles. The quadrants are numbered one to four counterclockwise (Fig. 5-2).

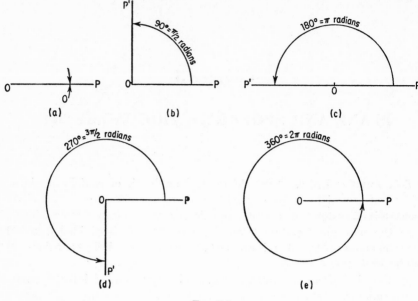

FIG. 5-1

The dividing lines XX' and YY' are considered a set of axes for cartesian coordinates; OX and OY are positive directions and OX' and OY' are negative directions.

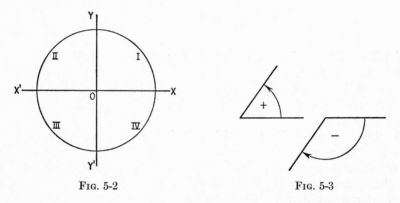

FIG. 5-2   FIG. 5-3

5-4. Sign of an Angle.   An angle is positive if its generatrix rotates about the vertex counterclockwise from one side, taken as the base, or axis, to the other side.   An angle is negative if the rotation is clockwise (Fig. 5-3).

5-5. Natural Trigonometric Functions.   Trigonometric functions of an angle may be defined in terms of a radius vector (Art. 5-2) and its

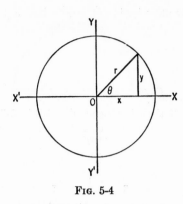

Fig. 5-4

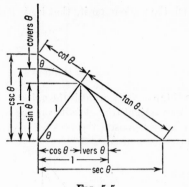

Fig. 5-5

cartesian coordinates (Art. 5-3). The radius vector is always taken as positive, but its components parallel to the axes are positive if they lie in the directions OX and OY and negative if in the directions OX' and OY' (Fig. 5-4).

$$\text{sine } \theta = \sin \theta = \frac{y}{r}$$

$$\text{cosine } \theta = \cos \theta = \frac{x}{r}$$

$$\text{tangent } \theta = \tan \theta = \frac{y}{x}$$

$$\text{cotangent } \theta = \cot \theta = \frac{x}{y}$$

$$\text{secant } \theta = \sec \theta = \frac{r}{x}$$

$$\text{cosecant } \theta = \csc \theta = \frac{r}{y}$$

$$\text{versine } \theta = \text{vers } \theta = \frac{r - x}{r}$$

$$\text{coversine } \theta = \text{covers } \theta = \frac{r - y}{r}$$

For graphical representation of these functions see Arts. 6-56 to 6-59.

If the radius vector is taken as unity, the natural trigonometric functions may be represented geometrically as in Fig. 5-5.

• If $\sin \theta = x$, what is the value of $\cos \theta$? $\tan \theta$?

Draw a right triangle with hypotenuse unity and one leg equal to x (Fig. 5-6).

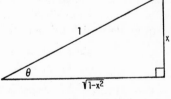

Fig. 5-6

Then, the angle opposite that leg is θ and the other leg is $\sqrt{1-x^2}$. Hence

$$\cos \theta = \frac{\sqrt{1-x^2}}{1} = \sqrt{1-x^2}$$

$$\tan \theta = \frac{x}{\sqrt{1-x^2}}$$

• If $\tan \varphi = \frac{3}{4}$, what is the value of $\sin \varphi$? $\cos \varphi$?
Draw a right triangle with one leg equal to 3, the other leg equal to 4 (Fig. 5-7). The angle opposite the first leg is φ and the hypotenuse equals 5. Therefore, $\sin \varphi = \frac{3}{5}$ and $\cos \varphi = \frac{4}{5}$.

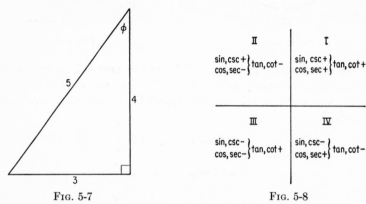

FIG. 5-7 FIG. 5-8

5-6. Sign of a Trigonometric Function. The sign of a trigonometric function of an angle depends on the quadrant in which the radius vector is located (Art. 5-3).

The sine and cosecant of an angle always have the same sign. So do the cosine and secant and the tangent and cotangent. The tangent is positive if the sine and cosine have like signs, negative if they have unlike signs.

Since the radius vector is always considered positive, the sign of a function depends only on the signs of the components of the radius vector (Fig. 5-4). The signs of the functions in each of the four quadrants are shown in Fig. 5-8.

5-7. Functions of Negative Angles. The natural trigonometric functions of a negative angle are equal to the same functions of a positive angle taken equal to 360° minus the absolute value of the angle in degrees or 2π minus the absolute value in radians. A comparison of the functions of an angle θ with those of $2\pi - \theta$, using Fig. 5-8, shows that:

$$\sin (-\theta) = -\sin \theta \qquad \csc (-\theta) = -\csc \theta$$
$$\cos (-\theta) = \cos \theta \qquad \sec (-\theta) = \sec \theta$$
$$\tan (-\theta) = -\tan \theta \qquad \cot (-\theta) = -\cot \theta$$

5-8. Values of Trigonometric Functions. Tables giving the values of trigonometric functions and their logarithms are available. A short table is presented here.   Also, some slide rules show trigonometric functions.   Electronic digital computers generally use subroutines that evaluate the functions by converging approximations.   Usually, when the functions are computed manually, a few terms of an infinite series are used (Art. 12-13).

SOME ANGLES AND THEIR FUNCTIONS

θ, degrees	θ, radians	$\sin \theta$	$\cos \theta$	$\tan \theta$	$\cot \theta$	$\sec \theta$	$\csc \theta$
0	0	0	1	0	∞	1	∞
30	$\dfrac{\pi}{6}$	$\dfrac{1}{2}$	$\dfrac{\sqrt{3}}{2}$	$\dfrac{\sqrt{3}}{3}$	$\sqrt{3}$	$\dfrac{2}{3}\sqrt{3}$	2
45	$\dfrac{\pi}{4}$	$\dfrac{\sqrt{2}}{2}$	$\dfrac{\sqrt{2}}{2}$	1	1	$\sqrt{2}$	$\sqrt{2}$
60	$\dfrac{\pi}{3}$	$\dfrac{\sqrt{3}}{2}$	$\dfrac{1}{2}$	$\sqrt{3}$	$\dfrac{\sqrt{3}}{3}$	2	$\dfrac{2}{3}\sqrt{3}$
90	$\dfrac{\pi}{2}$	1	0	∞	0	∞	1
120	$\dfrac{2\pi}{3}$	$\dfrac{\sqrt{3}}{2}$	$-\dfrac{1}{2}$	$-\sqrt{3}$	$-\dfrac{\sqrt{3}}{3}$	-2	$\dfrac{2}{3}\sqrt{3}$
135	$\dfrac{3\pi}{4}$	$\dfrac{\sqrt{2}}{2}$	$-\dfrac{\sqrt{2}}{2}$	-1	-1	$-\sqrt{2}$	$\sqrt{2}$
150	$\dfrac{5\pi}{6}$	$\dfrac{1}{2}$	$-\dfrac{\sqrt{3}}{2}$	$-\dfrac{\sqrt{3}}{3}$	$-\sqrt{3}$	$-\dfrac{2}{3}\sqrt{3}$	2
180	π	0	-1	0	∞	-1	∞
210	$\dfrac{7\pi}{6}$	$-\dfrac{1}{2}$	$-\dfrac{\sqrt{3}}{2}$	$\dfrac{\sqrt{3}}{3}$	$\sqrt{3}$	$-\dfrac{2}{3}\sqrt{3}$	-2
225	$\dfrac{5\pi}{4}$	$-\dfrac{\sqrt{2}}{2}$	$-\dfrac{\sqrt{2}}{2}$	1	1	$-\sqrt{2}$	$-\sqrt{2}$
240	$\dfrac{4\pi}{3}$	$-\dfrac{\sqrt{3}}{2}$	$-\dfrac{1}{2}$	$\sqrt{3}$	$\dfrac{\sqrt{3}}{3}$	-2	$-\dfrac{2}{3}\sqrt{3}$
270	$\dfrac{3\pi}{2}$	-1	0	∞	0	∞	-1
300	$\dfrac{5\pi}{3}$	$-\dfrac{\sqrt{3}}{2}$	$\dfrac{1}{2}$	$-\sqrt{3}$	$-\dfrac{\sqrt{3}}{3}$	2	$-\dfrac{2}{3}\sqrt{3}$
315	$\dfrac{7\pi}{4}$	$-\dfrac{\sqrt{2}}{2}$	$\dfrac{\sqrt{2}}{2}$	-1	-1	$\sqrt{2}$	$-\sqrt{2}$
330	$\dfrac{11\pi}{6}$	$-\dfrac{1}{2}$	$\dfrac{\sqrt{3}}{2}$	$-\dfrac{\sqrt{3}}{3}$	$-\sqrt{3}$	$\dfrac{2}{3}\sqrt{3}$	-2
360	2π	0	1	0	∞	1	∞

The sine scale S on a slide rule should be used with the A and B scales. For angles less than 30′, use the value of the angle in radians.   For cosines, use 90° minus the angle on the S scale.

The tangent scale should be used with the C and D scales.   For angles less than 6°, use the angle in radians.   For angles greater than 45°, use the reciprocal of the tangent of 90° minus the angle.

For cotangents of angles up to 45°, use the reciprocal of the tangent of the angle. For angles greater than 45°, use the tangent of 90° minus the angle.

5-9. Functions of Small Angles. When angle θ is small, $\sin \theta$ and $\tan \theta$ may be replaced by θ in radians.

$\sin \theta$ will be given accurately by θ to three decimal places up to about 8°, to four decimal places up to about 4°, and to five decimal places up to about 2°. If greater accuracy is needed, use $\sin \theta = \theta - \theta^3/6$.

$\tan \theta$ will be given accurately by θ to three decimal places up to about $6\frac{1}{2}°$, to four decimal places up to about 3°, and to five decimal places up to about 1°. If greater accuracy is needed, use $\tan \theta = \theta + \theta^3/3$.

$\cos \theta$ can be replaced by 1 with an accuracy of three decimal places up to 1° 50' and of four decimal places to 35'. If greater accuracy is needed, use $\cos \theta = 1 - \theta^2/2$, where θ is in radians. This is accurate to three decimal places up to about 19°, to four decimal places up to about $10\frac{1}{2}°$, and to five decimal places up to about 6°. If still greater accuracy is needed, use $\cos \theta = 1 - \theta^2/2 + \theta^4/24$.

5-10. Relationships between Complementary Angles. Since θ is complementary to $\pi/2 - \theta$:

$$\sin \theta = \cos \left(\frac{\pi}{2} - \theta \right)$$

$$\tan \theta = \cot \left(\frac{\pi}{2} - \theta \right)$$

$$\sec \theta = \csc \left(\frac{\pi}{2} - \theta \right)$$

5-11. Relationships between Trigonometric Functions.   (See Fig. 5-5.)

$\sin \theta =$	$\sin \theta$	$\sqrt{1 - \cos^2 \theta}$	$\dfrac{\tan \theta}{\sqrt{1 + \tan^2 \theta}}$	$\dfrac{1}{\sqrt{1 + \cot^2 \theta}}$	$\dfrac{\sqrt{\sec^2 \theta - 1}}{\sec \theta}$	$\dfrac{1}{\csc \theta}$
$\cos \theta =$	$\sqrt{1 - \sin^2 \theta}$	$\cos \theta$	$\dfrac{1}{\sqrt{1 + \tan^2 \theta}}$	$\dfrac{\cot \theta}{\sqrt{1 + \cot^2 \theta}}$	$\dfrac{1}{\sec \theta}$	$\dfrac{\sqrt{\csc^2 \theta - 1}}{\csc \theta}$
$\tan \theta =$	$\dfrac{\sin \theta}{\sqrt{1 - \sin^2 \theta}}$	$\dfrac{\sqrt{1 - \cos^2 \theta}}{\cos \theta}$	$\tan \theta$	$\dfrac{1}{\cot \theta}$	$\sqrt{\sec^2 \theta - 1}$	$\dfrac{1}{\sqrt{\csc^2 \theta - 1}}$
$\cot \theta =$	$\dfrac{\sqrt{1 - \sin^2 \theta}}{\sin \theta}$	$\dfrac{\cos \theta}{\sqrt{1 - \cos^2 \theta}}$	$\dfrac{1}{\tan \theta}$	$\cot \theta$	$\dfrac{1}{\sqrt{\sec^2 \theta - 1}}$	$\sqrt{\csc^2 \theta - 1}$
$\sec \theta =$	$\dfrac{1}{\sqrt{1 - \sin^2 \theta}}$	$\dfrac{1}{\cos \theta}$	$\sqrt{1 + \tan^2 \theta}$	$\dfrac{\sqrt{1 + \cot^2 \theta}}{\cot \theta}$	$\sec \theta$	$\dfrac{\csc \theta}{\sqrt{\csc^2 \theta - 1}}$
$\csc \theta =$	$\dfrac{1}{\sin \theta}$	$\dfrac{1}{\sqrt{1 - \cos^2 \theta}}$	$\dfrac{\sqrt{1 + \tan^2 \theta}}{\tan \theta}$	$\sqrt{1 + \cot^2 \theta}$	$\dfrac{\sec \theta}{\sqrt{\sec^2 \theta - 1}}$	$\csc \theta$

The relationships in this table are based on the assumption that θ is positive and in the first quadrant. For negative angles, the sign of θ should be determined as explained in Art. 5-7, and for angles other than those in the first quadrant, signs should be determined as indicated in Fig. 5-8.

5-12. Functions of Angles Larger than 360°. If the angle is in degrees, divide it by 360; if it is in radians, divide it by 2π. Find the required function of the remainder (see also Art. 5-13).

• What is the value of sin 750 deg?

$750/360 = 2 + 30/360$, so $\sin 750° = \sin 30° = \frac{1}{2}$.

• What is the value of $\cos 15\pi/4$?

$15\pi/4 \times 1/2\pi = 2 - (\pi/4)/2\pi$. Hence, $\cos 15\pi/4 = \cos(-\pi/4) = \cos \pi/4 = \sqrt{2}/2$.

5-13. Functions of Angles Larger than 90°. If the angle is greater than 360° or 2π radians, reduce it to an equivalent angle less than this (Art. 5-12). Determine the quadrant of the equivalent angle and from that the sign of the function. Write the equivalent (or the original) angle in the form $90n \pm \theta$ if it is in degrees, or $n\pi/2 \pm \theta$ if in radians, where n is an integer. If n is even, use the same function of θ as for the given angle; if n is odd, use the cofunction.

FUNCTIONS OF ANGLES IN TERMS OF FUNCTIONS OF AN ACUTE ANGLE

Function	$-\theta$	$\frac{\pi}{2} \pm \theta$	$\pi \pm \theta$	$\frac{3\pi}{2} \pm \theta$	$2k\pi \pm \theta$
$\theta =$	$-\theta$	$90° \pm \theta$	$180° \pm \theta$	$270° \pm \theta$	$k(360)° \pm \theta$
$\sin \theta =$	$-\sin \theta$	$+\cos \theta$	$\mp \sin \theta$	$-\cos \theta$	$\pm \sin \theta$
$\cos \theta =$	$+\cos \theta$	$\mp \sin \theta$	$-\cos \theta$	$\pm \sin \theta$	$+\cos \theta$
$\tan \theta =$	$-\tan \theta$	$\mp \cot \theta$	$\pm \tan \theta$	$\mp \cot \theta$	$\pm \tan \theta$
$\cot \theta =$	$-\cot \theta$	$\mp \tan \theta$	$\pm \cot \theta$	$\mp \tan \theta$	$\pm \cot \theta$
$\sec \theta =$	$+\sec \theta$	$\mp \csc \theta$	$-\sec \theta$	$\pm \csc \theta$	$+\sec \theta$
$\csc \theta =$	$-\csc \theta$	$+\sec \theta$	$\mp \csc \theta$	$-\sec \theta$	$\pm \csc \theta$

• What is the value of $\cos 15\pi/4$?

$15\pi/4 \times 1/2\pi = 1 + (7\pi/4)/2\pi$. Hence, $\cos 15\pi/4 = \cos 7\pi/4 = \cos(3\pi/2 + \pi/4)$. Since the coefficient of $\pi/2$ is 3, which is odd, the cofunction should be used. Note also that the angle is in the fourth quadrant, where the cosine is positive (Fig. 5-8), so that

$$\cos \frac{15\pi}{4} = \cos\left(\frac{3\pi}{2} + \frac{\pi}{4}\right) = \sin \frac{\pi}{4} = \frac{\sqrt{2}}{2}$$

• What is the value of sin 410°?

$410/360 = 1 + 150/360$. Hence, $\sin 410° = \sin 150° = \sin(2 \times 90° - 30°)$.

Here the coefficient of 90° is 2, which is even; the function remains the same. The angle is in the second quadrant, where the sine is positive.

$$\sin 410° = \sin (2 \times 90° - 30°) = \sin 30° = \tfrac{1}{2}$$

• What is the value of $\tan (-660°)$?

$^{660}\!/_{360} = 1 + {}^{300}\!/_{360}$, so $\tan (-660°) = \tan (-300°) = -\tan (3 \times 90° + 30°)$. The coefficient of 90° is odd, so the cofunction should be used. The angle is in the fourth quadrant, where the tangent is negative. Hence

$$\tan (-660°) = -\tan (3 \times 90° + 30°) = -(-\cot 30°) = \cot 30° = \sqrt{3}$$

5-14. Functions of the Sum of Two Angles.

$$\sin (\alpha + \beta) = \sin \alpha \cos \beta + \cos \alpha \sin \beta$$
$$\cos (\alpha + \beta) = \cos \alpha \cos \beta - \sin \alpha \sin \beta$$
$$\tan (\alpha + \beta) = \frac{\tan \alpha + \tan \beta}{1 - \tan \alpha \tan \beta}$$
$$\cot (\alpha + \beta) = \frac{\cot \alpha \cot \beta - 1}{\cot \alpha + \cot \beta}$$

5-15. Functions of the Difference of Two Angles

$$\sin (\alpha - \beta) = \sin \alpha \cos \beta - \cos \alpha \sin \beta$$
$$\cos (\alpha - \beta) = \cos \alpha \cos \beta + \sin \alpha \sin \beta$$
$$\tan (\alpha - \beta) = \frac{\tan \alpha - \tan \beta}{1 + \tan \alpha \tan \beta}$$
$$\cot (\alpha - \beta) = \frac{\cot \alpha \cot \beta + 1}{\cot \beta - \cot \alpha}$$

5-16. Functions of Multiples of an Angle

$$\sin 2\theta = 2 \sin \theta \cos \theta = \frac{2 \tan \theta}{1 + \tan^2 \theta}$$
$$\sin 3\theta = 3 \sin \theta - 4 \sin^3 \theta$$
$$\sin 4\theta = 4 \cos \theta(\sin \theta - 2 \sin^3 \theta)$$
$$\sin 5\theta = 5 \sin \theta - 20 \sin^3 \theta + 16 \sin^5 \theta$$
$$\sin 6\theta = \cos \theta(\sin \theta - 32 \sin^3 \theta + 32 \sin^5 \theta)$$
$$\sin n\theta = n \sin \theta - \frac{n(n^2 - 1^2)}{3!} \sin^3 \theta + \frac{n(n^2 - 1)(n^2 - 3^2)}{5!} \sin^5 \theta$$
$$\qquad - \frac{n(n^2 - 1^2)(n^2 - 3^2)(n^2 - 5^2)}{7!} \sin^7 \theta + \cdots \qquad n \text{ odd}$$
$$\sin n\theta = n \cos \theta \left[\sin \theta - \frac{(n^2 - 2^2)}{3!} \sin^3 \theta + \frac{(n^2 - 2^2)(n^2 - 4^2)}{5!} \sin^5 \theta \right.$$
$$\qquad \left. - \frac{(n^2 - 2^2)(n^2 - 4^2)(n^2 - 6^2)}{7!} \sin^7 \theta + \cdots \right] \qquad n \text{ even}$$
$$\cos 2\theta = \cos^2 \theta - \sin^2 \theta = 1 - 2 \sin^2 \theta = 2 \cos^2 \theta - 1$$
$$\cos 3\theta = 4 \cos^3 \theta - 3 \cos \theta$$

$\cos 4\theta = 8 \cos^4 \theta - 8 \cos^2 \theta + 1$

$\cos 5\theta = 16 \cos^5 \theta - 20 \cos^3 \theta + 5 \cos \theta$

$\cos 6\theta = 32 \cos^6 \theta - 48 \cos^4 \theta + 18 \cos^2 \theta - 1$

$\cos n\theta = 2^{n-1} \cos^n \theta - 2^{n-3}n \cos^{n-2} \theta + 2^{n-5} \dfrac{n(n-3)}{2!} \cos^{n-4} \theta$

$\qquad - 2^{n-7} \dfrac{n(n-4)(n-5)}{3!} \cos^{n-6} \theta + 2^{n-9} \dfrac{n(n-5)(n-6)(n-7)}{4!} \cos^{n-8} \theta$

$\qquad\qquad\qquad\qquad\qquad\qquad\qquad\qquad\qquad\qquad\qquad - \cdots$

$\tan 2\theta = \dfrac{2 \tan \theta}{1 - \tan^2 \theta}$

$\tan 3\theta = \dfrac{3 \tan \theta - \tan^3 \theta}{1 - 3 \tan^2 \theta}$

$\tan 4\theta = \dfrac{4 \tan \theta - 4 \tan^3 \theta}{1 - 6 \tan^2 \theta + \tan^4 \theta}$

$\cot 2\theta = \dfrac{\cot^2 \theta - 1}{2 \cot \theta}$

$\cot 3\theta = \dfrac{\cot^3 \theta - 3 \cot \theta}{3 \cot^2 \theta - 1}$

$\cot 4\theta = \dfrac{\cot^4 \theta - 6 \cot^2 \theta + 1}{4 \cot^3 \theta - 4 \cot \theta}$

5-17. Functions of Half an Angle

$$\sin \tfrac{1}{2}\theta = \pm \sqrt{\dfrac{1 - \cos \theta}{2}}$$

$$\cos \tfrac{1}{2}\theta = \pm \sqrt{\dfrac{1 + \cos \theta}{2}}$$

$$\tan \tfrac{1}{2}\theta = \pm \sqrt{\dfrac{1 - \cos \theta}{1 + \cos \theta}} = \dfrac{1 - \cos \theta}{\sin \theta} = \dfrac{\sin \theta}{1 + \cos \theta}$$

$$\cot \tfrac{1}{2}\theta = \pm \sqrt{\dfrac{1 + \cos \theta}{1 - \cos \theta}} = \dfrac{\sin \theta}{1 - \cos \theta} = \dfrac{1 + \cos \theta}{\sin \theta}$$

5-18. Sums and Differences of Functions.

The following formulas, expressing sums and differences of trigonometric functions as products, are suitable for logarithmic computations:

$$\sin \alpha + \sin \beta = 2 \sin \tfrac{1}{2}(\alpha + \beta) \cos \tfrac{1}{2}(\alpha - \beta)$$

$$\sin \alpha - \sin \beta = 2 \cos \tfrac{1}{2}(\alpha + \beta) \sin \tfrac{1}{2}(\alpha - \beta)$$

$$\cos \alpha + \cos \beta = 2 \cos \tfrac{1}{2}(\alpha + \beta) \cos \tfrac{1}{2}(\alpha - \beta)$$

$$\cos \alpha - \cos \beta = 2 \sin \tfrac{1}{2}(\alpha + \beta) \sin \tfrac{1}{2}(\beta - \alpha)$$

$$\sin^2 \alpha - \sin^2 \beta = \sin (\alpha + \beta) \sin (\alpha - \beta)$$

$$\cos^2 \alpha - \sin^2 \beta = \cos^2 \beta - \sin^2 \alpha = \cos (\alpha + \beta) \cos (\alpha - \beta)$$

$$\sin \theta + \cos \theta = \sqrt{2} \sin \left(\theta + \frac{\pi}{4} \right) = \sqrt{2} \cos \left(\theta - \frac{\pi}{4} \right)$$

$$\sec^2 \theta + \csc^2 \theta = \sec^2 \theta \csc^2 \theta = \frac{1}{\sin^2 \theta \cos^2 \theta}$$

$$a \sin \theta + b \cos \theta = \sqrt{a^2 + b^2} \sin (\theta + \varphi) = \sqrt{a^2 + b^2} \cos (\theta - \psi)$$

where

$$\varphi = \cos^{-1} \frac{a}{\sqrt{a^2 + b^2}} = \sin^{-1} \frac{b}{\sqrt{a^2 + b^2}}$$

$$\psi = \sin^{-1} \frac{a}{\sqrt{a^2 + b^2}} = \cos^{-1} \frac{b}{\sqrt{a^2 + b^2}}$$

$$\tan \alpha + \tan \beta = \frac{\sin (\alpha + \beta)}{\cos \alpha \cos \beta}$$

$$\tan \alpha - \tan \beta = \frac{\sin (\alpha - \beta)}{\cos \alpha \cos \beta}$$

$$\cot \alpha + \cot \beta = \frac{\sin (\alpha + \beta)}{\sin \alpha \sin \beta}$$

$$\cot \alpha - \cot \beta = \frac{\sin (\beta - \alpha)}{\sin \alpha \sin \beta}$$

$$\tan \alpha + \cot \beta = \frac{\cos (\alpha - \beta)}{\cos \alpha \sin \beta}$$

$$\cot \alpha - \tan \beta = \frac{\cos (\alpha + \beta)}{\sin \alpha \cos \beta}$$

5-19. Products of Functions

$$\sin \alpha \cos \beta = \tfrac{1}{2} \sin (\alpha + \beta) + \tfrac{1}{2} \sin (\alpha - \beta)$$
$$\cos \alpha \sin \beta = \tfrac{1}{2} \sin (\alpha + \beta) - \tfrac{1}{2} \sin (\alpha - \beta)$$
$$\cos \alpha \cos \beta = \tfrac{1}{2} \cos (\alpha + \beta) + \tfrac{1}{2} \cos (\alpha - \beta)$$
$$\sin \alpha \sin \beta = -\tfrac{1}{2} \cos (\alpha + \beta) + \tfrac{1}{2} \cos (\alpha - \beta)$$

5-20. Powers of Functions

$$\sin^2 \theta = \tfrac{1}{2}(-\cos 2\theta + 1)$$
$$\sin^3 \theta = \tfrac{1}{4}(-\sin 3\theta + 3 \sin \theta)$$
$$\sin^4 \theta = \tfrac{1}{8}(\cos 4\theta - 4 \cos 2\theta + 3)$$
$$\sin^5 \theta = \tfrac{1}{16}(\sin 5\theta - 5 \sin 3\theta + 10 \sin \theta)$$
$$\sin^6 \theta = \tfrac{1}{32}(-\cos 6\theta + 6 \cos 4\theta - 15 \cos 2\theta + 10)$$
$$\cos^2 \theta = \tfrac{1}{2}(\cos 2\theta + 1)$$
$$\cos^3 \theta = \tfrac{1}{4}(\cos 3\theta + 3 \cos \theta)$$
$$\cos^4 \theta = \tfrac{1}{8}(\cos 4\theta + 4 \cos 2\theta + 3)$$
$$\cos^5 \theta = \tfrac{1}{16}(\cos 5\theta + 5 \cos 3\theta + 10 \cos \theta)$$
$$\cos^6 \theta = \tfrac{1}{32}(\cos 6\theta + 6 \cos 4\theta + 15 \cos 2\theta + 10)$$

5-21. Law of Sines. In any triangle, the sides are proportional to the sines of the opposite angles. In triangle ABC (Fig. 5-9), for example,

$$\frac{a}{\sin A} = \frac{b}{\sin B} = \frac{c}{\sin C} = d$$

The constant d equals the diameter of the circumscribed circle.

• If a triangle has one angle equal to 45° and another equal to 60°, and the included side is 100 ft long, how long are the other two sides?

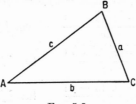

Fig. 5-9

In triangle ABC (Fig. 5-9), let $A = 45°$, $B = 60°$, and $c = 100$. Then $C = 180° - 45° - 60° = 75°$.

$$\frac{a}{\sin 45} = \frac{b}{\sin 60} = \frac{100}{\sin 75}$$

$$a = \frac{\sin 45}{\sin 75} \times 100 = \frac{0.707}{0.966} \times 100 = 73.2 \text{ ft}$$

$$b = \frac{\sin 60}{\sin 75} \times 100 = \frac{0.866}{0.966} \times 100 = 89.7 \text{ ft}$$

5-22. Law of Cosines. In any triangle, the square of a side equals the sum of the squares of the other sides minus twice the product of these sides and the cosine of their included angle. In triangle ABC (Fig. 5-9):

$$a^2 = b^2 + c^2 - 2bc \cos A$$

• If angle C of triangle ABC is 75°, $a = 73.2$ ft, and $b = 89.7$ ft, how long is side c?

$$c^2 = a^2 + b^2 - 2ab \cos C = (73.2)^2 + (89.7)^2 - 2(73.2)(89.7) \cos 75°$$
$$= 5,360 + 8,040 - 3,400 = 10,000$$
$$c = 100 \text{ ft}$$

5-23. Law of Tangents. In any triangle, the sum of any two sides is to their difference as the tangent of half the sum of the opposite angles is to the tangent of half the difference of these angles. In triangle ABC (Fig. 5-9):

$$\frac{a + c}{a - c} = \frac{\tan \frac{1}{2}(A + C)}{\tan \frac{1}{2}(A - C)}$$

• Given $a = 206$ ft, $c = 111$ ft, and $B = 31°22'$, compute A and C.

$A + C = 180° - B = 180° - 31°22' = 148°38'$.
$\frac{1}{2}(A + C) = 74°19'$

$$\tan \frac{1}{2}(A - C) = \frac{a - c}{a + c} \tan \frac{1}{2}(A + C) = \frac{206 - 111}{206 + 111} \tan 74°19' = 1.067$$

$\frac{1}{2}(A - C) = 46°52'$
$A = \frac{1}{2}(A + C) + \frac{1}{2}(A - C) = 74°19' + 46°52' = 121°11'$.
$C = \frac{1}{2}(A + C) - \frac{1}{2}(A - C) = 74°19' - 46°52' = 27°27'$.

5-24. Functions of an Angle in Terms of Three Sides of a Triangle.
In triangle ABC (Fig. 5-9), let $s = \frac{1}{2}(a + b + c)$ and let the radius r of the inscribed circle equal $\sqrt{(s - a)(s - b)(s - c)/s}$. Then

$$\sin \tfrac{1}{2}A = \sqrt{\frac{(s - b)(s - c)}{bc}} \qquad \cos \tfrac{1}{2}C = \sqrt{\frac{s(s - c)}{ab}}$$

$$\sin \tfrac{1}{2}B = \sqrt{\frac{(s - a)(s - c)}{ac}} \qquad \tan \tfrac{1}{2}A = \frac{r}{s - a}$$

$$\sin \tfrac{1}{2}C = \sqrt{\frac{(s - a)(s - b)}{ab}} \qquad \tan \tfrac{1}{2}B = \frac{r}{s - b}$$

$$\cos \tfrac{1}{2}A = \sqrt{\frac{s(s - a)}{bc}} \qquad \tan \tfrac{1}{2}C = \frac{r}{s - c}$$

$$\cos \tfrac{1}{2}B = \sqrt{\frac{s(s - b)}{ac}}$$

(See also Art. 5-26.5.)

5-25. Mollweide's Equations. These equations, useful for checking solutions of triangles and for logarithmic computations, involve the three sides and angles of a triangle.

$$\frac{a - b}{c} = \frac{\sin \tfrac{1}{2}(A - B)}{\cos \tfrac{1}{2}C}$$
$$\frac{a + b}{c} = \frac{\cos \tfrac{1}{2}(A - B)}{\sin \tfrac{1}{2}C}$$

• Check the solution of triangle ABC in Art. 5-21.
Given: $A = 45°$, $B = 60°$, $C = 75°$, $a = 73.2$, $b = 89.7$, and $c = 100$.

$$A - B = 45 - 60 = -15 \qquad \tfrac{1}{2}(A - B) = -15 \times \tfrac{1}{2} = -7.5$$
$$\tfrac{1}{2}C = \tfrac{1}{2} \times 75 = 37.5$$

From the first Mollweide equation,

$$\frac{73.2 - 89.7}{100} = \frac{\sin(-7.5)}{\cos(75/2)} = \frac{-\sin 7.5}{\cos 37.5}$$
$$-0.165 = -0.165$$

From the second Mollweide equation,

$$\frac{73.2 + 89.7}{100} = \frac{\cos(-7.5)}{\sin 37.5} = \frac{\cos 7.5}{\sin 37.5}$$
$$1.629 = 1.629$$

5-26. Solution of Plane Triangles. Sketch the triangle. List the given parts and those to be computed. Select the trigonometric formulas to be used, and solve them for the required parts. Substitute numerical data in the formulas and check the answers by formulas not used in the solution. It is desirable not to use a computed value in the calculation of another value.

1. Given two sides or an angle and a side of a right triangle. If two sides are given, use the Pythagorean theorem (Art. 4-24). If an angle and a side are known, the unknown angle can be found by subtracting the given angle from 90°. The unknown side can be found by solving a trigonometric function of one of the angles expressed as the ratio of the unknown side to a known side. Angles can be determined, when sides are given, from the trigonometric functions represented by the ratios of the sides (Art. 5-5).

• If a leg of a right triangle is 325 ft long and the angle opposite it is 47°30′, find the other acute angle, the other leg, and the hypotenuse.

In triangle ABC (Fig. 5-10), let $a = 325$ and $A = 47°30′$; find b, c, and B.

$$b = a \cot A = 325 \cot 47°30′ = 298 \text{ ft}$$

FIG. 5-10

(This solution could be obtained by slide rule, using D and T scales with the complement of angle A, or by looking up $\cot 47°30′ = 0.91633$ in a table of trigonometric functions. It also can be computed by logarithms.)

$$c = \frac{a}{\sin A} = \frac{325}{\sin 47°30′} = 441 \text{ ft}$$
$$B = 90° - A = 90° - 47°30′ = 42°30′$$

Check: $b = c \sin B = 441 \sin 42°30′ = 298 \text{ ft}$

• If one angle of a right triangle is equal to ½° and the adjacent leg is 7,163 ft long, what is the length of the other leg to four significant figures?

In triangle ABC (Fig. 5-10), let $A = $ ½° and $b = 7,163$; find a.

$$a = b \tan A = 7,163 \tan \text{½}°$$

Since A is small, substitute its value in radians for its tangent (Art. 5-9).

$$a = 7,163 \times \text{½} \times 0.01745 = 62.5 \text{ ft.}$$

2. Given two angles and a side of an oblique triangle. Find the third angle by subtracting the sum of the given angles from 180°. Then use the law of sines (Art. 5-21) to determine the sides.

• In triangle ABC, if $B = 60°$, $C = 75°$, and $a = 73.2$ ft, how long are b and c?
$A = 180° - 60° - 75° = 45°$.

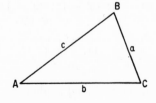

$$\frac{b}{\sin B} = \frac{c}{\sin C} = \frac{a}{\sin A}$$

$$\frac{b}{\sin 60} = \frac{c}{\sin 75} = \frac{73.2}{\sin 45}$$

$$b = \frac{\sin 60}{\sin 45}\, 73.2 = 89.7 \text{ ft}$$

$$c = \frac{\sin 75}{\sin 45}\, 73.2 = 100 \text{ ft}$$

Check with a Mollweide equation (Art. 5-25).

$$\frac{a - b}{c} = \frac{73.2 - 89.7}{100} = \frac{\sin \tfrac{1}{2}(A - B)}{\cos \tfrac{1}{2}C} = \frac{\sin \tfrac{1}{2}(45 - 60)}{\cos \left(7\tfrac{5}{2}\right)}$$
$$-0.165 = -0.165$$

3. Given two sides of an oblique triangle and an angle opposite one side. Construct the triangle to scale to determine if there is one solution, two solutions, or no solution (Art. 4-52.5). Solve by the law of sines (Art. 5-21).

• In triangle ABC, if $a = 117.5$ ft, $b = 156.1$ ft, and $A = 34°15'$, what are the values of c, B, and C?

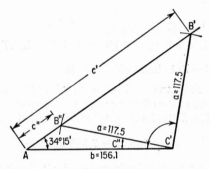

FIG. 5-11

A sketch to scale (Fig. 5-11) shows that two solutions are possible. (A protractor was used to lay off angle A.)

$$\frac{a}{\sin A} = \frac{b}{\sin B} \qquad \sin B = \frac{b}{a} \sin A = \frac{156.1}{117.5} \sin 34°15' = 0.749$$

Two angles whose sine equals 0.749 are $B' = 48°30'$ and $B'' = 180° - 48°30' = 131°30'$. Then

$$C' = 180° - 34°15' - 48°30' = 97°15'$$
$$C'' = 180° - 34°15' - 131°30' = 14°15'$$
$$c' = \frac{a \sin C'}{\sin A} = \frac{117.5 \sin 97°15'}{\sin 34°15'} = 207 \text{ ft}$$
$$c'' = \frac{a \sin C''}{\sin A} = \frac{117.5 \sin 14°15'}{\sin 34°15'} = 51.4 \text{ ft}$$

4. Given two sides and the included angle of an oblique triangle. One

method is to use the law of cosines (Art. 5-22) to find the third side, then the law of sines (Art. 5-21) to compute the two unknown angles.   Another method is to employ the law of tangents (Art. 5-23) to find the two angles.

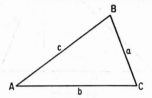

• In triangle ABC, if $a = 206.3$ ft, $b = 125.5$ ft, and $C = 27°30'$, what are the values of c, A, and B?

$c^2 = a^2 + b^2 - 2ab \cos C = (206.3)^2 + (125.5)^2 - 2(206.3)(125.5) \cos 27°30'$
$\quad = 42{,}500 + 15{,}800 - 46{,}000 = 12{,}300$
$c = 111$ ft
$$A + B = 180° - C = 180° - 27°30' = 152°30'$$
$$\tfrac{1}{2}(A + B) = 76°15'$$

From the law of tangents:

$$\tan \tfrac{1}{2}(A - B) = \frac{a - b}{a + b} \tan \tfrac{1}{2}(A + B) = \frac{206.3 - 125.5}{206.3 + 125.5} \tan 76°15' = 0.996$$
$$\tfrac{1}{2}(A - B) = 44°53'$$
$$A = \tfrac{1}{2}(A + B) + \tfrac{1}{2}(A - B) = 76°15' + 44°53' = 121°8'$$
$$B = \tfrac{1}{2}(A + B) - \tfrac{1}{2}(A - B) = 76°15' - 44°53' = 31°22'$$

Check with a Mollweide equation (Art. 5-25).

$$\frac{a - b}{c} = \frac{206.3 - 125.5}{111} = \frac{\sin \tfrac{1}{2}(A - B)}{\cos \tfrac{1}{2}C} = \frac{\sin 44°53'}{\cos 13°45'}$$
$$0.727 = 0.727$$

5. Given three sides of an oblique triangle.   One method is to use the law of cosines (Art. 5-22) to solve for the three angles.   Another method, better with logarithms, is to use the tangent functions in Art. 5-24.

• In triangle ABC, given $a = 100$, $b = 120$, and $c = 140$, find A, B, and C.
$s = \tfrac{1}{2}(a + b + c) = \tfrac{1}{2}(100 + 120 + 140) = 180$.

$$s - a = 180 - 100 = 80$$
$$s - b = 180 - 120 = 60$$
$$s - c = 180 - 140 = 40$$
$$r = \sqrt{\frac{(s - a)(s - b)(s - c)}{s}} = \sqrt{\frac{80 \times 60 \times 40}{180}} = 32.66$$

$$\tan \tfrac{1}{2}A = \frac{r}{s - a} = \frac{32.66}{80} = 0.4082 \qquad \tfrac{1}{2}A = 22°12' \qquad A = 44°24'$$

$$\tan \tfrac{1}{2}B = \frac{r}{s - b} = \frac{32.66}{60} = 0.5443 \qquad \tfrac{1}{2}B = 28°34' \qquad B = 57°8'$$

$$\tan \tfrac{1}{2}C = \frac{r}{s - c} = \frac{32.66}{40} = 0.8165 \qquad \tfrac{1}{2}C = 39°14' \qquad C = 78°28'$$

Check: $A + B + C = 44°24' + 57°8' + 78°28' = 180°$

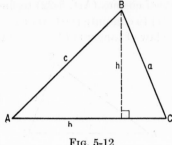

FIG. 5-12

5-27. Area of a Triangle. If K is the area of a triangle such as ABC (Fig. 5-12),

$$K = \tfrac{1}{2}bh$$

where b = base
h = altitude

Also,

$$K = \frac{b^2 \sin A \sin C}{2 \sin B} = \frac{a^2 \sin B \sin C}{2 \sin A} = \frac{c^2 \sin A \sin B}{2 \sin C}$$
$$= \tfrac{1}{2}ab \sin C = \tfrac{1}{2}bc \sin A = \tfrac{1}{2}ac \sin B$$
$$= \sqrt{s(s - a)(s - b)(s - c)}$$

where $s = \tfrac{1}{2}(a + b + c)$

5-28. Azimuth. The direction of a line, measured by the angle between a fixed reference line (called a meridian) and the line, in a clockwise direction (Fig. 5-13).

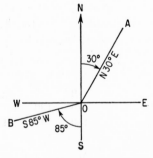

FIG. 5-13   FIG. 5-14

In astronomical observations, azimuths are measured from true south. Surveyors reckon azimuths either from north or south.

Azimuths may be true, magnetic, or assumed, depending on whether the meridian is true, magnetic, or assumed.

5-29. Bearing. The direction of any line with respect to a fixed reference line (called a meridian), measured by the acute angle the line makes with the meridian (Fig. 5-14).

The bearing of a line depends on the quadrant in which it lies. In Fig. 5-14, the bearing of OA, in the northeast quadrant, is given by N30°E (north 30° east). The bearing of OB, in the southwest quadrant, is given by S85°W (south 85° west).

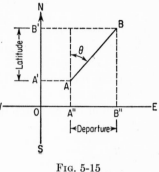

FIG. 5-15

5-30. Latitudes and Departures. The latitude of a line is its projection on the meridian, or fixed reference line. In Fig. 5-15, the latitude of line AB is $A'B'$.

The departure of a line is its projection on a parallel (a reference line normal to the meridian). In Fig. 5-15, the departure of AB is $A''B''$.

If θ is the bearing angle (Fig. 5-15), then

Latitude = length of line times cosine bearing angle
Departure = length of line times sine bearing angle

Thus, $A'B' = AB \cos \theta$; $A''B'' = AB \sin \theta$.

In a closed traverse, or polygon, the sum of the north latitudes must equal the sum of the south latitudes. If the north latitudes are considered positive and the south latitudes are considered negative, the sum of the latitudes of a traverse must be zero. Similarly, the sum of the east departures must equal the sum of the west departures. If the east departures are taken positive and the west departures are taken negative, the sum of the departures of a traverse must be zero.

If the sums are not zero, the error of closure is given by

$$e = \sqrt{(\Sigma L)^2 + (\Sigma D)^2}$$

where ΣL = algebraic sum of latitudes
ΣD = algebraic sum of departures
The direction of the error (Fig. 5-16) is given by

$$\epsilon = \tan^{-1} \frac{-\Sigma D}{-\Sigma L}$$

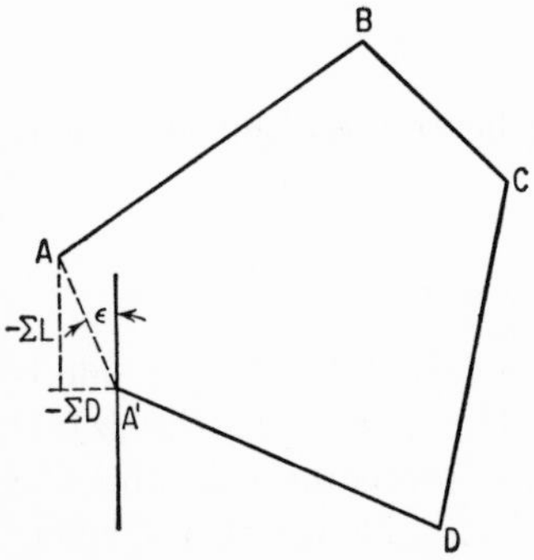

FIG. 5-16

5-31. Computation of Area by Double Meridian Distances. To determine the area of a closed traverse or polygon, assume a reference meridian through one vertex (Fig. 5-17). (Usually, the most westerly vertex will be the most convenient choice.) Compute double the areas of the triangles and trapezoids formed by projecting the sides of the polygon on the meridian. The algebraic sum of these double areas is twice the area of the polygon.

The perpendicular distance from any point of the polygon to the meridian is the meridian distance of the point. The meridian distance is positive if measured to the east of the meridian and negative if measured to the west.

The sum of the meridian distances of the two extremities of a line is the double meridian distance of the line.

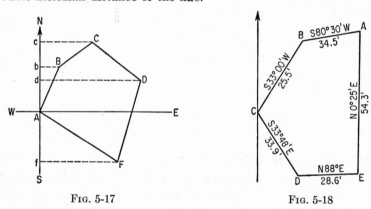

FIG. 5-17   FIG. 5-18

The double area of any triangle or trapezoid formed by a side of a polygon, its projection on a meridian, and the projection lines is given by

$$2a = DMD \times L$$

where DMD = double meridian distance of side
L = latitude of side
Latitudes are positive if the line has a north bearing and negative if it has a south bearing.

If A is the area of the polygon, then

$$A = \tfrac{1}{2}\Sigma\, 2a = \tfrac{1}{2}\Sigma\, DMD \times L$$

If the meridian passes through a vertex, the DMD of the first side is equal to its departure. The DMD of any other side equals the DMD of the preceding side plus the departure of the preceding side plus the departure of the side itself. The DMD of the last side equals its departure with sign changed.

• What is the area of the polygon in Fig. 5-18?

The calculation is summarized in the table below. The sides and their bearings and lengths are given at the left of the table. Latitudes and departures, given next, were computed by multiplying each length by the cosine and sine, respectively, of the bearing angle (Art. 5-30).

AREA OF $ABCDE$ BY DOUBLE MERIDIAN DISTANCES

Side	Bearing	Length	Latitude	Departure	DMD	Double areas +	Double areas −
AB	S80°30′W	34.5	−5.7	−34.0	61.8	.....	353
BC	S33°00′W	25.5	−21.4	−13.9	13.9	......	298
CD	S33°48′E	33.9	−28.2	+18.9	18.9	......	532
DE	N88°00′E	28.6	+1.0	+28.6	66.4	66	
EA	N0°25′E	54.3	+54.3	+0.4	95.4	5,181	
						5,247	1,183

Sum of double areas = 5,247 − 1,183 = 4,064

$$\text{Area of polygon} = \frac{4,064}{2} = 2,032 \text{ sq ft}$$

A meridian for the area computation was selected north and south through the most westerly vertex C, so that all DMD would be positive. Computation of DMD started with CD. Its DMD is its departure, 18.9. The DMD for DE then equals the preceding DMD, 18.9, plus the preceding departure, 18.9, plus its own departure, 28.6, for a total of 66.4. The DMD of $EA = 66.4 + 28.6 + 0.4 = 95.4$, etc. As a check, note that the DMD of the last side BC is the negative of its departure.

The double areas at the right of the table were obtained by multiplying each DMD by the latitude of the side, taking note of the signs of the latitude and DMD. Thus, the double area for AB is $−5.7 \times 61.8 = −353$. Algebraic summation of the double areas and division by 2 yield the area of the polygon, 2,032 ft².

5-32. Inverse Trigonometric Functions. If $x = \sin\theta$, then θ is the angle whose sine equals x. The latter relationship usually is written

$$\theta = \sin^{-1} x$$

and read antisine x, inverse sine x, or arcsine x. [NOTE: $1/\sin x$ is written $(\sin x)^{-1}$, though the square of $\sin x$ is written $\sin^2 x$ and its nth power is written $\sin^n x$.] Similarly, if $y = \cos\varphi$, $\varphi = \cos^{-1} y$.

The inverse functions are multivalued. Principal values are defined as follows:

$$\sin^{-1} x \text{ and } \tan^{-1} x \qquad -\frac{\pi}{2} \text{ to } \frac{\pi}{2}$$

$$\cos^{-1} x \text{ and } \cot^{-1} x \qquad 0 \text{ to } \pi$$

If θ_1 is a principal value of an inverse function, then $\theta = (-1)^n\theta_1 + n\pi$

is the general solution of $\theta = \sin^{-1} x$. The general solution of $\theta = \cos^{-1} x$ is $\theta = 2\pi n \pm \theta_1$; the general solution of $\theta = \tan^{-1} x$ is $\theta = \theta_1 + n\pi$.

$$\sin^{-1} x = \cos^{-1} \sqrt{1 - x^2} = \tan^{-1} \frac{x}{\sqrt{1 - x^2}} = \frac{\pi}{2} - \cos^{-1} x$$

$$\cos^{-1} x = \sin^{-1} \sqrt{1 - x^2} = \tan^{-1} \frac{\sqrt{1 - x^2}}{x} = \frac{\pi}{2} - \sin^{-1} x$$

$$\tan^{-1} x = \sin^{-1} \frac{x}{\sqrt{1 + x^2}} = \cos^{-1} \frac{1}{\sqrt{1 + x^2}} = \frac{\pi}{2} - \tan^{-1} \frac{1}{x}$$

$$\cot^{-1} x = \sin^{-1} \frac{1}{\sqrt{1 + x^2}} = \cos^{-1} \frac{x}{\sqrt{1 + x^2}} = \frac{\pi}{2} - \cot^{-1} \frac{1}{x}$$

5-33. Trigonometric Equations. The solutions of a trigonometric equation are the angles that satisfy the equation.

• Solve $\sin \theta = \frac{1}{2} \sqrt{2}$ for θ.

The value of θ that satisfies this equation can be obtained from the table in Art. 5-8 or, since $\sqrt{2}/2 = 0.70711$, from a five-place table of trigonometric functions. $\theta = 45°$.

In general, a trigonometric equation will involve more than one function of an angle. To solve it, first transform all terms so that only one function is involved (see Arts. 5-7 and 5-10 to 5-20).

• Solve $2 \tan^2 \theta - \sec^2 \theta = 1$ for $\theta < 90°$.

To make both functions in this equation the same, substitute $1 + \tan^2 \theta$ for $\sec^2 \theta$.

$$2 \tan^2 \theta - (1 + \tan^2 \theta) = \tan^2 \theta - 1 = 1$$
$$\tan^2 \theta = 2$$
$$\tan \theta = \sqrt{2} = 1.4142$$

$\theta = 54°44'$, determined from a five-place table of functions.

When the answers may be greater than 90°, find the acute angles φ that satisfy the equation. Then, disregarding signs, the general solutions are $360° - \varphi$ or $180° \pm \varphi$. Select the angles in the proper quadrants to satisfy the given equation.

• Solve $\sqrt{3} \sin \theta + \cos \theta = \sqrt{3}$.

According to Art. 5-18, $a \sin \theta + b \cos \theta = \sqrt{a^2 + b^2} \sin (\theta + \varphi)$, where $\varphi = \cos^{-1} a/\sqrt{a^2 + b^2}$. In the given equation, $a = \sqrt{3}$ and $b = 1$. Hence, $\sqrt{a^2 + b^2} = \sqrt{3 + 1} = 2$, and $\varphi = \cos^{-1} (\sqrt{3}/2) = 30°$. The given equation can be transformed to

$$2 \sin (\theta + 30°) = \sqrt{3}$$
Then $\qquad \sin (\theta + 30°) = \sqrt{3}/2$
$$\theta + 30° = 60°, 120°$$
$$\theta = 30°, 90°$$

Trigonometric equations may also be solved graphically. If only one function of the unknown is involved, transform the equation so that zero is on the right-hand side; plot the function on the left-hand side and find the intersections of the curve with the abscissa axis. If there is more than one function on the left when zero is on the right, divide them into two groups; plot each group and find the intersection of the curves.

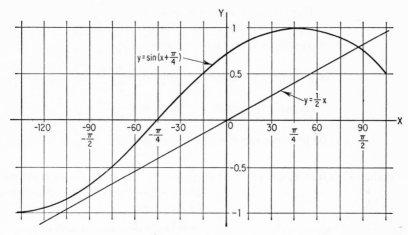

FIG. 5-19

• Solve $\frac{1}{2}x - \sin(x + \pi/4) = 0$, where x is an angle in radians.

Plot $y = \frac{1}{2}x$ and $y = \sin(x + \pi/4)$, as shown in Fig. 5-19. The intersection of the curves is the solution: 86°. The plot shows that only one solution is possible because there is only one intersection.

5-34. Hyperbolic Functions. Functions similar to the natural trigonometric functions may be defined with respect to the equilateral hyperbola $x^2 - y^2 = a^2$. Just as the cartesian coordinates (x,y) of the circle of unit radius can be expressed as the cosine and sine, respectively, of a parameter θ, the coordinates of a point P on the hyperbola $x^2 - y^2 = 1$ can be given as the hyperbolic cosine and hyperbolic sine of a parameter u (Fig. 5-20).

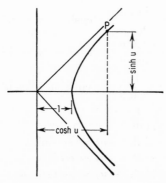

FIG. 5-20

The hyperbolic functions are expressed in terms of $e = 2.718 \cdots$ as follows:

$$\text{Hyperbolic sine } u = \sinh u = \frac{e^u - e^{-u}}{2}$$

$$\text{Hyperbolic cosine } u = \cosh u = \frac{e^u + e^{-u}}{2}$$

$$\text{Hyperbolic tangent } u = \tanh u = \frac{e^u - e^{-u}}{e^u + e^{-u}}$$

$$\text{Hyperbolic cotangent } u = \coth u = \frac{e^u + e^{-u}}{e^u - e^{-u}}$$

$$\text{Hyperbolic secant } u = \operatorname{sech} u = \frac{2}{e^u + e^{-u}}$$

$$\text{Hyperbolic cosecant } u = \operatorname{csch} u = \frac{2}{e^u - e^{-u}}$$

Note that $\sinh 0 = 0$, $\cosh 0 = 1$, $\tanh 0 = 0$. For graphical representation of the hyperbolic functions, see Art. 6-63.

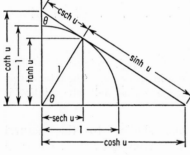

FIG. 5-21

• What is the value of $\sinh 3$?

If a table of hyperbolic functions is available, the value can be obtained directly. Otherwise, it can be computed from an infinite series (Art. 12-14) or by calculation from the definition.

$$\sinh 3 = \frac{e^3 - e^{-3}}{2}$$

$$= \frac{20.09 - 0.05}{2} = 10.02$$

5-35. Relationships between Hyperbolic Functions. Figure 5-21 shows the geometric relationships between hyperbolic functions of the same argument.

$$\tanh u = \frac{\sinh u}{\cosh u} \qquad \cosh^2 u - \sinh^2 u = 1$$

$$\coth u = \frac{\cosh u}{\sinh u} \qquad \tanh^2 u + \operatorname{sech}^2 u = 1$$

$$\operatorname{sech} u = \frac{1}{\cosh u} \qquad \coth^2 u - \operatorname{csch}^2 u = 1$$

$$\operatorname{csch} u = \frac{1}{\sinh u}$$

$$\sinh (-u) = -\sinh u$$
$$\cosh (-u) = \cosh u$$
$$\tanh (-u) = -\tanh u$$

$$\cosh u + \sinh u = e^u$$

$$\cosh u - \sinh u = e^{-u} = \frac{1}{\sinh u + \cosh u}$$

$$\sinh (u \pm v) = \sinh u \cosh v \pm \cosh u \sinh v$$

$$\cosh (u \pm v) = \cosh u \cosh v \pm \sinh u \sinh v$$

$$\tanh (u \pm v) = \frac{\tanh u \pm \tanh v}{1 \pm \tanh u \tanh v}$$

$$(\sinh u + \cosh u)^n = \sinh nu + \cosh nu$$

5-36. Lambda Function and Gudermannian. The angle θ in Fig. 5-21 is a function of the hyperbolic argument of u (Arts. 5-34 and 5-35). This function is called the gudermannian of u and is written

$$\theta = \text{gd } u = \tan^{-1} \sinh u = \text{amh } u$$

The inverse of a gudermannian is called the lambda function.

$$u = \text{gd}^{-1} \theta = \lambda(\theta)$$

From Fig. 5-21,

$$\sinh u = \tan \theta$$
$$\tanh u = \sin \theta$$
$$\cosh u = \sec \theta$$

5-37. Relationships between Trigonometric and Hyperbolic Functions. If $i = \sqrt{-1}$, then if x is a real or complex variable,

$$\sinh ix = i \sin x \qquad \sin ix = i \sinh x$$
$$\cosh ix = \cos x \qquad \cos ix = \cosh x$$
$$\tanh ix = i \tan x \qquad \tan ix = i \tanh x$$

$$\sinh (x \pm iy) = \sinh x \cos y \pm i \cosh x \sin y$$
$$\cosh (x \pm iy) = \cosh x \cos y \pm i \sinh x \sin y$$
$$\tanh (x \pm iy) = \frac{\sinh 2x \pm i \sin 2y}{\cosh 2x + \cos 2y}$$
$$\coth (x \pm iy) = \frac{\sinh 2x \mp i \sin 2y}{\cosh 2x - \cos 2y}$$

5-38. Inverse Hyperbolic Functions. If $x = \sinh \theta$, then θ is a function (not necessarily an angle as in natural trigonometric functions) whose hyperbolic sine is x. That is,

$$\theta = \text{argsinh } x = \sinh^{-1} x$$

read θ is the argument whose hyperbolic sine is x. The inverses of the other hyperbolic functions are similarly defined.

Since the hyperbolic functions also are functions of $e = 2.718 \cdots$, the inverses can be expressed in terms of logarithms to the base e (written ln).

$$\text{argsinh } x = \ln (x + \sqrt{x^2 + 1})$$

$$\text{argcosh } x = \ln (x \pm \sqrt{x^2 - 1}) \qquad x \geq 1$$

$$\text{argtanh } x = \tfrac{1}{2} \ln \frac{1 + x}{1 - x} \qquad -1 < x < +1$$

$$\text{argcoth } x = \tfrac{1}{2} \ln \frac{x + 1}{x - 1} \qquad x < -1 \text{ and } x > 1$$

$$\text{argsech } x = \ln \frac{1 \pm \sqrt{1 - x^2}}{x} \qquad 0 < x \leq 1$$

$$\text{argsech } x = \ln \frac{1 + \sqrt{x^2 + 1}}{x} \qquad x > 0$$

$$\text{argcsch } x = \ln \frac{1 - \sqrt{x^2 + 1}}{x} \qquad x < 0$$

5-39. Law of Cosines for Spherical Triangles.   This law, similar to the law of cosines for plane triangles, is useful for finding the third side of a spherical triangle ABC (Fig. 5-22) when two sides and the included angle are given, or for finding the third angle when two angles and the included side are given.

Law of cosines for sides:

$$\cos a = \cos b \cos c + \sin b \sin c \cos A$$
$$\cos b = \cos a \cos c + \sin a \sin c \cos B$$
$$\cos c = \cos a \cos b + \sin a \sin b \cos C$$

Law of cosines for angles:

$$\cos A = - \cos B \cos C + \sin B \sin C \cos a$$
$$\cos B = - \cos A \cos C + \sin A \sin C \cos b$$
$$\cos C = - \cos A \cos B + \sin A \sin B \cos C$$

• Given, in spherical triangle ABC, $a = 60°$, $b = 120°$, and $C = 150°$. Compute the other parts.

$$\cos c = \cos a \cos b + \sin a \sin b \cos C = \cos 60 \cos 120$$
$$+ \sin 60 \sin 120 \cos 150 = -0.899$$

$$c = 180° - 26° = 154°$$

Cos B can be computed from one of the equations for sides.

$$\cos B = \frac{\cos b - \cos a \cos c}{\sin a \sin c} = \frac{\cos 120 - \cos 60 \cos 154}{\sin 60 \sin 154} = -0.133$$

$$B = 180° - 82° = 98°$$

Similarly,

$$\cos A = \frac{\cos a - \cos b \cos c}{\sin b \sin c} = \frac{\cos 60 - \cos 120 \cos 154}{\sin 120 \sin 154} = 0.133$$

$$A = 82°$$

5-40. Law of Sines for Spherical Triangles. In spherical triangle ABC (Fig. 5-22),

$$\frac{\sin A}{\sin a} = \frac{\sin B}{\sin b} = \frac{\sin C}{\sin c}$$

• Given, in spherical triangle ABC, $a = 60°$, $b = 120°$, $C = 150°$, and $A = 82°$. Find B.

$$\sin B = \frac{\sin A}{\sin a} \sin b = \frac{\sin 82}{\sin 60} \sin 120 = 0.990$$
$$B = 82°, 98°$$

Check by the law of cosines for angles (Art. 5-39).

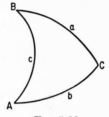

$$\cos B = -\cos A \cos C + \sin A \sin C \cos b$$
$$= -\cos 82 \cos 150 + \sin 82 \sin 150 \cos 120$$
$$= -0.133$$
$$B = 180° - 82° = 98° \qquad 180° + 82° = 262°$$
$$B = 98° \text{ is the only solution.}$$

Fig. 5-22

5-41. Functions of an Angle in Terms of Three Sides of a Spherical Triangle. In spherical triangle ABC (Fig. 5-22), let $s = \frac{1}{2}(a + b + c)$ and $r = \sqrt{\sin(s-a)\sin(s-b)\sin(s-c)/\sin s}$.

$$\sin \tfrac{1}{2}A = \sqrt{\frac{\sin(s-b)\sin(s-c)}{\sin b \sin c}}$$

$$\sin \tfrac{1}{2}B = \sqrt{\frac{\sin(s-a)\sin(s-c)}{\sin a \sin c}}$$

$$\sin \tfrac{1}{2}C = \sqrt{\frac{\sin(s-a)\sin(s-b)}{\sin a \sin b}}$$

$$\cos \tfrac{1}{2}A = \sqrt{\frac{\sin s \sin(s-a)}{\sin b \sin c}}$$

$$\cos \tfrac{1}{2}B = \sqrt{\frac{\sin s \sin(s-b)}{\sin a \sin c}}$$

$$\cos \tfrac{1}{2}C = \sqrt{\frac{\sin s \sin(s-c)}{\sin a \sin b}}$$

$$\tan \tfrac{1}{2}A = \frac{r}{\sin(s-a)}$$

$$\tan \tfrac{1}{2}B = \frac{r}{\sin(s-b)}$$

$$\tan \tfrac{1}{2}C = \frac{r}{\sin(s-c)}$$

(See also Art. 5-46.1.)

5-42. Functions of a Side of a Spherical Triangle in Terms of Three Angles. In spherical triangle ABC (Fig. 5-22) let $S = \frac{1}{2}(A + B + C)$ and $R = \sqrt{-\cos S / \cos(S - A)\cos(S - B)\cos(S - C)}$.

$$\cos \tfrac{1}{2}a = \sqrt{\frac{\cos(S - B)\cos(S - C)}{\sin B \sin C}}$$

$$\cos \tfrac{1}{2}b = \sqrt{\frac{\cos(S - A)\cos(S - C)}{\sin A \sin C}}$$

$$\cos \tfrac{1}{2}c = \sqrt{\frac{\cos(S - A)\cos(S - B)}{\sin A \sin B}}$$

$$\sin \tfrac{1}{2}a = \sqrt{\frac{\cos S \cos(S - A)}{\sin B \sin C}}$$

$$\sin \tfrac{1}{2}b = \sqrt{\frac{\cos S \cos(S - B)}{\sin A \sin C}}$$

$$\sin \tfrac{1}{2}c = \sqrt{\frac{\cos S \cos(S - C)}{\sin A \sin B}}$$

$$\tan \tfrac{1}{2}a = R \cos(S - A)$$
$$\tan \tfrac{1}{2}b = R \cos(S - B)$$
$$\tan \tfrac{1}{2}c = R \cos(S - C)$$

(See also Art. 5-46.2.)

5-43. Napier's Analogies for Spherical Triangles. In spherical triangle ABC

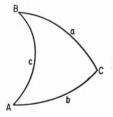

$$\frac{\tan \tfrac{1}{2}(a - b)}{\tan \tfrac{1}{2}c} = \frac{\sin \tfrac{1}{2}(A - B)}{\sin \tfrac{1}{2}(A + B)}$$

$$\frac{\tan \tfrac{1}{2}(A - B)}{\cot \tfrac{1}{2}C} = \frac{\sin \tfrac{1}{2}(a - b)}{\sin \tfrac{1}{2}(a + b)}$$

$$\frac{\tan \tfrac{1}{2}(a + b)}{\tan \tfrac{1}{2}c} = \frac{\cos \tfrac{1}{2}(A - B)}{\cos \tfrac{1}{2}(A + B)}$$

$$\frac{\tan \tfrac{1}{2}(A + B)}{\cot \tfrac{1}{2}C} = \frac{\cos \tfrac{1}{2}(a - b)}{\cos \tfrac{1}{2}(a + b)}$$

5-44. Gauss' Equations (Delambre's Analogies) for Spherical Triangles. In spherical triangle ABC,

$$\cos \tfrac{1}{2}c \sin \tfrac{1}{2}(A + B) = \cos \tfrac{1}{2}C \cos \tfrac{1}{2}(a - b)$$
$$\cos \tfrac{1}{2}c \cos \tfrac{1}{2}(A + B) = \sin \tfrac{1}{2}C \cos \tfrac{1}{2}(a + b)$$
$$\sin \tfrac{1}{2}c \sin \tfrac{1}{2}(A - B) = \cos \tfrac{1}{2}C \sin \tfrac{1}{2}(a - b)$$
$$\sin \tfrac{1}{2}c \cos \tfrac{1}{2}(A - B) = \sin \tfrac{1}{2}C \sin \tfrac{1}{2}(a + b)$$

5-45. Solution of Right and Quadrantal Spherical Triangles. (See also Art. 5-46.) A quadrantal triangle has one side equal to 90°. Its polar triangle, therefore, has an angle equal to 90° (Art. 4-44). Hence, a quadrantal triangle can be solved by first solving its polar right triangle and then finding the supplements of the polar triangle's sides and angles.

These supplements equal the opposite angles and sides, respectively, of the given triangle.

In a right spherical triangle, an oblique angle and its opposite side always belong to the same quadrant. If the angle is less than 90°, for example, the side opposite it also is less than 90°.

If the hypotenuse is less than 90°, the other two angles and legs all belong to the same quadrant. If the hypotenuse is larger than 90°, the other two angles belong to different quadrants, as do the two sides. That is, one angle or side is in the first quadrant, and the other is in the second quadrant.

When any two parts of a spherical triangle ABC with a 90° angle at C (Fig. 5-23) are given, the other parts can be computed from

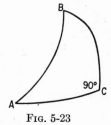

FIG. 5-23

$$\cos c = \cos a \cos b = \cot A \cot B$$
$$\sin a = \sin A \sin c = \tan b \cot B$$
$$\sin b = \sin B \sin c = \tan a \cot A$$
$$\cos A = \sin B \cos a = \tan b \cot c$$
$$\cos B = \sin A \cos b = \tan a \cot c$$

• Given spherical triangle ABC (Fig. 5-23) with $A = 31°$, $B = 72°$, and $C = 90°$. Find the three sides.

$$\cos b = \frac{\cos B}{\sin A} = \frac{\cos 72}{\sin 31} = 0.600$$
$$b = 53°7'$$

(not 180° − 53°7′, because b must belong to the same quadrant as $B = 72°$)

$$\cos a = \frac{\cos A}{\sin B} = \frac{\cos 31}{\sin 72} = 0.900$$
$$a = 25°50'$$

(not 180° − 25°50′, because a must belong to the same quadrant as $A = 31°$)

$$\cos c = \cot A \cot B = \cot 31 \cot 72 = 0.541$$
$$c = 57°15'$$

(not 180° − 57°15′, because c must be less than $a + b$)

$$\textit{Check: } \cos b = \frac{\tan a \cot c}{\sin A} = \frac{\tan 25°50' \cot 57°15'}{\sin 31°} = 0.600$$

• Given spherical triangle ABC with $A = 32°$, $C = 90°$, and $a = 24°$. Find b, c, and B.

$$\sin B = \frac{\cos A}{\cos a} = \frac{\cos 32}{\cos 24} = 0.927$$
$$B = 68° \quad \text{or} \quad 180 - 68° = 112°$$
$$\sin b = \tan a \cot A = \tan 24 \cot 32 = 0.711$$
$$b = 45°20' \quad \text{or} \quad 180° - 45°20' = 134°40'$$
$$\sin c = \frac{\sin a}{\sin A} = \frac{\sin 24}{\sin 32} = 0.768$$
$$c = 50°10' \quad \text{or} \quad 180° - 50°10' = 129°50'$$

Since $(a + b + c) < 360°$ and the pairs a,A and b,B must belong to the same quadrant, the solutions are

$$a = 24° \qquad b = 45°20' \qquad c = 50°10' \qquad A = 32° \qquad B = 68° \qquad C = 90°$$

(From $\cot c = \sin A \cos b / \tan a$, it can be seen that $\cot c$ must be positive; therefore, $c \neq 129°50'$ with the above values of a, b, and A.)

$$a = 24° \qquad b = 134°40' \qquad c = 129°50' \qquad A = 32° \qquad B = 112° \qquad C = 90°$$

(With these values of a, b, and A, $\cot c$ is negative; therefore, $c \neq 50°10'$.)

The two solutions represent two triangles with common side a, forming a lune (Fig. 5-24).

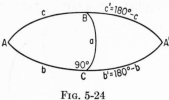

FIG. 5-24

• Given spherical triangle ABC with $a = 75°$, $b = 85°30'$, and $c = 90°$. Find the three angles.

The polar triangle $A'B'C'$ of ABC is a right triangle with $C' = 90°$, $B' = 180° - b = 180° - 85°30' = 94°30'$, and $A' = 180° - a = 180° - 75° = 105°$. The unknown parts of ABC can be found by first finding the unknown parts of $A'B'C'$.

$$\cos a' = \frac{\cos A'}{\sin B'} = \frac{\cos 105°}{\sin 94°30'} = -0.260$$
$$a' = 180° - 74°55'$$
$$\cos b' = \frac{\cos B'}{\sin A'} = \frac{\cos 94°30'}{\sin 105°} = -0.0813$$
$$b' = 180° - 85°20'$$
$$\cos c' = \cot A' \cot B' = \cot 105° \cot 94°30' = 0.0211$$
$$c' = 88°47'$$

Hence, in triangle ABC,
$$A = 180° - a' = 74°55'$$
$$B = 180° - b' = 85°20'$$
$$C = 180° - c' = 91°13'$$

Check by Gauss' equation (Art. 5-44).

$$\cos \tfrac{1}{2}c = \frac{\cos \tfrac{1}{2}(a + b)}{\cos \tfrac{1}{2}(A + B)} \sin \tfrac{1}{2}C = \frac{\cos 80°15'}{\cos 80°8'} \sin \frac{91°13'}{2} = 0.707$$
$$\tfrac{1}{2}c = 45° \text{ and } c = 90°$$

5-46. Solution of Oblique Spherical Triangles. List the given parts and those to be found. Select the formulas to be used (preferably those involving only given parts); solve these for the required parts. Substitute numerical data in the formulas. Check the answers by formulas not used in the solution. Note also that each angle of a spherical triangle is less than 180°.

If a side (or an angle) differs from 90° by a larger number of degrees than another side (or angle) in a spherical triangle, then the first side and its opposite angle belong to the same quadrant.

Half the sum of two sides and half the sum of the two angles opposite them belong to the same quadrant.

• If, in spherical triangle ABC, $a = 120°$, $b = 70°$, and $c = 130°$, to what quadrants do A, B, and C belong?

Since $(a - 90° = 120° - 90°) > (90° - b = 90° - 70°)$, A and a are both in the same quadrant, the second.

$(c - 90° = 130° - 90°) > (90° - b = 90° - 70°)$, so C and c are both in the second quadrant.

B can be determined from the fact that $\frac{1}{2}(b + a) = 95°$ and $\frac{1}{2}(b + c) = 100°$ both lie in the second quadrant; $\frac{1}{2}(B + A)$ and $\frac{1}{2}(B + C)$ also must lie in the second quadrant. Thus, B's quadrant depends on the valves computed for A and C.

1. Given three sides. The three angles can be found readily from the tangent formulas in Art. 5-41.

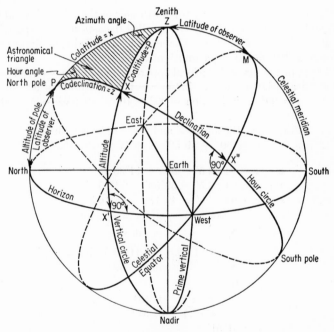

Fig. 5-25

• What is the azimuth of the sun and the local apparent time in New York City (latitude 40°43′ north) when the altitude of the sun is 30°10′ bearing west and its declination is 10° north?

The celestial sphere is shown in Fig. 5-25. Two spherical coordinate systems are indicated. One is based on the observer's horizon, the great circle cut on the celestial sphere by a plane perpendicular to the axis joining the observer's zenith and nadir. In this system, points on the celestial sphere are located by giving their azimuths and altitudes. The azimuth of a point is measured by the arc of the horizon from the north point eastward and terminating at the vertical circle through the point. The altitude is measured by the arc of the vertical circle from the horizon to the point. If, in Fig. 5-25, the sun is assumed to be at X, its azimuth would be given by the arc north-east-south-west-X' and its altitude by $X'X$.

The second coordinate system is based on the celestial equator, the great circle cut by the plane through the earth's equator. In this system, points on the celestial sphere are located by giving the meridian angles or local hour angles and the declinations. The meridian angle is measured by the arc of the equator from the celestial meridian of the observer east or west to the hour circle through the point. The local hour angle is measured by the arc of the equator from the celestial meridian westward to the hour circle through the point. The declination is measured by the arc of the hour circle from the equator to the point. In Fig. 5-25, the local hour angle of the sun at X is given by MX'' and the declination by $X''X$.

The spherical triangle formed by the north pole of the celestial equator P, the zenith Z, and a point X on the sphere is called the astronomical triangle. The angle at P equals the hour angle. The angle at Z, called the azimuth angle, in this case equals $360°$ minus the azimuth of X.

The azimuth of the sun and the local apparent time can be obtained by solving the astronomical triangle ZPX. $ZP = x = 90° - \text{latitude} = 90° - 40°43' = 49°17'; PX = z = 90° - \text{declination} = 90° - 10° = 80°; XZ = p = 90° - \text{altitude} = 90° - 30°10' = 59°50'$.

From these three sides, the angles Z and P are to be computed.
$s = \frac{1}{2}(z + p + x) = \frac{1}{2}(80° + 59°50' + 49°17') = 94°33.5'$.

$$s - z = 94°33.5' - 80° = 14°33.5'$$
$$s - p = 94°33.5' - 59°50' = 34°43.5'$$
$$s - x = 94°33.5' - 49°17' = 45°16.5'$$
$$r = \sqrt{\frac{\sin (s - z) \sin (s - p) \sin (s - x)}{\sin s}}$$
$$= \sqrt{\frac{\sin 14°33.5' \sin 34°43.5' \sin 45°16.5'}{\sin 94°33.5'}}$$
$$= 0.320$$
$$\tan \tfrac{1}{2}Z = \frac{r}{\sin (s - z)} = \frac{0.320}{0.2514} = 1.273$$
$$\tfrac{1}{2}Z = 51°51'$$
$$Z = 103°42'$$

The azimuth $= 360° - Z = 360° - 103°42' = 256°18'$.

$$\tan \tfrac{1}{2}P = \frac{r}{\sin (s - p)} = \frac{0.320}{0.570} = 0.562$$
$$\tfrac{1}{2}P = 29°20'$$
$$P = 58°40'$$

Local time $= P/15° = 58°40'/15° = 3$ hours 54 min 40 sec.

2. Given three angles. The three sides can be found from the tangent formulas in Art. 5-42.

• If, in spherical triangle ABC, $A = 107°33'$, $B = 127°22'$, and $C = 128°42'$, find a, b, and c.

$S = \frac{1}{2}(A + B + C) = \frac{1}{2}(107°33' + 127°22' + 128°42') = 181°48.5'$.

$$S - A = 181°48.5' - 107°33' = 74°15.5'$$
$$S - B = 181°48.5' - 127°22' = 54°26.5'$$
$$S - C = 181°48.5' - 128°42' = 53°6.5'$$

$$R = \sqrt{\frac{- \cos S}{\cos (S - A) \cos (S - B) \cos (S - C)}}$$

$$= \sqrt{\frac{- \cos 181°48.5'}{\cos 74°15.5' \cos 54°26.5' \cos 53°6.5'}}$$

$$= 3.244$$

$$\tan \tfrac{1}{2}a = R \cos (S - A) = 3.244 \cos 74°15.5' = 0.880$$
$$\tfrac{1}{2}a = 41°21'$$
$$a = 82°42'$$
$$\tan \tfrac{1}{2}b = R \cos (S - B) = 3.244 \cos 54°26.5' = 1.890$$
$$\tfrac{1}{2}b = 62°7'$$
$$b = 124°14'$$
$$\tan \tfrac{1}{2}c = R \cos (S - C) = 3.244 \cos 53°6.5' = 1.947$$
$$\tfrac{1}{2}c = 62°49'$$
$$c = 125°38'$$

Check with Gauss' equation.

$$\cos \tfrac{1}{2}c = \frac{\cos \frac{1}{2}(a + b)}{\cos \frac{1}{2}(A + B)} \sin \tfrac{1}{2}C = \frac{\cos 103°28'}{\cos 117°27.5'} \sin 64°21' = 0.456$$
$$\tfrac{1}{2}c = 62°51'$$
$$c = 125°42'$$

3. Given two sides and the included angle. The sum and difference of the two unknown angles, and hence the angles, can be found from Napier's analogies (Art. 5-43), as can the third side.

• Find the initial course and great-circle distance from Annapolis, Md. (latitude 38°58' north, longitude 76°30' west), to Los Angeles, Calif. (latitude 34°5' north, longitude 118°12' west).

In spherical triangle ABC in Fig. 5-26, B represents the north pole of the earth, A represents Annapolis, and C represents Los Angeles. Angle $B = 118°12' - 76°30' = 41°42'$; $c = 90° - 38°58' = 51°2'$; $a = 90° - 34°5' = 55°55'$.

$$\tan \tfrac{1}{2}(A - C) = \frac{\sin \frac{1}{2}(a - c)}{\sin \frac{1}{2}(a + c)} \cot \tfrac{1}{2}B = \frac{\sin \frac{1}{2}(55°55' - 51°2')}{\sin \frac{1}{2}(55°55' + 51°2')} \cot \frac{41°42'}{2}$$
$$= 0.1396$$
$$\tfrac{1}{2}(A - C) = 8°1'$$
$$\tan \tfrac{1}{2}(A + C) = \frac{\cos \frac{1}{2}(a - c)}{\cos \frac{1}{2}(a + c)} \cot \tfrac{1}{2}B = \frac{\cos 2°26.5'}{\cos 53°28.5'} \cot 20°51' = 4.42$$
$$\tfrac{1}{2}(A + C) = 77°15'$$
$$A = \tfrac{1}{2}(A + C) + \tfrac{1}{2}(A - C) = 77°15' + 8°1' = 85°16'$$

The initial course angle A then is N85°16'W, and the true course, measured from the north to the east, is $360° - 85°16' = 274°44'$.

$$\tan \tfrac{1}{2}b = \frac{\cos \tfrac{1}{2}(A + C)}{\cos \tfrac{1}{2}(A - C)} \tan \tfrac{1}{2}(a + c) = \frac{\cos 77°15'}{\cos 8°1'} \tan 53°28.5' = 0.301$$
$$\tfrac{1}{2}b = 16°45'$$
$$b = 33°30' = 33.5°$$

If the radius of the earth is taken as 3,960 statute miles, the distance from Annapolis to Los Angeles is $3,960 \times 33.5/57.3 = 2,320$ miles $= 2,010$ nautical miles. (Division by 57.3 converts the angle in degrees to radians.)

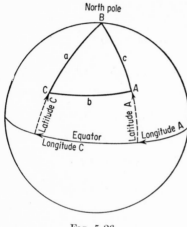

FIG. 5-26

4. Given two angles and the included side. The sum and difference of the two unknown sides, and hence the sides, can be found from Napier's analogies (Art. 5-43). The third angle can be computed from these equations or from the law of cosines (Arts. 5-39 and 5-43). The computation is similar to the case in which two sides and the included angle are given.

5. Given two sides and the angle opposite one of them (two solutions are possible). Find the angle opposite the second side by the sine law. Napier's analogies (Art. 5-43) can be used to determine the unknown side and the third angle.

• In spherical triangle ABC, given $a = 46°$, $b = 65°$, and $A = 40°$, find c, B, and C.

$$\sin B = \frac{\sin A}{\sin a} \sin b = \frac{\sin 40°}{\sin 46°} \sin 65° = 0.810$$
$$B = 54°6' \quad \text{or} \quad 180° - 54°6' = 125°54'$$

Since $(90° - a) > (90° - b)$, a and A belong to the same quadrant. But both values of B are possible. Also, $\tfrac{1}{2}(a + b) = \tfrac{1}{2}(46° + 65°) = 55.5°$ is in the first quadrant, so $\tfrac{1}{2}(A + B)$ must also be in the first quadrant; both values of B satisfy this equation.

$$\tan \tfrac{1}{2}c' = \frac{\sin \tfrac{1}{2}(A + B')}{\sin \tfrac{1}{2}(A - B')} \tan \tfrac{1}{2}(a - b)$$
$$= \frac{\sin \tfrac{1}{2}(40° + 54°6')}{\sin \tfrac{1}{2}(40° - 54°6')} \tan \tfrac{1}{2}(46° - 65°)$$
$$= 1.000$$
$$\tfrac{1}{2}c' = 45°$$
$$c' = 90°$$

$$\tan \tfrac{1}{2}c'' = \frac{\sin \tfrac{1}{2}(A + B'')}{\sin \tfrac{1}{2}(A - B'')} \tan \tfrac{1}{2}(a - b)$$

$$= \frac{\sin \tfrac{1}{2}(40° + 125°54')}{\sin \tfrac{1}{2}(40° - 125°54')} \tan \tfrac{1}{2}(46° - 65°)$$

$$\tfrac{1}{2}c'' = 13°43'$$
$$c'' = 27°26'$$

Similarly, C is found to be $116°38'$ or $24°16'$. Hence, the solutions are

$B' = 54°6'$	$C' = 116°38'$	$c' = 90°$
$B'' = 125°54'$	$C'' = 24°16'$	$c'' = 27°26'$

6. Given two angles and the side opposite one of them. Use the same formulas as for the case when two sides and the angle opposite one of them are given.

5-47. L'Huilier's Formula. The spherical excess of a triangle is the number of degrees the sum of the angles of the triangle exceeds 180°. It is equal to the area of the triangle in degrees (Art. 4-44). The area in square units equals $\pi r^2 E/180$, where E is the spherical excess.

In spherical triangle ABC, with three sides given and $s = \tfrac{1}{2}(a + b + c)$, the spherical excess can be computed from

$$\tan \tfrac{1}{4}E = \sqrt{\tan \tfrac{1}{2}s \tan \tfrac{1}{2}(s - a) \tan \tfrac{1}{2}(s - b) \tan \tfrac{1}{2}(s - c)}$$

• Given spherical triangle ABC with $a = 47°$, $b = 53°$, and $c = 64°$ on a sphere 200 ft in diameter. Find the area.

$$s = \tfrac{1}{2}(47° + 53° + 64°) = 82°; \ \tfrac{1}{2}s = 41°.$$
$$\tfrac{1}{2}(s - a) = 17°30'$$
$$\tfrac{1}{2}(s - b) = 14°30'$$
$$\tfrac{1}{2}(s - c) = 9°$$
$$\tan \tfrac{1}{4}E = \sqrt{\tan 41° \tan 17°30' \tan 14°30' \tan 9°} = 0.1062$$
$$\tfrac{1}{4}E = 6°4'$$
$$E = 24°16' = 24.27°$$
$$\text{Area} = \pi r^2 E/180 = \pi(100)^2 24.27/180 = 4,230 \text{ sq ft}$$

BIBLIOGRAPHY

Brooks, M., A. C. Schock, and A. I. Oliver: "Trigonometry for Today," 2d ed., McGraw-Hill Book Company, Inc., New York, 1959.

Bruhns, C.: "A New Manual of Logarithms to Seven Places of Decimals," D. Van Nostrand Company, Inc., Princeton, N.J., 1941.

Burington, R. S.: "Handbook of Mathematical Tables and Formulas," 3d ed., McGraw-Hill Book Company, Inc., New York, 1948.

Ives, H. C.: "Natural Trigonometric Functions to Seven Decimal Places for Every Ten Seconds of Arc," 2d ed., John Wiley & Sons, Inc., New York, 1942.

Kells, L. M., W. F. Kern, and J. R. Bland: "Plane and Spherical Trigonometry," 3d ed., McGraw-Hill Book Company, Inc., New York, 1951.

Palmer, C. I., C. W. Leigh, and S. Kimball: "Plane and Spherical Trigonometry," 5th ed., McGraw-Hill Book Company, Inc., New York, 1950.

Chapter 6

ANALYTIC GEOMETRY

6-1. Rectangular (Cartesian) Coordinates in a Plane. Every point in a plane can be located by its distances from two perpendicular reference lines, called axes. The distances are called the cartesian or rectangular coordinates of the point.

The intersection of the axes is called the origin (O in Fig. 6-1). The segments of the axes on one side of the origin (OX and OY) are assigned positive values; the positive directions are indicated by arrows on the axes. Coordinates measured in those directions also are positive. The parts of the axes on the other side of the origin (OX' and OY') and coordinates measured in those directions are negative.

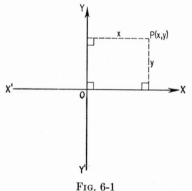

FIG. 6-1

One axis (XX', usually drawn horizontal and associated with the variable x) is called the abscissa axis and coordinates along or parallel to it are called abscissas. The other axis (YY', usually drawn vertical and associated with the variable y) is called the ordinate axis; coordinates measured along or parallel to it are called ordinates.

A point is denoted by the number pair (x,y), where x is its abscissa and y its ordinate. The origin is $(0,0)$.

6-2. Skew (Oblique) Coordinates in a Plane. Points in a plane can be located by specifying two coordinates with respect to two oblique reference lines, or axes. The coordinates are measured parallel to the respective axes (Fig. 6-2).

6-3. Curvilinear Coordinates in a Plane. Coordinates of points in a plane need not be measured along straight lines. Two intersecting curves can be used to locate a point (Fig. 6-3). The coordinates determine the specific curves of a family that pass through the point.

144

If the curvilinear system consists of two families of curves u_1 and u_2, then in cartesian coordinates

$$u_1 = f_1(x,y)$$
$$u_2 = f_2(x,y)$$

The most useful systems, in general, are those in which one family of curves is normal to the other family. Such systems are said to be orthogonal. Cartesian and polar coordinates are special cases of orthogonal curvilinear coordinates. For cartesian coordinates, u_1 and u_2 are the

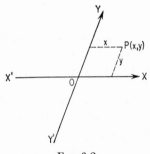

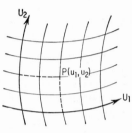

FIG. 6-2   FIG. 6-3

straight lines $f_1(x,y) = x = c_1$ and $f_2(x,y) = y = c_2$. For polar coordinates, u_1 curves are the circles $f_1(x,y) = x^2 + y^2 = \rho^2$ and u_2 curves are the straight lines $f_2(x,y) = y/x = \tan \theta$.

6-4. Polar Coordinates in a Plane. The position of a point in a plane is fixed by its coordinates with respect to a fixed line OX (Fig. 6-4), called the polar axis, and a fixed point O on the line, called the pole. One coordinate is ρ, the length of the line from the origin to the point, and the second coordinate is θ, the angle from the polar axis to that line.

The angle θ is called the vectorial angle and is considered positive when measured counterclockwise; ρ, called the radius vector, is usually taken positive at all times (but it is feasible to consider

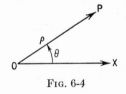

FIG. 6-4

it positive or negative, depending on whether P lies on the terminal line of θ or on that line produced through the pole).

Polar coordinates are related to cartesian coordinates with origin coincident with the pole and abscissa axis coincident with the polar axis by the equations

$$x = \rho \cos \theta$$
$$y = \rho \sin \theta$$
$$\rho^2 = x^2 + y^2$$
$$\theta = \tan^{-1} \frac{y}{x}$$

6-5. Rectangular (Cartesian) Coordinates in Space.

A point in space can be located by its distance from three mutually perpendicular reference planes. The distances are called the coordinates of the point and the planes are called coordinate planes. The perpendicular intersections of the planes are called the coordinate-axes and the intersection of the axes is called the origin. In Fig. 6-5, O is the origin; OX, OY, and OZ are the axes.

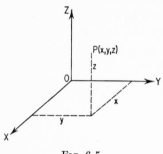

FIG. 6-5

Rectangular coordinate systems are said to be right-handed when the positive directions of the axes are chosen in such a way that if the x axis is rotated into the y axis, a right-handed screw similarly rotated would move in the positive z direction.

A point in space is denoted by a number triplet (x,y,z), where x, y, and z are the coordinates of the point measured along or parallel to the x, y, and z axes, respectively. The origin is $(0,0,0)$.

6-6. Polar Coordinates in Space.

A point in space can be located by the radius vector ρ (from the origin to the point) and the direction angles α, β, and γ of ρ (Fig. 6-6).

Polar coordinates are related to cartesian coordinates with origin at the pole by

$$x = \rho \cos \alpha \qquad y = \rho \cos \beta \qquad z = \rho \cos \gamma$$
$$\rho^2 = x^2 + y^2 + z^2$$

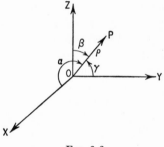

FIG. 6-6

6-7. Curvilinear Coordinates in Space.

The location of a point in space is determined by the intersection of three surfaces. The coordinates determine the specific surfaces of a family that pass through the point.

If the curvilinear coordinates are u_1, u_2, and u_3, then in cartesian coordinates,

$$u_1 = f_1(x,y,z)$$
$$u_2 = f_2(x,y,z)$$
$$u_3 = f_3(x,y,z)$$

The most useful systems, in general, are those in which the families of surfaces are normal to each other. Such systems are called orthogonal.

Special cases of orthogonal systems include cartesian, spherical, cylindrical, and ellipsoidal coordinates. For cartesian coordinates, u_1, u_2, and u_3 are the planes $u_1 = c_1$, $u_2 = c_2$, and $u_3 = c_3$ (Art. 6-5). For spherical coordinates, the surfaces are the spheres $u_1 = \rho$, a family of cones $u_2 = \theta$, and a family of planes through the z axis, $u_3 = \varphi$ (Art. 6-8). For cylindrical coordinates, the surfaces are a family of coaxial cylinders $u_1 = x^2 + y^2 = \rho^2$, a family of planes through the z axis, $u_2 = \varphi$, and a family of planes perpendicular to the z axis, $u_3 = c$ (Art. 6-9). Ellipsoidal coordinates are defined in Art. 6-10.

6-8. Spherical Coordinates. This system of coordinates for locating a point in space, similar to polar coordinates, is based on the radius vector ρ from the origin to the point (Arts. 6-4 and 6-6). A point is completely determined by its distance OP from the origin O (Fig. 6-7), the angle θ which the radius vector makes with the z axis, and the angle φ which the plane through ρ and the z axis makes with the xz coordinate plane.

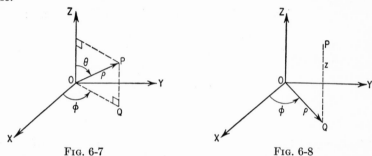

FIG. 6-7   FIG. 6-8

Cartesian coordinates can be transformed into spherical coordinates by the equations

$$x = \rho \sin \theta \cos \varphi$$
$$y = \rho \sin \theta \sin \varphi$$
$$z = \rho \cos \theta$$

6-9. Cylindrical Coordinates. Another system for locating a point in space, similar to polar coordinates in a plane (Art. 6-4), is based on the radius vector ρ to the projection of the point on the xy plane. The third coordinate is the same as the z coordinate in the cartesian system (Art. 6-5). A point P is determined by the polar coordinates ρ and φ (Fig. 6-8) of the projection Q of P on the xy plane and the distance QP of P from that plane.

Cartesian coordinates can be transformed into cylindrical coordinates by the equations

$$x = \rho \cos \varphi$$
$$y = \rho \sin \varphi$$
$$z = z$$

6-10. Ellipsoidal Coordinates. A point in space can be located at the intersection of an ellipsoid (Art. 6-68), a hyperboloid of one sheet (Art.

6-69), and a hyperboloid of two sheets (Art. 6-70). These three surfaces in the ellipsoidal system are confocal. Their equations in cartesian coordinates are:

$$\frac{x^2}{a^2 + u_1} + \frac{y^2}{b^2 + u_1} + \frac{z^2}{c^2 + u_1} = 1 \qquad u_1 > -c^2$$

$$\frac{x^2}{a^2 + u_2} + \frac{y^2}{b^2 + u_2} + \frac{z^2}{c^2 + u_2} = 1 \qquad -c^2 > u_2 > -b^2$$

$$\frac{x^2}{a^2 + u_3} + \frac{y^2}{b^2 + u_3} + \frac{z^2}{c^2 + u_3} = 1 \qquad -b^2 > u_3 > -a^2$$

where a, b, and c are half the major, mean, and minor axes of the ellipsoid (first equation) with $a > b > c$, and u_1, u_2, and u_3 are ellipsoidal coordinates.

The transformation equations from cartesian to ellipsoidal are:

$$x^2 = \frac{(a^2 + u_1)(a^2 + u_2)(a^2 + u_3)}{(a^2 - b^2)(a^2 - c^2)}$$

$$y^2 = \frac{(b^2 + u_1)(b^2 + u_2)(b^2 + u_3)}{(b^2 - c^2)(b^2 - a^2)}$$

$$z^2 = \frac{(c^2 + u_1)(c^2 + u_2)(c^2 + u_3)}{(c^2 - a^2)(c^2 - b^2)}$$

6-11. Parametric Equations.   Coordinates of curves or surfaces may be expressed as functions of other variables, or parameters. If t is a parameter, for example, the equations of a curve in plane coordinates may have the form $x = f_1(t)$, $y = f_2(t)$. A curve or surface may be defined by any of several possible parametric equations.

• Express $x = 2/t$, $y = 4t$ in rectangular coordinates.
Eliminate the parameter t from the equations.

$$t = \frac{y}{4} = \frac{2}{x}$$

Hence, the required equation is the hyperbola $xy = 8$.

• Given $x^2y^2 - 4x^2 + 9y^2 = 0$, write parametric equations in terms of trigonometric functions of an angle θ.
Let $y = 2 \sin \theta$ and substitute in the equation. Thus,

$$4x^2 \sin^2 \theta - 4x^2 + 36 \sin^2 \theta = x^2(\sin^2 \theta - 1) + 9 \sin^2 \theta = 0$$

Substitute $-\cos^2 \theta$ for $\sin^2 \theta - 1$ and transpose.

$$x^2 \cos^2 \theta = 9 \sin^2 \theta$$

$$x = \frac{3 \sin \theta}{\cos \theta} = 3 \tan \theta$$

Thus the parametric equations could be $x = 3 \tan \theta$, $y = 2 \sin \theta$

6-12. Translation of Rectangular Axes.

Assume that in a plane a point P (Fig. 6-9) has coordinates (x,y). If the axes are shifted without rotation so that the origin is translated to the point (x_0,y_0), the equations of transformation are:

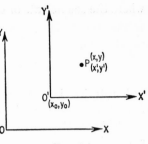

FIG. 6-9

$$x = x' + x_0$$
$$y = y' + y_0$$

where (x',y') are the new coordinates of P.

Similarly, the transformation equations for translation of the origin of cartesian coordinates in space to (x_0,y_0,z_0) are:

$$x = x' + x_0$$
$$y = y' + y_0$$
$$z = z' + z_0$$

• Simplify the equation $x^2 + 2x + y^2 - 6y = 6$.

Collect the terms in x and y and complete the squares by adding $1 + 9 = 10$ to both sides of the equation.

$$(x^2 + 2x + 1) + (y^2 - 6y + 9) = 6 + 10$$
$$(x + 1)^2 + (y - 3)^2 = 16$$

Set $x + 1 = x'$ and $y - 3 = y'$ by translating the origin to $(-1,3)$. The simplified equation is

$$x'^2 + y'^2 = 16$$

6-13. Rotation of Rectangular Axes.

If a point P in a plane (Fig. 6-10) has coordinates (x,y), and if the axes are rotated through an angle θ about the origin, the transformation equations are:

$$x = x' \cos \theta - y' \sin \theta$$
$$y = x' \sin \theta + y' \cos \theta$$

where (x',y') are the new coordinates of P.

The rotation of cartesian axes in space can be expressed as successive rotations about each of the three axes in turn.

FIG. 6-10

Any rotation of two axes about the third is called a planar rotation.

The transformation equations for a general rotation of axes in space have the form:

$$x = a_{11}x' + a_{12}y' + a_{13}z'$$
$$y = a_{21}x' + a_{22}y' + a_{23}z'$$
$$z = a_{31}x' + a_{32}y' + a_{33}z'$$

where the coefficients of the variables are constants and their determinant

$$\begin{vmatrix} a_{11} & a_{12} & a_{13} \\ a_{21} & a_{22} & a_{23} \\ a_{31} & a_{32} & a_{33} \end{vmatrix} = 1$$

A rotation is one type of orthogonal transformation of coordinates (Art. 6-18) and a special case of affine transformation (Art. 6-16).

A shift of the origin to (x_0, y_0, z_0) can be carried out at the same time as a rotation by adding to the preceding transformation equations x_0, y_0, and z_0, respectively (see also Art. 6-12).

When the product xy of two variables appears in quadratic equations, it can be removed by rotation of the axes through an angle θ determined by

$$\tan 2\theta = \frac{B}{A - C}$$

where B = coefficient of xy term
A = coefficient of x^2
C = coefficient of y^2

• Remove the xy term from $x^2 - 2xy + y^2 = 12$.
Here, $B = -2$, $A = 1$, and $C = 1$.

$$\tan 2\theta = \frac{B}{A - C} = \frac{-2}{1 - 1} = \infty$$
$$2\theta = 90°$$
$$\theta = 45°$$

Hence, the transformation equations are

$$x = x' \cos 45° - y' \sin 45° = \frac{\sqrt{2}}{2} x' - \frac{\sqrt{2}}{2} y'$$

$$y = x' \sin 45° + y' \cos 45° = \frac{\sqrt{2}}{2} x' + \frac{\sqrt{2}}{2} y'$$

Substitution of these values in the original equation yields

$$\left(\frac{\sqrt{2}}{2} x' - \frac{\sqrt{2}}{2} y' \right)^2 - 2 \left(\frac{\sqrt{2}}{2} x' - \frac{\sqrt{2}}{2} y' \right) \left(\frac{\sqrt{2}}{2} x' + \frac{\sqrt{2}}{2} y' \right)$$
$$+ \left(\frac{\sqrt{2}}{2} x' + \frac{\sqrt{2}}{2} y' \right)^2 = 12$$
$$\tfrac{1}{2}x'^2 - x'y' + \tfrac{1}{2}y'^2 - x'^2 + y'^2 + \tfrac{1}{2}x'^2 + x'y' + \tfrac{1}{2}y'^2 = 12$$
$$y'^2 = 6$$

Thus, after rotation of the axes through 45°, $x^2 - 2xy + y^2 = 12$ becomes $y'^2 = 6$.

6-14. Inversion (Reflection). A transformation of coordinates effected by changing the direction of one coordinate axis.

If, for example, the x axis is inverted, the transformation equations are

$$x = -x'$$
$$y = y$$
$$z = z$$

Successive inversion of two axes is equivalent to a rotation of the two axes 180° about the third.

6-15. Perversion. A transformation of coordinates effected by changing the direction of a coordinate axis, then rotating the axes. It is one type of orthogonal transformation (Art. 6-18).

The transformation equations have the form

$$x = a_{11}x' + a_{12}y' + a_{13}z'$$
$$y = a_{21}x' + a_{22}y' + a_{23}z'$$
$$z = a_{31}x' + a_{32}y' + a_{33}z'$$

where the coefficients of the variables are constants and their determinant

$$\begin{vmatrix} a_{11} & a_{12} & a_{13} \\ a_{21} & a_{22} & a_{23} \\ a_{31} & a_{32} & a_{33} \end{vmatrix} = -1$$

6-16. Linear (Affine) Transformation. A transformation of coordinates such that a point $P(x,y,z)$, in effect, is shifted to $Q(x',y',z')$, and all points in the original coordinate system that were finite remain finite, infinite points remain infinite, straight lines remain straight, and parallel lines remain parallel.

The transformation equations, when the origin is unchanged, are:

$$x = a_{11}x' + a_{12}y' + a_{13}z'$$
$$y = a_{21}x' + a_{22}y' + a_{23}z'$$
$$z = a_{31}x' + a_{32}y' + a_{33}z'$$

where the coefficients of the variables are constants.

Orthogonal transformations and rotations of axes are special cases of linear transformation (Arts. 6-18 and 6-13), as are inversions or reflections (Art. 6-14) and perversions (Art. 6-15).

6-17. Successive Transformations. Two successively applied linear (affine) transformations (Art. 6-16) are equivalent to a single affine transformation. If the two transformations are given by

$$x = a_{11}x' + a_{12}y' + a_{13}z' \qquad x' = b_{11}x'' + b_{12}y'' + b_{13}z''$$
$$y = a_{21}x' + a_{22}y' + a_{23}z' \qquad y' = b_{21}x'' + b_{22}y'' + b_{23}z''$$
$$z = z_{31}x' + a_{32}y' + a_{33}z' \qquad z' = b_{31}x'' + b_{32}y'' + b_{33}z''$$

then the single equivalent transformation is given by

$$x = c_{11}x'' + c_{12}y'' + c_{13}z''$$
$$y = c_{21}x'' + c_{22}y'' + c_{23}z''$$
$$z = c_{31}x'' + c_{32}y'' + c_{33}z''$$

The c coefficients can be obtained by premultiplying the matrix of the a coefficients by the matrix of the b coefficients (Art. 13-1). The single equivalent transformation is called the product of the two transformations.

6-18. Orthogonal Transformations. A special case of linear transformation (Art. 6-16) in which the determinant of the coefficients of the variables in the transformation equations equals ± 1.

$$\begin{vmatrix} a_{11} & a_{12} & a_{13} \\ a_{21} & a_{22} & a_{23} \\ a_{31} & a_{32} & a_{33} \end{vmatrix} = \pm 1$$

If the determinant has the value $+1$, the transformation equations represent a rotation (Art. 6-13). If the determinant equals -1, the equations represent a perversion (Art. 6-15).

Under an orthogonal transformation, lengths and distances remain unchanged. The general characteristics of a linear transformation also hold: finite points remain finite, infinite points remain infinite, straight lines remain straight, and parallel lines remain parallel. If (x,y,z) are the coordinates of a point before an orthogonal transformation and (x',y',z') are its coordinates after the transformation, then

$$x^2 + y^2 + z^2 = x'^2 + y'^2 + z'^2$$

When the transformation equations are solved for x', y', and z',

$$x' = a_{11}x + a_{21}y + a_{31}z$$
$$y' = a_{12}x + a_{22}y + a_{32}z$$
$$z' = a_{13}x + a_{23}y + a_{33}z$$

the same nine a coefficients appear as before, but now those that were in the same row appear in the same column.

The coefficients are related by the equations

$$a_{i1}^2 + a_{i2}^2 + a_{i3}^2 = 1 \qquad i = 1,2,3$$
$$a_{i1}a_{j1} + a_{i2}a_{j2} + a_{i3}a_{j3} = 0 \qquad i = 1,2,3; j = 1,2,3; i \neq j$$

6-19. Intersection of Curves. If two curves meet, the coordinates of the points of intersection can be found by simultaneous solution of their equations. Discard imaginary and complex solutions.

• At what points does $y^2 = x^3$ intersect $x^2 + y^2 - 6x = 0$?
Substitute $y^2 = x^3$ in the second equation and solve for x.

$$x^2 + x^3 - 6x = 0$$
$$x(x^2 + x - 6) = x(x - 2)(x + 3) = 0$$
$$x = 0, 2, -3$$

The root $x = -3$ gives an imaginary value of y and must be excluded.

$$y = \pm x^{\frac{3}{2}} = 0, \pm \sqrt{8}$$

The points of intersection are $(0,0)$, $(2,\sqrt{8})$, and $(2,-\sqrt{8})$.

6-20. Length of a Straight Line. The length of a straight line in a plane joining two points $P_1(x_1,y_1)$ and $P_2(x_2,y_2)$ is given by

$$l = \sqrt{(x_1 - x_2)^2 + (y_1 - y_2)^2}$$

Similarly, the length of a line in space between (x_1,y_1,z_1) and (x_2,y_2,z_2) is equal to

$$l = \sqrt{(x_1 - x_2)^2 + (y_1 - y_2)^2 + (z_1 - z_2)^2}$$

In polar coordinates, the length of a line in a plane between $P_1(\rho_1,\theta_1)$ and $P_2(\rho_2,\theta_2)$ is given by

$$l = \sqrt{\rho_1{}^2 + \rho_2{}^2 - 2\rho_1\rho_2 \cos (\theta_2 - \theta_1)}$$

• What is the distance between $(4,8)$ and $(1,4)$?

Let $x_1 = 4$, $y_1 = 8$; $x_2 = 1$, $y_2 = 4$. Then the distance (length of line) is

$$l = \sqrt{(4 - 1)^2 + (8 - 4)^2} = \sqrt{(3)^2 + (4)^2} = 5$$

6-21. Division of a Line. The straight line between $P_1(x_1,y_1,z_1)$ and $P_2(x_2,y_2,z_2)$ is divided by point $P(x,y,z)$ in the ratio $P_1P/PP_2 = r$ when

$$x = \frac{x_1 + rx_2}{1 + r} \qquad y = \frac{y_1 + ry_2}{1 + r} \qquad z = \frac{z_1 + rz_2}{1 + r}$$

If P is the midpoint, $r = 1$ and

$$x = \tfrac{1}{2}(x_1 + x_2) \qquad y = \tfrac{1}{2}(y_1 + y_2) \qquad z = \tfrac{1}{2}(z_1 + z_2)$$

• Show that the diagonals of a parallelogram bisect each other.

Place the parallelogram $OABC$ with one vertex at the origin and one side along the x axis. Assign coordinates to the vertices as shown in Fig. 6-11, taking into account the fact that opposite sides of the parallelogram are equal and parallel.

The coordinates of the midpoint of diagonal AC are

$$x = \tfrac{1}{2}(a + c) \qquad y = \tfrac{1}{2}(b + 0) = \frac{b}{2}$$

The coordinates of the midpoint of diagonal OB are

$$x = \tfrac{1}{2}(a + c + 0) = \tfrac{1}{2}(a + c)$$

$$y = \tfrac{1}{2}(b + 0) = \frac{b}{2}$$

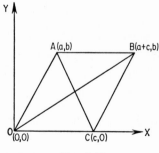

FIG. 6-11

Since the coordinates of both midpoints are identical, they represent the same point.   Therefore, the diagonals bisect each other.

6-22. Slope of a Straight Line.   The inclination θ of a line is the angle between it and the x axis, measured counterclockwise from the x axis (Fig. 6-12).   The slope of a line is the tangent of its inclination.

Lines parallel to the x axis have zero slope; lines parallel to the y axis have infinitely large slope.

If the coordinates (x_1, y_1) and (x_2, y_2) of two points on a line are known, the slope is given by

$$m = \frac{y_2 - y_1}{x_2 - x_1}$$

• What are the slope and inclination of a line passing through $(-1, -2)$ and $(3, -4)$?

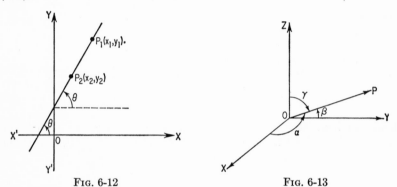

FIG. 6-12   FIG. 6-13

Here, $x_1 = -1$, $y_1 = -2$; $x_2 = 3$, $y_2 = -4$.   Then

$$m = \frac{-4 - (-2)}{3 - (-1)} = \frac{-2}{4} = -\frac{1}{2}$$

The inclination is $\theta = \tan^{-1} m = \tan^{-1}(-\tfrac{1}{2}) = 153°26'$.

6-23. Direction Angles of a Line.   The direction angles of a directed straight line in space are the angles between the line (or a line parallel to it through the origin) and each of the coordinate axes.   In Fig. 6-13, α, β, and γ are the direction angles of OP.

6-24. Direction Cosines of a Line.   These are the cosines of the direction angles (α, β, and γ in Fig. 6-13).

The sum of the squares of the direction cosines equals unity.

$$\cos^2 \alpha + \cos^2 \beta + \cos^2 \gamma = 1$$

If $P_1(x_1, y_1, z_1)$ and $P_2(x_2, y_2, z_2)$ are any two points on a line in space and the distance between them is

$$l = \sqrt{(x_2 - x_1)^2 + (y_2 - y_1)^2 + (z_2 - z_1)^2}$$

then the direction cosines of the line, when the positive direction is from P_1 to P_2, are

$$\cos \alpha = \frac{x_2 - x_1}{l} \qquad \cos \beta = \frac{y_2 - y_1}{l} \qquad \cos \gamma = \frac{z_2 - z_1}{l}$$

• Find the direction cosines of the line directed from $P_1(2,-1,3)$ to $P_2(5,3,15)$. The length of the segment P_1P_2 is

$$l = \sqrt{(5-2)^2 + (3+1)^2 + (15-3)^2} = 13$$

$$\cos \alpha = \frac{5-2}{13} = \frac{3}{13}$$

$$\cos \beta = \frac{3+1}{13} = \frac{4}{13}$$

$$\cos \gamma = \frac{15-3}{13} = \frac{12}{13}$$

6-25. Direction Numbers of a Line. The cartesian coordinates of any point on a line through the origin parallel to the given line and similarly directed.

These numbers are proportional to the direction cosines of the line and thus determine its direction in space. They are also proportional to the difference between the corresponding coordinates of any two points on the line.

If a, b, and c are the direction numbers of a line, the direction cosines are

$$\cos \alpha = \frac{a}{\sqrt{a^2 + b^2 + c^2}} \qquad \cos \beta = \frac{b}{\sqrt{a^2 + b^2 + c^2}} \qquad \cos \gamma = \frac{c}{\sqrt{a^2 + b^2 + c^2}}$$

If $P_1(x_1,y_1,z_1)$ and $P_2(x_2,y_2,z_2)$ are two points on a line directed from P_1 to P_2, then

$$\frac{x_2 - x_1}{a} = \frac{y_2 - y_1}{b} = \frac{z_2 - z_1}{c}$$

If a direction number is zero, the line is perpendicular to the corresponding coordinate axis.

• Find a set of direction numbers and direction cosines of a line directed from $P_1(3,-2,4)$ to $P_2(6,2,16)$.

The direction numbers are proportional to the differences between corresponding coordinates.

$$\frac{6-3}{a} = \frac{2+2}{b} = \frac{16-4}{c} \qquad \text{or} \qquad \frac{3}{a} = \frac{4}{b} = \frac{12}{c}$$

Thus, 3, 4, and 12 are a set of direction numbers.

$$\sqrt{(3)^2 + (4)^2 + (12)^2} = 13, \text{ and}$$

$$\cos \alpha = \tfrac{3}{13} \qquad \cos \beta = \tfrac{4}{13} \qquad \cos \gamma = \tfrac{12}{13}$$

6-26. Angle between Two Lines. In a plane coordinate system, the angle between two directed lines with slopes m_1 and m_2 ($m_1 > m_2$) is given by

$$\tan \theta = \frac{m_1 - m_2}{1 + m_1 m_2}$$

• What is the angle between the line from $P_1(3,3)$ to $P_2(5,6)$ and the line from $P_3(-3,1)$ to $P_4(9,7)$?

From the slope formula in Art. 6-22,

$$m_1 = \frac{y_2 - y_1}{x_2 - x_1} = \frac{6 - 3}{5 - 3} = \frac{3}{2} = 1.5$$

$$m_2 = \frac{y_4 - y_3}{x_4 - x_3} = \frac{7 - 1}{9 + 3} = \frac{6}{12} = 0.5$$

$$\tan \theta = \frac{1.5 - 0.5}{1 + 1.5 \times 0.5} = 0.57143$$

$$\theta = 29°45'$$

The angle between two lines in space is defined as the angle between two intersecting lines parallel to the given lines and similarly directed. If α_1, β_1, and γ_1 and α_2, β_2, and γ_2 are the direction angles of the two lines, the angle between them can be computed from

$$\cos \theta = \cos \alpha_1 \cos \alpha_2 + \cos \beta_1 \cos \beta_2 + \cos \gamma_1 \cos \gamma_2$$

• What is the angle between the line whose direction cosines are 0.30, 0.60, and 0.74 and the line whose direction cosines are 0.85, 0.25, and 0.46?

$$\cos \theta = 0.30 \times 0.85 + 0.60 \times 0.25 + 0.74 \times 0.46 = 0.745$$

$$\theta = 41°50'$$

6-27. Angle between Two Plane Curves. If two plane curves intersect, the angle between them is either of two supplementary angles between the tangents at the point of intersection. The slopes m_1 and m_2 of the tangents are the values of the derivatives at the point. The angle of intersection can be obtained from

$$\theta = \tan^{-1} \frac{m_2 - m_1}{1 + m_1 m_2}$$

• At what angles do the parabolas $y^2 = ax$ and $x^2 = ay$ intersect?

The curves meet at $(0,0)$ and (a,a). If m_1 is the slope of the tangent to $y^2 = ax$ and m_2 is the slope of the tangent to $x^2 = ay$, then

$$m_1 = \frac{dy}{dx} = \frac{a}{2y} = \infty, \tfrac{1}{2}$$

$$m_2 = \frac{dy}{dx} = \frac{2x}{a} = 0, 2$$

$$\tan \theta = \frac{m_2 - m_1}{1 + m_1 m_2} = \frac{m_2/m_1 - 1}{1/m_1 + m_2} = \infty, \tfrac{3}{4}$$

$$\theta = 90°, 36°52'$$

6-28. Angle between a Line and a Plane. This is defined as the angle between the line and its projection on the plane (see also Art. 6-26). It can be computed from

$$\sin \theta = \cos \alpha_1 \cos \alpha_2 + \cos \beta_1 \cos \beta_2 + \cos \gamma_1 \cos \gamma_2$$

where α_1, β_1, and γ_1 are the direction cosines of the line and α_2, β_2, and γ_2 are the direction cosines of the normal to the plane (Arts. 6-24 and 6-45).

• Find the angle between the line whose direction cosines are 0.30, 0.60, and 0.74 and the plane $3.4x + y + 1.85z = 40$.

To obtain the direction cosines of the normal to the plane, put the plane equation in the normal form by dividing through by $\sqrt{(3.4)^2 + (1)^2 + (1.85)^2} = 4$.

$$0.85x + 0.25y + 0.46z = 10$$

Hence, the direction cosines of the normal to the plane are 0.85, 0.25, and 0.46.

$$\sin \theta = 0.30 \times 0.85 + 0.60 \times 0.25 + 0.74 \times 0.46 = 0.745$$
$$\theta = 48°10'$$

6-29. Angle between Two Planes. This is equal to the angle between two intersecting normals to the planes. If α_1, β_1, and γ_1 and α_2, β_2, and γ_2 are the direction cosines of the normals (Art. 6-45), the angle between the planes can be computed from

$$\cos \theta = \cos \alpha_1 \cos \alpha_2 + \cos \beta_1 \cos \beta_2 + \cos \gamma_1 \cos \gamma_2$$

• Find the angle between the planes $x + 2y + 2.47z = 6.66$ and $3.4x + y + 1.85z = 16$.

Divide the first equation by $\sqrt{(1)^2 + (2)^2 + (2.47)^2} = 3.33$ and the second by $\sqrt{(3.4)^2 + (1)^2 + (1.85)^2} = 4$ to put them in normal form (Art. 6-45). The equations then are

$$0.30x + 0.60y + 0.74z = 2$$
$$0.85x + 0.25y + 0.46z = 4$$

Hence, the direction cosines of the two normals to the planes are, respectively, 0.30, 0.60, 0.74 and 0.85, 0.25, 0.46. The angle between the planes is given by

$$\cos \theta = 0.30 \times 0.85 + 0.60 \times 0.25 + 0.74 \times 0.46 = 0.745$$
$$\theta = 41°50'$$

6-30. Tests for Parallel Lines. In a plane coordinate system, if lines are parallel, their slopes are equal.

Two lines in space are parallel and in the same direction if their direction angles or direction cosines are equal or if their direction numbers are proportional. They will be parallel but oppositely directed if the direction angles are supplementary.

• Show that the line from $P_1(1,2,3)$ to $P_2(4,6,15)$ and the line from $P_3(-1,-2,3)$ to $P_4(5,6,27)$ are parallel.

Direction numbers for P_1P_2 (Art. 6-25) are:

$$a_1 = x_2 - x_1 = 4 - 1 = 3$$
$$b_1 = y_2 - y_1 = 6 - 2 = 4$$
$$c_1 = z_2 - z_1 = 15 - 3 = 12$$

Direction numbers for P_3P_4 are

$$a_2 = x_4 - x_3 = 5 + 1 = 6$$
$$b_2 = y_4 - y_3 = 6 + 2 = 8$$
$$c_2 = z_4 - z_3 = 27 - 3 = 24$$

Since $a_1/a_2 = b_1/b_2 = c_1/c_2$, the lines are parallel.

6-31. Tests for a Parallel to a Plane. A line is parallel to a plane if the sum of the products of the direction numbers of the line (Art. 6-25) and of a normal to the plane (Art. 6-45) is zero.

• Is the plane $4x + 3y - 2z = 24$ parallel to the line

$$8x + 3y - 3z = 5$$
$$16x - 3y - 3z = 1$$

To obtain the direction numbers of the line, put its equations in symmetric form (Art. 6-44). First eliminate y from the equations, then eliminate z. Addition of the equations gives $24x - 6z = 6$; subtraction yields $8x - 6y = -4$. Solution of both equations for x results in

$$\frac{x}{1} = \frac{y - \frac{2}{3}}{\frac{4}{3}} = \frac{z + 1}{4}$$

The denominators are a set of direction numbers. The direction numbers of a normal to the plane are the coefficients in its equation, 4, 3, and -2. The sum of the products of the direction numbers is

$$4 \times 1 + 3 \times \tfrac{4}{3} - 2 \times 4 = 0$$

so the line is parallel to the plane.

6-32. Tests for Perpendicularity. In a plane coordinate system, two lines are perpendicular if their slopes are negative reciprocals.

$$m_1 m_2 = -1$$

• Show that the line from $P_1(-2,-1)$ to $P_2(4,1)$ is perpendicular to the line from $P_3(3,4)$ to P_2.

$$m_1 = \frac{y_2 - y_1}{x_2 - x_1} = \frac{1 - (-1)}{4 - (-2)} = \frac{1}{3}$$
$$m_2 = \frac{y_2 - y_3}{x_2 - x_3} = \frac{1 - 4}{4 - 3} = -3$$

Since $m_1 m_2 = -3 \times \tfrac{1}{3} = -1$, the lines are perpendicular.

Two lines in space are perpendicular if the sum of the products of their direction numbers (Art. 6-25) is zero.

• If line AB has direction numbers 2, $\frac{3}{2}$, and -1, and CD has direction numbers 3, 4, and 12, show that AB is perpendicular to CD.

Since $2 \times 3 + \frac{3}{2} \times 4 - 1 \times 12 = 0$, the lines are perpendicular.

A line is perpendicular to a plane if the direction numbers of the line are proportional to the direction numbers of a normal to the plane (Art. 6-45).

• Show that the plane $3x + 4y + 12z = 24$ is perpendicular to the line

$$8x + 3y - 3z = 5$$
$$16x - 3y - 3z = 1$$

To obtain the direction numbers of the line, put the equation in symmetric form (Art. 6-44). First, eliminate y from the equations, then eliminate z. Addition of the equations gives $24x - 6z = 6$; subtraction yields $8x - 6y = -4$. Solution of both equations for x results in

$$\frac{x}{1} = \frac{y - \frac{2}{3}}{\frac{4}{3}} = \frac{z + 1}{4}$$

The denominators are a set of direction numbers. The direction numbers of a normal to the plane are the coefficients in its equation, 3, 4, and 12. Since these numbers are proportional to those for the line ($\frac{3}{1} = 4/(\frac{4}{3}) = 1\frac{2}{4}$), the line and the plane are perpendicular.

Two planes are perpendicular if the sum of the products of the direction numbers of a normal to each is zero.

• Show that the plane $3x + 4y + 12z = 24$ is perpendicular to the plane $4x + 3y - 2z = 17$.

Direction numbers of a normal to the first plane are the coefficients 3, 4, and 12; direction numbers of a normal to the second plane are 4, 3, and -2. The sum of the products of the direction numbers is $3 \times 4 + 4 \times 3 - 12 \times 2 = 0$. Hence the planes are perpendicular.

6-33. Orthogonality of Curves. Two curves are orthogonal at their intersection if the tangents to the curves there are perpendicular. In plane coordinates, the curves are orthogonal if the slopes of the tangents are negative reciprocals; in space coordinates, the curves are orthogonal if the sum of the products of the direction numbers of the tangents is zero.

• Show that the circles $x^2 + y^2 = 2x$ and $x^2 + y^2 = 4y$ are orthogonal.

The curves intersect at $(0,0)$ and $(\frac{8}{5}, \frac{4}{5})$. The tangents to the first circle at the intersections have the slopes ∞ and $-\frac{3}{4}$, respectively (see also Art. 6-27). The tangents to the second circle at the same points have the slopes 0 and $\frac{4}{3}$, respectively. Since one set of tangents are perpendicular and parallel to the x axis and the product of the slopes of the other set is -1, the tangents are perpendicular and the curves are orthogonal.

6-34. Distance in Rectangular Coordinates. The distance from any point in space $P(x,y,z)$ to the origin is

$$\rho = \sqrt{x^2 + y^2 + z^2}$$

The distance to the x axis is $\sqrt{y^2 + z^2}$, the distance to the y axis is $\sqrt{x^2 + z^2}$, and the distance to the z axis is $\sqrt{x^2 + y^2}$.

The distance to any other point $P_1(x_1,y_1,z_1)$ is

$$l = \sqrt{(x_1 - x)^2 + (y_1 - y)^2 + (z_1 - z)^2}$$

The distance from a point $P_1(x_1,y_1,z_1)$ to any line

$$\frac{x - x_0}{\cos \alpha} = \frac{y - y_0}{\cos \beta} = \frac{z - z_0}{\cos \gamma}$$

where α, β, and γ are direction cosines, is

$$l = \{(x_0 - x_1)^2 + (y_0 - y_1)^2 + (z_0 - z_1)^2$$
$$- [(x_0 - x_1) \cos \alpha + (y_0 - y_1) \cos \beta + (z_0 - z_1) \cos \gamma]^2\}^{\frac{1}{2}}$$

In plane coordinates, the distance to any line $x \cos \theta + y \sin \theta = p$ (Art. 6-43) is

$$l = x_1 \cos \theta + y_1 \sin \theta - p$$

If the computation gives $l > 0$, P_1 and the origin are on opposite sides of the line. If $l < 0$, P_1 and the origin are on the same side of the line.

Similarly, the distance from a plane $x \cos \alpha + y \cos \beta + z \cos \gamma = p$ to a point $P_1(x_1,y_1,z_1)$ is $l = x_1 \cos \alpha + y_1 \cos \beta + z_1 \cos \gamma - p$.

• What is the perpendicular distance from the line $4x - y + 2 = 0$ to the point $P(3,2)$?

Put the equation in normal form by dividing by $- \sqrt{(4)^2 + (-1)^2} = - \sqrt{17}$; then substitute (3,2) for (x,y).

$$l = - \frac{4}{\sqrt{17}} 3 + \frac{1}{\sqrt{17}} 2 - \frac{2}{\sqrt{17}} = - \frac{12}{17} \sqrt{17}$$

• Compute the distance between $(6,7,2)$ and the line $(x - 2)/4 = (y + 1)/3 = (z - 2)/12$.

The denominators in the line equations are direction numbers. Divide them by $\sqrt{(4)^2 + (3)^2 + (12)^2} = 13$ to convert them to the direction cosines: $\cos \alpha = \frac{4}{13}$; $\cos \beta = \frac{3}{13}$; $\cos \gamma = \frac{12}{13}$. In the line equations $(x_0,y_0,z_0) = (2,-1,2)$. Hence, the required distance is

$$l = \sqrt{(2 - 6)^2 + (-1 - 7)^2 + (2 - 2)^2}$$
$$- [(2 - 6)\tfrac{4}{13} + (-1 - 7)\tfrac{3}{13} + (2 - 2)\tfrac{12}{13}]^2 = 8.40$$

• Find the distance between $(5,4,6)$ and the plane $2x + 2y + z = 15$.

To put the plane equation in the normal form, transpose 15 to the left-hand side and divide the equation by $\sqrt{(2)^2 + (2)^2 + (1)^2} = 3$. Then substitute $(5,4,6)$ for (x,y,z).

$$l = \tfrac{2}{3}x_1 + \tfrac{2}{3}y_1 + \tfrac{1}{3}z_1 - 5 = \tfrac{2}{3}5 + \tfrac{2}{3}4 + \tfrac{1}{3}6 - 5 = 3$$

6-35. Computation of Area by Coordinates. (See also Art. 5-31.) Given the coordinates of the vertices of an n-sided polygon, $P_1(x_1,y_1)$, $P_2(x_2,y_2)$, . . . , $P_n(x_n,y_n)$, the area can be computed from

$$A = \tfrac{1}{2}[y_1(x_n - x_2) + y_2(x_1 - x_3) + y_3(x_2 - x_4) + \cdots + y_n(x_{n-1} - x_1)]$$

$$= \tfrac{1}{2} \sum_1^n y_m(x_{m-1} - x_{m+1})$$

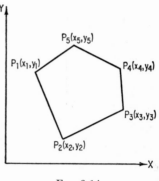

where $x_0 = x_n$

$x_{n+1} = x_1$

(See Fig. 6-14.)

Rule: Deal with the vertices of the polygon consecutively, keeping the area on the left (counterclockwise). Multiply each ordinate by the difference obtained by subtracting the abscissa of the following vertex from the abscissa of the preceding vertex. Add these products and divide by 2.

FIG. 6-14

• Find the area of the quadrilateral whose vertices are $(-2,3)$, $(-3,-4)$, $(5,-1)$, and $(2,2)$.

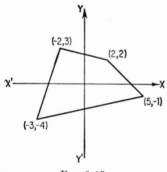

FIG. 6-15

Sketch the quadrilateral (Fig. 6-15) and proceed from vertex to vertex counterclockwise, starting with $(-2,3)$.

$$A = \tfrac{1}{2}[3(2 + 3) - 4(-2 - 5) - 1(-3 - 2) + 2(5 + 2)] = 31$$

6-36. Area of a Triangle by Determinants. Given the coordinates $P_1(x_1,y_1)$, $P_2(x_2,y_2)$, and $P_3(x_3,y_3)$ of the vertices of a triangle, numbered consecutively counterclockwise, the area is

$$A = \tfrac{1}{2} \begin{vmatrix} x_1 & y_1 & 1 \\ x_2 & y_2 & 1 \\ x_3 & y_3 & 1 \end{vmatrix}$$

(See Art. 13-13.)

• Compute the area of a triangle with vertices at (5,5), (−6,7), and (−7,−2). Sketch the triangle (Fig. 6-16) and proceed from vertex to vertex counterclockwise, starting with (5,5).

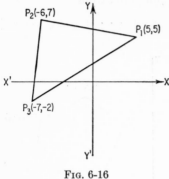

$$A = \tfrac{1}{2} \begin{vmatrix} 5 & 5 & 1 \\ -6 & 7 & 1 \\ -7 & -2 & 1 \end{vmatrix} = 10\tfrac{1}{2} = 50.5$$

FIG. 6-16

6-37. Asymptote. If the distance between a plane curve and a straight line approaches zero as the coordinates of the curve approach infinity, the line is an asymptote of the curve.

Asymptotes parallel to the x axis are called horizontal; those parallel to the y axis are vertical.

To find the vertical asymptotes of a curve, solve the equation for y and determine values of x that make the denominator, but not the numerator, zero. If it is not convenient to solve for y, write the equation as $f(y) = 0$, treating x as a constant. Then equate the coefficient of the highest power of y to zero and solve for x. (If this coefficient does not contain x, there is no vertical asymptote.) To find the horizontal asymptotes, follow the same procedure with x and y interchanged.

• Determine the vertical and horizontal asymptotes of $xy - y = 3x + 1$. Solve for y.

$$y = \frac{3x + 1}{x - 1}$$

As x approaches 1, the denominator approaches 0 and y becomes infinitely large. Therefore, $x = 1$ is a vertical asymptote. Now solve for x.

$$x = \frac{y + 1}{y - 3}$$

As y approaches 3, x becomes infinitely large, so $y = 3$ is a horizontal asymptote.

• Determine the horizontal and vertical asymptotes of $3x^2 - 2y^2 + x^2y^2 + 3x - 2y + 1 = 0$.

The highest power of y is the second; the coefficient of y^2 is $-2 + x^2$. Equate this to zero to yield $x = \pm \sqrt{2}$ as the vertical asymptotes.

The highest power of x is the second; the coefficient of x^2 is $3 + y^2$. But the equation $3 + y^2 = 0$ has no real roots. Hence there are no horizontal asymptotes.

6-38. Equations of Tangents. A tangent to a curve at a point P is the limiting position of a line through P and a second point Q on the curve, as Q approaches P along the curve.

The slope of a plane curve at a point is given by the slope of the tangent at the point. The slope of the tangent is equal to $\dfrac{dy}{dx}$, the derivative of the ordinate with respect to the abscissa, evaluated at the point.

The equation of a tangent to a curve at a point $P_1(x_1,y_1)$ on the curve is

$$m_1 = \left.\frac{dy}{dx}\right]_{P_1} = \frac{y - y_1}{x - x_1}$$

The length of a tangent at $P_1(x_1,y_1)$ is defined as the segment included between the point of tangency and the x axis. It is equal to

$$t = \left|\frac{y_1}{m_1}\right| \sqrt{1 + m_1{}^2}$$

where $m_1 = \left.\dfrac{dy}{dx}\right]_{P_1}$ and the vertical bars denote absolute value.

The projection of the tangent on the x axis is called the subtangent. It is equal to y_1/m_1.

• Find the equation of the tangent, its length, and the length of the subtangent of the parabola $y^2 - 6y - 8x - 31 = 0$ at $(-3,-1)$.

To obtain the derivative of y with respect to x, which gives the slope of the tangent at any point, differentiate both sides of the equation of the parabola.

$$2y\frac{dy}{dx} - 6\frac{dy}{dx} - 8 = 0$$

$$\frac{dy}{dx} = \frac{8}{2y - 6} = \frac{4}{y - 3}$$

At the point of tangency $(-3,-1)$, this derivative is equal to $m_1 = -1$. Hence, the equation of the tangent is

$$-1 = \frac{y + 1}{x + 3} \text{ or } x + y + 4 = 0$$

The length of the subtangent is $y_1/m_1 = -1/-1 = 1$.

The length of the tangent is $|y_1/m_1| \sqrt{1 + m_1{}^2} = 1 \sqrt{1 + 1} = \sqrt{2}$.

• Find the equations of the tangents from $(3,-1)$ to the circle $x^2 + y^2 = 5$.

Differentiate both sides of the equation to determine $\dfrac{dy}{dx}$.

$$2x + 2y\frac{dy}{dx} = 0$$

$$\frac{dy}{dx} = -\frac{x}{y}$$

The tangent must have this slope at the point of tangency $P_1(x_1,y_1)$ and also pass through $P_2(3,-1)$. Hence, the equation of the tangent is

$$m_1 = \frac{y - y_1}{x - x_1} = -\frac{x_1}{y_1} \quad \text{or} \quad x_1x - x_1{}^2 + yy_1 - y_1{}^2 = 0$$

and $x = 3$, $y = -1$ must satisfy it. Thus,

$$3x_1 - x_1^2 - y_1 - y_1^2 = 0$$

Also, (x_1,y_1) satisfies $x_1^2 + y_1^2 = 5$, since the point of tangency lies on the circle. Addition of this equation to the preceding one yields

$$3x_1 - y_1 = 5$$
$$y_1 = 3x_1 - 5$$

Substitution of this value of y_1 in $x_1^2 + y_1^2 = 5$ and solution first for x_1 and then for y_1 gives the points of tangency as $(2,1)$ and $(1,-2)$. Then the equations of the tangents are $2x + y - 5 = 0$ and $x - 2y - 5 = 0$.

The equations of the line tangent at $P_1(x_1,y_1,z_1)$ to the skew (space) curve whose equations are $x = f_1(t)$, $y = f_2(t)$, and $z = f_3(t)$ are

$$\frac{x - x_1}{\dfrac{dx}{dt}\bigg]_{P_1}} = \frac{y - y_1}{\dfrac{dy}{dt}\bigg]_{P_1}} = \frac{z - z_1}{\dfrac{dz}{dt}\bigg]_{P_1}}$$

where the total derivatives are to be evaluated at P_1.

If the skew curve is the intersection of two surfaces $F(x,y,z) = 0$ and $G(x,y,z) = 0$, the equations of the tangent line at $P_1(x_1,y_1,z_1)$ are

$$\frac{x - x_1}{A} = \frac{y - y_1}{B} = \frac{z - z_1}{C}$$

$$A = \frac{\partial F}{\partial y}\frac{\partial G}{\partial z} - \frac{\partial F}{\partial z}\frac{\partial G}{\partial y}$$

$$B = \frac{\partial F}{\partial z}\frac{\partial G}{\partial x} - \frac{\partial F}{\partial x}\frac{\partial G}{\partial z}$$

$$C = \frac{\partial F}{\partial x}\frac{\partial G}{\partial y} - \frac{\partial F}{\partial y}\frac{\partial G}{\partial x}$$

where all the partial derivatives are to be evaluated at P_1.

A straight line is tangent to a surface at a point P if it is the limiting position of a secant through P and a neighboring point Q, as Q approaches P along a curve on the surface. All lines tangent to a surface at a given point lie on a plane, called the tangent plane. Its equation is

$$\frac{\partial F}{\partial x}\bigg]_{P_1} (x - x_1) + \frac{\partial F}{\partial y}\bigg]_{P_1} (y - y_1) + \frac{\partial F}{\partial z}\bigg]_{P_1} (z - z_1) = 0$$

where the partial derivatives are to be evaluated at the point of tangency $P_1(x_1,y_1,z_1)$ and $F(x,y,z) = 0$ is the equation of the surface. The partial derivatives give the slopes of the lines tangent to the curves of intersection of a surface and planes parallel to the coordinate planes. For example, the slope of the tangent to $z = f(x,y)$ in the plane $y = k$ is equal to $\dfrac{\partial z}{\partial x}$. Similarly, the slope in the plane $x = c$ is given by $\dfrac{\partial z}{\partial y}$.

6-39. Equations of Normals to Curves. The normal at any point of a plane curve is the line perpendicular to the tangent at the point. Its slope is the negative reciprocal of the slope of the tangent (Art. 6-38). Hence, the equation of the normal to a curve at a point $P_1(x_1,y_1)$ is

$$x - x_1 = \frac{dy}{dx}\bigg]_{P_1} (y_1 - y) = m_1(y_1 - y)$$

where the derivative of y with respect to x (equal to m_1, the slope of the tangent) is to be evaluated at P_1.

The length of the normal is the segment included between the point of contact with the curve and the x axis. It is equal to

$$n = |y_1| \sqrt{1 + m_1{}^2}$$

where m_1 is the derivative of y with respect to x evaluated at the point of contact and the vertical lines denote absolute value.

The subnormal is the projection of the normal on the x axis and is equal to $m_1 y_1$.

• Find the equation of the normal, its length, and the length of the subnormal of the parabola $y^2 - 6y - 8x - 31 = 0$ at $(-3,-1)$.

Differentiate both sides of the equation to determine $\dfrac{dy}{dx}$.

$$2y \frac{dy}{dx} - 6 \frac{dy}{dx} - 8 = 0$$

$$\frac{dy}{dx} = \frac{8}{2y - 6} = \frac{4}{y - 3}$$

At the point of contact $(-3,-1)$, this derivative is equal to $m_1 = -1$. Hence, the equation of the normal is $x + 3 = -1(-1 - y)$ or $x - y + 2 = 0$.

The length of the subnormal is $m_1 y_1 = -1(-1) = 1$.

The length of the normal is $|y_1| \sqrt{1 + m_1{}^2} = 1 \sqrt{1 + 1} = \sqrt{2}$.

The normal plane of a skew (space) curve at a point $P_1(x_1,y_1,z_1)$ is the plane through P_1 perpendicular to the tangent at that point. The equation of the plane normal to the skew curve whose equations are $x = f_1(t)$, $y = f_2(t)$, and $z = f_3(t)$ is

$$\frac{dx}{dt}\bigg]_{P_1} (x - x_1) + \frac{dy}{dt}\bigg]_{P_1} (y - y_1) + \frac{dz}{dt}\bigg]_{P_1} (z - z_1) = 0$$

where all the total derivatives are to be evaluated at the point of contact P_1.

If the skew curve is the intersection of two surfaces $F(x,y,z) = 0$ and

$G(x,y,z) = 0$, the equation of the normal plane is

$$A(x - x_1) + B(y - y_1) + C(z - z_1) = 0$$

$$A = \frac{\partial F}{\partial y}\frac{\partial G}{\partial z} - \frac{\partial F}{\partial z}\frac{\partial G}{\partial y}$$

$$B = \frac{\partial F}{\partial z}\frac{\partial G}{\partial x} - \frac{\partial F}{\partial x}\frac{\partial G}{\partial z}$$

$$C = \frac{\partial F}{\partial x}\frac{\partial G}{\partial y} - \frac{\partial F}{\partial y}\frac{\partial G}{\partial x}$$

where all the partial derivatives are to be evaluated at the point of contact P_1.

The line normal to a surface at a given point $P_1(x_1,y_1,z_1)$ is perpendicular to the tangent plane at the point. The equations of the normal line are

$$\frac{x - x_1}{\dfrac{\partial F}{\partial x}\bigg]_{P_1}} = \frac{y - y_1}{\dfrac{\partial F}{\partial y}\bigg]_{P_1}} = \frac{z - z_1}{\dfrac{\partial F}{\partial z}\bigg]_{P_1}}$$

where $F(x,y,z) = 0$ is the equation of the surface and all the partial derivatives are to be evaluated at P_1.

6-40. Curvature of a Plane Curve. Curvature is the rate of change of direction of a curve. In cartesian plane coordinates, it is equal to

$$K = \frac{y''}{(1 + y'^2)^{3/2}} = \frac{-x''}{(1 + x'^2)^{3/2}}$$

where y',y'' = first and second derivatives of y with respect to x
x',x'' = first and second derivatives of x with respect to y
If ds is the differential length of curve,

$$K = \sqrt{\left(\frac{d^2x}{ds^2}\right)^2 + \left(\frac{d^2y}{ds^2}\right)^2}$$

When the equation of the curve is in parametric form $x = f_1(t)$, $y = f_2(t)$, the curvature can be computed from

$$K = \frac{x'y'' - y'x''}{(x'^2 + y'^2)^{3/2}}$$

where x',y' = first derivatives of x,y with respect to t
x'',y'' = second derivatives with respect to t
In polar coordinates, curvature is given by

$$K = \frac{\rho^2 + 2\rho'^2 - \rho\rho''}{(\rho^2 + \rho'^2)^{3/2}}$$

where ρ' and ρ'' are, respectively, the first and second derivatives of the radius vector ρ with respect to the vectorial angle θ.

When the curvature is positive, the curve is concave upward (tangents are below the curve). When the curvature is negative, the curve is concave downward (tangents are above the curve).

The radius of curvature is the reciprocal of the curvature.

$$R = \frac{1}{K} = \frac{(1 + y'^2)^{3/2}}{y''}$$

• Find the radius of curvature of the parabola $y = x^2$ at its vertex.

At the vertex (0,0), $y' = \dfrac{dy}{dx} = 2x = 0$. Also, $y'' = \dfrac{d}{dx}(2x) = 2$.

$$R = \frac{(1 + 0)^{3/2}}{2} = \frac{1}{2}$$

6-41. Analysis of Curve Equations. Much can be learned about the characteristics of a curve before plotting its equation. An analysis should include a study of (1) intercepts on the axes, (2) symmetry, (3) excluded values of the variables and extent of the curve, (4) asymptotes, (5) slopes of tangents and maximum and minimum values of the variables, (6) inflection points (changes in sign of the slopes of the tangents), and (7) curvature.

1. Intercepts. Find the points where the curve crosses the x axis by setting $y = 0$ and solving for x. Find the y intercepts by setting $x = 0$ and solving for y. Discard imaginary and complex roots.

2. Symmetry. Two points are symmetric about a line if the line is the perpendicular bisector of a line drawn between the points. Two points are symmetric about a point if it is at the midpoint of a line drawn between the two points.

A plane curve is symmetric about the x axis if its equation is unchanged when $-y$ is substituted for y; it is symmetric about the y axis if it is unaffected by substitution of $-x$ for x; it is symmetric about the origin if it is not altered by changing both y to $-y$ and x to $-x$.

3. Extent of a curve. Values of a variable that require imaginary or complex values of the other variable must be excluded. From the values that are permissible, determine whether the curve is closed or open to infinity.

4. Asymptotes. Determine the horizontal and vertical asymptotes, if any exist (Art. 6-37).

5. Slopes, maxima, and minima. Determine the slope of the curve at critical points by computing the value of the first derivative of y with respect to x. A function increases with increasing values of x when the first derivative is positive; it decreases when the derivative is negative.

In particular, find the points where the derivative is zero.   These points are maxima if the second derivative is negative, minima if it is positive. (If the second derivative also vanishes at these points, compute the third, fourth, fifth derivatives, etc., until one is found that does not vanish. If the nonvanishing derivative is an even derivative, the point is a maximum or minimum.)

6. Inflection points.   A curve is concave upward (tangents are below the curve) if the second derivative of y with respect to x is positive; it is concave downward (tangents are above the curve) if this derivative is negative.   It changes concavity where the second derivative is zero; such points are called inflection points.

7. Curvature.   Determine the radius of curvature at important points (Art. 6-40).

• Analyze the equation $x^2 + 4y^2 = 16$.
1. Intercepts.   If $y = 0$, $x = \pm 4$; if $x = 0$, $y = \pm 2$.
2. Symmetry.   The equation is unchanged by substitution of $-y$ for y and $-x$ for x.   Hence, the curve is symmetric about both axes and the origin.
3. Extent.   Solve for both variables.

$$x = \pm 2 \sqrt{4 - y^2}$$
$$y = \pm \tfrac{1}{2} \sqrt{16 - x^2}$$

Examination of the radicals indicates that x will be imaginary if y is larger than 2 or smaller than -2.   Also, y will be imaginary if x is larger than 4 or smaller than -4.   Hence, the curve must lie within the rectangle $x = \pm 4$, $y = \pm 2$ and must be closed.

4. Asymptotes.   There are no horizontal or vertical asymptotes.
5. Slopes, maxima, and minima.   The slope at any point is

$$\frac{dy}{dx} = -\frac{x}{4y}$$

The slope is negative where x and y have the same sign, positive where signs differ.   The slope is zero where $x = 0$; the tangents at those points are parallel to the x axis.   The second derivative shows whether these points are maximum or minimum points.

The slope is infinite where y is zero; the tangents at those points are parallel to the y axis.

6. Inflection points

$$\frac{d^2y}{dx^2} = -\frac{1}{y^3}$$

The second derivative cannot be zero; hence there are no inflection points.   The curve is concave upward if y is negative, downward if y is positive.   Since the second derivative is negative at $(0,2)$, where the first derivative is zero, the point is a maximum.   There is a minimum at $(0,-2)$.

7. Curvature. The radius of curvature at $y = \pm 2$ is

$$R = \frac{(1 + y'^2)^{3/2}}{y''} = \frac{(1 + x^2/16y^2)^{3/2}}{-1/y^3} = \pm 8$$

The radius at $y = 0$ is ± 1. Thus, the curvature is much sharper at the x axis than at the y axis. The equation represents an ellipse (Fig. 6-17).

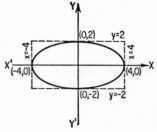

6-42. Analysis of Surface Equations. Just as a great deal can be learned from analysis of a plane curve before it is plotted, so can a surface be profitably analyzed. The analysis is similar to that for a plane curve (Art. 6-41).

FIG. 6-17

1. Intercepts. Find the points where the coordinate axes cut the surface. Set each pair of variables, in turn, equal to zero and solve for real values of the remaining variable.

2. Traces. Find the curves of intersection with the coordinate planes. In succession, set x, y, and z equal to zero.

3. Symmetry. If the surface is unaffected by a change in sign of one variable, the surface is symmetric about the coordinate plane from which that variable is measured. If the signs of two variables can be changed, the surface is symmetric about the axis of the third variable. If the signs of all three variables can be changed, the surface is symmetric about the origin.

4. Plane sections. Take sections parallel to the coordinate planes or in any other direction that might yield important information, and examine the curve of intersection.

• Analyze the equation $x^2 + 4y^2 + 9z^2 = 36$.
1. Intercepts. With $y = z = 0$, $x = \pm 6$. With $x = z = 0$, $y = \pm 3$. With $x = y = 0$, $z = \pm 2$.
2. Traces. When $x = 0$, the trace on the yz plane is the ellipse

$$\frac{y^2}{9} + \frac{z^2}{4} = 1$$

When $y = 0$, the trace is the ellipse

$$\frac{x^2}{36} + \frac{z^2}{4} = 1$$

When $z = 0$, the trace also is an ellipse,

$$\frac{x^2}{36} + \frac{y^2}{9} = 1$$

The surface, therefore, intersects all three coordinate planes in ellipses.

3. **Symmetry.** The surface is symmetrical about the three coordinate planes, the three axes, and the origin, since the equation is unaffected by any sign changes in the variables.

4. **Plane sections.** The planes $x = k_1$, where $k_1 > 6$ or $k_1 < -6$ do not intersect the surface; neither do the planes $y = k_2$ and $z = k_3$, where $k_2 > 3$ or $k_2 < -3$ and $k_3 > 2$ or $k_3 < -2$. Therefore, it is a closed surface, bounded by the rectangular parallelepiped $x = \pm 6, y = \pm 3$, and $z = \pm 2$.

A section taken at $x = \pm k$, where $|k| < 6$, intersects the surface in an ellipse $4y^2 + 9z^2 = 36 - k^2$, the axes of which are smaller than those of the trace on the yz plane. When $x = \pm 6$, the intersection is a point ellipse. Similar results are obtained by taking sections parallel to the other coordinate planes. The surface is an ellipsoid (Fig. 6-18).

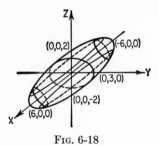

FIG. 6-18

6-43. Line Equations in Plane Coordinates.

In the general form, the equation of a straight line is the linear equation

$$Ax + By + C = 0$$

where A, B, and C are constants. The slope is $-A/B$ and the y intercept is $-C/B$.

1. Point-slope form

$$y - y_1 = m(x - x_1)$$

where m = slope

(x_1, y_1) = a point on the line

2. Slope-intercept form

$$y = mx + b$$

where m = slope

b = intercept on y axis

3. Two-point form

$$\frac{y - y_1}{x - x_1} = \frac{y_2 - y_1}{x_2 - x_1} = m$$

where (x_1, y_1), (x_2, y_2) = points on line

4. Intercept form

$$\frac{x}{a} + \frac{y}{b} = 1$$

where a = intercept on x axis

b = intercept on y axis

5. Normal form

$$x \cos \theta + y \sin \theta - p = 0$$

where θ = angle from x axis (measured counterclockwise) to normal to line through origin

p = perpendicular distance from origin to line

To put in the normal form an equation given in the general form

$$Ax + By + C = 0$$

divide it by $\sqrt{A^2 + B^2}$ with sign opposite to that of C.

6. Polar coordinates.   If, in the general form, $C \neq 0$, so that the line does not pass through the origin, the polar equation is

$$\rho = -\frac{C}{A \cos \theta + B \sin \theta}$$

If the line passes through the origin, its equation can be written as $\theta = k$, where k is a constant.   The equation of a line through two points (ρ_1, θ_1) and (ρ_2, θ_2) is

$$\rho\rho_1 \sin (\theta - \theta_1) + \rho_1\rho_2 \sin (\theta_1 - \theta_2) + \rho\rho_2 \sin (\theta_2 - \theta) = 0$$

6-44. Line Equations in Space Coordinates.   In the general form, a line is defined by two simultaneous linear equations, each representing a plane; the line is their intersection.

$$A_1x + B_1y + C_1z = 0$$
$$A_2x + B_2y + C_2z = 0$$

where A_1, B_1, C_1 and A_2, B_2, C_2 are constants.   If α, β, and γ are the direction angles of the line, the direction cosines can be obtained from

$$\frac{\cos \alpha}{B_1C_2 - B_2C_1} = \frac{\cos \beta}{C_1A_2 - C_2A_1} = \frac{\cos \gamma}{A_1B_2 - A_2B_1}$$

1. Parametric form

$$x = x_1 + t \cos \alpha \qquad y = y_1 + t \cos \beta \qquad z = z_1 + t \cos \gamma$$

where t = parameter
(x_1, y_1, z_1) = point on line

2. Symmetric form

$$\frac{x - x_1}{\cos \alpha} = \frac{y - y_1}{\cos \beta} = \frac{z - z_1}{\cos \gamma}$$

or

$$\frac{x - x_1}{a} = \frac{y - y_1}{b} = \frac{z - z_1}{c}$$

where α, β, γ = direction cosines
a, b, c = direction numbers
(x_1, y_1, z_1) = point on line

3. Two-point form

$$\frac{x - x_1}{x_2 - x_1} = \frac{y - y_1}{y_2 - y_1} = \frac{z - z_1}{z_2 - z_1}$$

where (x_1, y_1, z_1) and (x_2, y_2, z_2) are two points on the line.

4. Projection form

$$x = mz + p \qquad y = nz + q$$

where $(p,q,0)$ = point where line cuts xy plane

$m,n,1$ = direction numbers of line

6-45. Plane Equations.   The general form of the equation of a plane is the linear equation

$$Ax + By + Cz + D = 0$$

where A, B, C, and D are constants. A, B, and C are direction numbers of a line normal to the plane.

If A, B, or C is zero, the plane is perpendicular to the coordinate plane corresponding to the two remaining variables.   If two coefficients are zero, the plane is perpendicular to the axis corresponding to the remaining variable.

1. Normal form

$$x \cos \alpha + y \cos \beta + z \cos \gamma - p = 0$$

where p = perpendicular distance from origin to plane

α, β, γ = direction cosines of normal

To reduce the general form $Ax + By + Cz + D = 0$ to the normal form, divide by $\sqrt{A^2 + B^2 + C^2}$ with sign opposite that of D.

2. Intercept form

$$\frac{x}{a} + \frac{y}{b} + \frac{z}{c} = 1$$

where a, b, and c are the intercepts on the x, y, and z axes, respectively.

6-46. Circle Equations.   1. Standard form.   The equation of a circle with center at (h,k) and radius r (Fig. 6-19) is

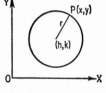

$$(x - h)^2 + (y - k)^2 = r^2$$

When the center is at the origin,

$$x^2 + y^2 = r^2$$

FIG. 6-19

2. General form

$$Ax^2 + Ay^2 + Dx + Ey + F = 0$$

The equation is second degree in x and y.   The coefficients of x^2 and y^2 are equal, and there is no xy term.

3. Polar coordinates

$$\rho^2 + \rho(D \cos \theta + E \sin \theta) + F = 0$$

where D, E, and F are constants.

If the center is at the origin, the equation of the circle is $\rho = r$.

If the pole is on the circumference and the polar axis is a diameter,

$$\rho = 2r \cos \theta$$

If the circle touches the polar axis only at the pole and extends above the polar axis,

$$\rho = 2r \sin \theta$$

6-47. Sphere Equations. The equation of a sphere with center at (h,k,l) and radius r is

$$(x - h)^2 + (y - k)^2 + (z - l)^2 = r^2$$

When the center is at the origin,

$$x^2 + y^2 + z^2 = r^2$$

6-48. Cone Equations. The equation of a cone with vertex at the origin and lying along the z axis (Fig. 6-20) is

$$\frac{x^2}{a^2} + \frac{y^2}{b^2} - \frac{z^2}{c^2} = 0$$

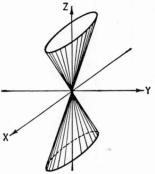

If the minus sign appears before the x^2 term, the cone lies along the x axis; if the minus sign appears before y^2, the cone lies along the y axis.

6-49. Conic Sections. When the cone in Fig. 6-20 (Art. 6-48) is cut by a plane $z = k$, the intersection is the ellipse

$$\frac{x^2}{a^2} + \frac{y^2}{b^2} = \frac{k^2}{c^2}$$

FIG. 6-20

When the cone is cut by a plane $x = k$ or $y = k$, the intersections are the hyperbolas

$$\frac{z^2}{c^2} - \frac{y^2}{b^2} = \frac{k^2}{a^2} \quad \text{or} \quad \frac{z^2}{c^2} - \frac{x^2}{a^2} = \frac{k^2}{b^2}$$

The intersection of the cone with the plane $y - (b/c)z = k$ is the parabola

$$\frac{2k}{bc} z = -\frac{x^2}{a^2} - \frac{k^2}{b^2}$$

If a plane passes through the vertex, it intersects the cone in two straight lines or a point.

A conic section is described by a point moving so that its distances from a given fixed point (focus) and a given fixed line (directrix) are in a constant ratio (eccentricity) (see also Art. 6-50).

6-50. Curves Represented by Second-degree Equations. In general, a second-degree equation

$$Ax^2 + Bxy + Cy^2 + Dx + Ey + F = 0$$

is a circle if $A = C$ and $B = 0$, a parabola if $B^2 - 4AC = 0$, an ellipse if $B^2 - 4AC$ is negative, or a hyperbola if $B^2 - 4AC$ is positive.

If the equation can be factored into two factors of the first degree in x and y, each factor, set equal to zero, represents the equation of a line or point.

If a second-degree equation is written in terms of eccentricity e (Art. 6-49), with y axis as directrix and focus at $(p,0)$,

$$(1 - e^2)x^2 + y^2 - 2px + p^2 = 0$$

the conic is a parabola if $e = 1$, an ellipse if $e < 1$, or a hyperbola if $e > 1$.

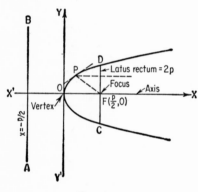

FIG. 6-21

In polar coordinates, the equation of a conic with pole at a focus and polar axis along the x axis or the directrix has the form

$$\rho = \frac{ep}{1 \pm e \cos \theta}$$

6-51. Parabola Equations. A parabola is the locus of points equidistant from a fixed point (focus F) and a fixed line (directrix AB) (see Fig. 6-21). It also has a focusing property: if, from a point on the curve, a line is drawn parallel to the axis of the curve and another is drawn to the focus, these lines make equal angles with the tangent at the point (see also Arts. 6-49 and 6-50).

The equation of a parabola with axis along the x axis, vertex at the origin, and focus at $(p/2,0)$ is

$$y^2 = 2px$$

If p is positive, the curve lies completely on the positive side of the y axis; if p is negative, the curve lies completely on the negative side.

Similarly, if the axis of the parabola lies along the y axis, with vertex at the origin and focus at $(0,p/2)$, the equation is

$$x^2 = 2py$$

If p is positive, the curve lies above the x axis; if p is negative, the curve lies below the x axis.

An equation of the form

$$y^2 + Dx + Ey + F = 0$$

is a parabola. It can also be written as

$$(y - k)^2 = 2p(x - h)$$

The vertex of the parabola is at (h,k) and the axis is parallel to the x axis (see also Art. 6-50).

The equation of a parabola with axis along the polar axis and pole at the focus is, in polar coordinates,

$$\rho = \frac{p}{1 - \cos \theta}$$

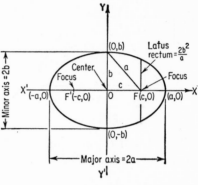

FIG. 6-22

6-52. Ellipse Equations. An ellipse is the locus of a point moving so that the sum of its distances from two fixed points (foci) is a constant (see also Art. 6-49). Its eccentricity equals c/a, where c is the distance from a focus to the center of the ellipse and a is half the constant. If the eccentricity is zero, the curve is a circle; if the eccentricity is unity, the curve is a straight line.

The equation of an ellipse with center at the origin and foci at $(\pm c,0)$ is

$$\frac{x^2}{a^2} + \frac{y^2}{b^2} = 1$$

where $2a$ is the major axis, $2b$ is the minor axis, and $a^2 - b^2 = c^2$ (Fig. 6-22).

An equation of the form

$$Ax^2 + Cy^2 + Dx + Ey + F = 0$$

where A and C have the same sign, is an ellipse. It can be written as

$$\frac{(x - h)^2}{a^2} + \frac{(y - k)^2}{b^2} = 1$$

(In special cases, the equation may represent a point or have no locus.) The center of the ellipse is at (h,k).

6-53. Hyperbola Equations. A hyperbola is the locus of a point moving so that the difference between its distances from two fixed points (foci F and F') is constant ($2a$). The curve has two symmetrical branches, both of which have the same two asymptotes (Fig. 6-23). Its eccentricity is c/a, where $2c$ is the distance between foci and $2a$ is the difference between the distances of any point on the curve from the two foci.

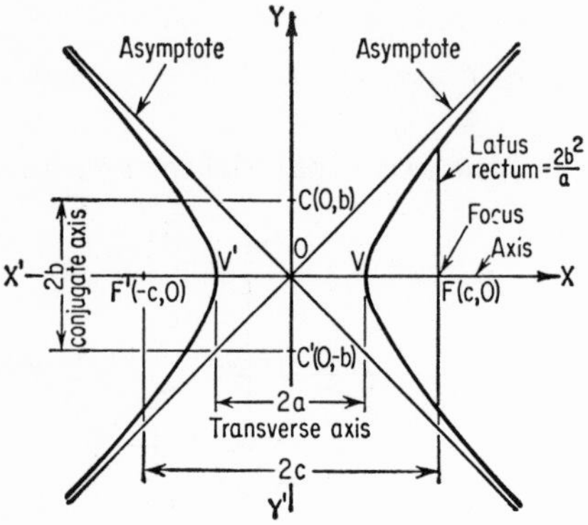

FIG. 6-23

The equation of a hyperbola with center at the origin and foci at $(\pm c,0)$ is

$$\frac{x^2}{a^2} - \frac{y^2}{b^2} = 1$$

VV' is the transverse axis, with length $2a$; CC' is the conjugate axis, with length $2b$; $b^2 = c^2 - a^2$. The asymptotes are $y = \pm (b/a)x$.

An equation of the form

$$Ax^2 + Cy^2 + Dx + Ey + F = 0$$

where A and C have opposite signs, is a hyperbola or two intersecting lines (see also Art. 6-50). It can be written as

$$\frac{(x - h)^2}{a^2} - \frac{(y - k)^2}{b^2} = 1 \quad \text{or} \quad \frac{(y - k)^2}{a^2} - \frac{(x - h)^2}{b^2} = 1$$

The center of the hyperbola is at (h,k).

6-54. Equilateral Hyperbola. The equation of an equilateral hyperbola is $x^2 - y^2 = a^2$. It has perpendicular asymptotes $y = \pm x$. If the axes are rotated -45 deg to coincide with the asymptotes, the equation becomes

$$2xy = a^2$$

6-55. Periodic Functions. A function is periodic with period k if $f(x + k) = f(x)$. Each repetition is called a cycle.

Sine and cosine curves are examples of periodic curves, since

$$\sin (x + 2\pi) = \sin x \quad \text{and} \quad \cos (x + 2\pi) = \cos x$$

6-56. Sine Curves. The curve $y = \sin x$ is a periodic function with period 2π. [Sin $(x + 2\pi) = \sin x$.] The maximum or minimum value of y without sign is called the amplitude. The maximum value is 1 and occurs at $x = \pi/2, \pi/2 + 2\pi, \ldots, -2\pi + \pi/2$, etc. The minimum value is -1 and occurs at $x = 3\pi/2, 3\pi/2 + 2\pi, \ldots, -\pi/2, -\pi/2 - 2\pi$, etc. The curve crosses the x axis at the origin and at $x = \pm\pi, \pm2\pi, \pm3\pi$, etc. It is symmetric about the origin (see Fig. 6-24).

The equation $y = a \sin x$ represents a similar curve with amplitude a. The curve $y = a \sin nx$ has amplitude a and period $2\pi/n$. In a length of 2π radians along the x axis, there are n cycles. Thus, $y = 3 \sin 2x$ has amplitude 3 and period π. As x varies from zero to 2π, this curve goes through two cycles, whereas $\sin x$ goes through only one.

The curve $y = a \sin (nx + \theta)$, where θ is a constant, has the same shape as $y = a \sin nx$ but is displaced a distance θ/n along the x axis. The displacement is to the left if θ is positive, to the right if θ is negative.

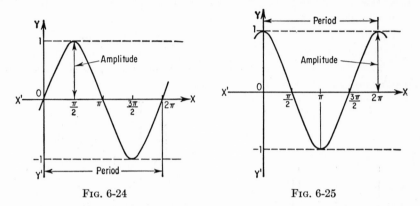

FIG. 6-24   FIG. 6-25

A sine curve $y = \sin x$ is equivalent to a cosine curve $y = \cos x$ displaced a distance $\pi/2$ to the right along the x axis.

$$y = \sin x = \cos \left(x - \frac{\pi}{2} \right)$$

6-57. Cosine Curves. The curve $y = \cos x$ is a periodic function with period 2π. [Cos $(x + 2\pi) = \cos x$.] The maximum value of y (amplitude) is 1 and occurs at $x = 0$, $\pm 2\pi$, $\pm 4\pi$, etc. The minimum value is -1 and occurs at $x = \pm \pi$, $\pm 3\pi$, $\pm 5\pi$, etc. The curve crosses the x axis at $\pm \pi/2$, $\pm 3\pi/2$, $\pm 5\pi/2$, etc. It is symmetric about the y axis (Fig. 6-25).

The equation $y = a \cos x$ represents a similar curve with amplitude a.

The curve $y = a \cos nx$ has amplitude a and period $2\pi/n$. In a length of 2π radians along the x axis, there are n cycles. Thus, $y = 5 \cos 3x$ has amplitude 5 and period $2\pi/3$. As x varies from zero to 2π, the curve goes through three cycles, whereas $\cos x$ goes through only one.

The curve $y = a \cos (nx + \theta)$ has the same shape as $y = a \cos nx$ but is displaced a distance θ/n along the x axis, to the left if θ is positive and to the right if θ is negative.

A cosine curve $y = \cos x$ is equivalent to a sine curve $y = \sin x$ displaced a distance $\pi/2$ to the left along the x axis.

$$y = \cos x = \sin \left(x + \frac{\pi}{2} \right)$$

6-58. Tangent and Cotangent Curves. The curves $y = \tan x$ and $y = \cot x$ are periodic curves with period π. [Tan $(x + \pi) = \tan x$; cot $(x + \pi) = \cot x$.] The tangent curves, symmetric about the origin, cross the x axis at the origin and at $x = \pm\pi$, $\pm 2\pi$, etc. They are asymptotic to $x = \pm\pi/2$, $\pm 3\pi/2$, $\pm 5\pi/2$, etc. (Fig. 6-26).

The cotangent curves are similar but are displaced a distance $\pi/2$ along the x axis and the signs are changed (the curves are reflected about the x axis).

The curves $y = \tan nx$ and $y = \cot nx$ have period π/n.

The curves $y = \tan (nx + \theta)$ and $y = \cot (nx + \theta)$ also have period π/n but are displaced a distance θ/n along the x axis, to the left if θ is positive and to the right if θ is negative.

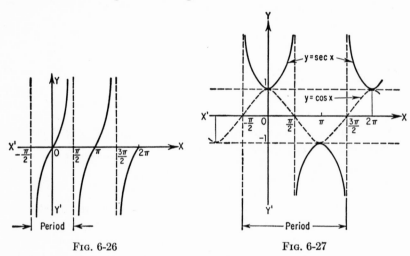

FIG. 6-26   FIG. 6-27

6-59. Secant and Cosecant Curves. The curves $y = \sec x = 1/\cos x$ and $y = \csc x = 1/\sin x$ are periodic functions with period 2π.

$$\sec (x + 2\pi) = \sec x \qquad \csc (x + 2\pi) = \csc x$$

Both curves have two branches per cycle; each branch is the reflection of the other about the x axis, but is displaced a distance π along the x axis. Neither curve crosses the x axis.

The secant curve, symmetric about the y axis, has a minimum at $y = 1$ when $x = 0$, $\pm 2\pi$, $\pm 4\pi$, etc., and a maximum at $y = -1$ when $x = \pm\pi$, $\pm 3\pi$, etc. (Fig. 6-27). The branches have asymptotes at $x = \pm\pi/2$, $\pm 3\pi/2$, $\pm 5\pi/2$, etc. The cosecant curve, symmetric about the origin, is a secant curve displaced a distance $\pi/2$ to the right along the x axis.

The curves $y = \sec nx$ and $y = \csc nx$ have period $2\pi/n$.

The curves $y = \sec (nx + \theta)$ and $y = \csc (nx + \theta)$ also have period

$2\pi/n$ but are displaced a distance θ/n along the x axis. The displacement is to the left if θ is positive and to the right if θ is negative.

6-60. Curves for Inverse Trigonometric Functions. These are the same as for the corresponding trigonometric functions except that the x and y axes are interchanged.

Thus, $y = \arcsin x = \sin^{-1} x$ is the same as $x = \sin y$. (See Art. 5-32.)

6-61. Exponential Curves. The curve $y = a^x$, where a is a positive number other than unity, is asymptotic to the x axis and lies completely above it. If $a > 1$, y increases with increasing x; if $a < 1$, y decreases as x increases (Fig. 6-28). All exponential curves pass through $(0,1)$. The curve $y = -a^x$ is the reflection of $y = a^x$ about the x axis.

In particular, if $a = e = 2.7183 \cdots$, the slope of the curve at any point equals the value of y at that point.

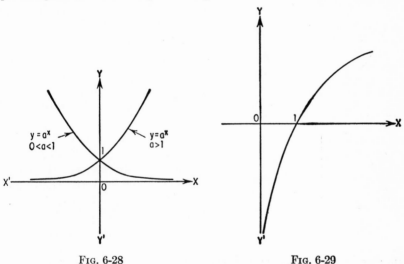

FIG. 6-28   FIG. 6-29

6-62. Logarithmic Curves. The curve for $y = \log_a x$ (logarithm to the base a of x) is the same as for the exponential equation $y = a^x$ with x and y axes interchanged. Logarithmic curves are asymptotic to the y axis and intersect the x axis at $x = 1$ (Fig. 6-29).

If $P_1(x_1,y_1)$ and $P_2(x_2,y_2)$ are any two points on $y = \log_a x$, then $y_1 - y_2 = \log_a x_1 - \log_a x_2 = \log_a x_1/x_2$. Hence, if the ratio x_1/x_2 is a constant, the change in y is a constant. Thus, the change in y is the same for the interval 1 to 10 as it is for the intervals 10 to 100, 100 to 1,000, and 0.1 to 0.01.

6-63. Curves for Hyperbolic Functions. The curve

$$y = \sinh x = \tfrac{1}{2}(e^x - e^{-x})$$

is symmetric about the origin and passes through it (Fig. 6-30). For large positive values of x, it approaches the exponential curve $y = \frac{1}{2}e^x$, and for large negative values it approaches $y = -\frac{1}{2}e^{-x}$ (see Art. 6-61).

The curve $y = \cosh x = \frac{1}{2}(e^x + e^{-x})$ is symmetric about the y axis and has a minimum at (0,1). For large positive values of x, the curve approaches the exponential curve $y = \frac{1}{2}e^x$ and therefore also $y = \sinh x$ (Fig. 6-30).

The curve $y = \tanh x = (e^x - e^{-x})/(e^x + e^{-x})$ is symmetric about the origin, passes through it, and is asymptotic to $y = \pm 1$.

The curve $y = \coth x = (e^x + e^{-x})/(e^x - e^{-x})$ does not cross either the x or y axis. Symmetric about the origin, it is asymptotic to the y axis and to $y = \pm 1$.

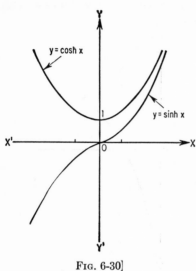

FIG. 6-30]

6-64. Skew (Space) Curves. A curve or straight line in space is defined by two simultaneous equations. The curve is the intersection of the two surfaces defined by the equations. Since many surfaces may be passed through a curve, it can be defined by a multitude of pairs of equations.

An important group of defining surfaces are the *projecting cylinders*, whose elements intersect the curve and are parallel to the coordinate axes. One of the three variables x, y, and z does not appear in each of the equations of these cylinders.

Analysis of a skew curve is similar to that of a plane curve (Art. 6-41), but additional characteristics, such as torsional curvature, may be investigated. (See also Art. 14-20.)

• What are the equations of the coordinate axes?

$$x \text{ axis: } y = 0 \qquad z = 0$$
$$y \text{ axis: } x = 0 \qquad z = 0$$
$$z \text{ axis: } x = 0 \qquad y = 0$$

• Find the equations of the projecting cylinders of $x^2 + y^2 + z^2 = 25$ and $x^2 + 4y^2 - z^2 = 0$.

Eliminate x, y, and z in turn. Add to eliminate z.

$$2x^2 + 5y^2 = 25$$

This is the equation of an elliptical cylinder along the z axis. Subtract to eliminate x.

$$3y^2 - 2z^2 = -25$$

This is the equation of a hyperbolic cylinder lying along the x axis. Multiply the first equation by 4 and subtract the second from it.

$$3x^2 + 5z^2 = 100$$

This is the equation of an elliptical cylinder along the y axis. Any two of these projecting cylinders define the same curve as the original pair of equations.

A space curve also may be defined by three parametric equations.

$$x = f_1(t) \qquad y = f_2(t) \qquad z = f_3(t)$$

6-65. Cylinder Equations. If one variable does not appear in an equation, the surface is a cylinder with elements parallel to the axis of that variable. The given equation, considered as a plane locus in the corresponding coordinate plane, represents the directrix.

• What surface is represented by $x^2 + y^2 = 16$?
The equation does not contain z. Therefore, it is a cylinder with elements parallel to the z axis. Its trace on the xy plane is the circle $x^2 + y^2 = 16$, with radius 4.

If the elements of a cylinder are not parallel to a coordinate axis, the equation contains all three variables. It represents a cylinder if it defines a ruled surface generated by parallel lines.

• Show that $x^2 + 2xy + y^2 + z^2 = 1$ is a cylinder.
Write the equation in the form $(x + y)^2 = 1 - z^2$, or $x + y = \pm \sqrt{1 - z^2}$. Then it can be seen that the lines $z = k$, $x + y = \pm \sqrt{1 - k^2}$ are the generators of the surface, because elimination of k in these equations gives the original equation. The lines are parallel, because their direction numbers are 1, -1, and 0 and are independent of k (Art. 6-30).

6-66. Surfaces of Revolution. These are generated by revolving a plane curve about a line in its plane (axis of revolution). All sections cut by planes through the axis of revolution are identical, and all sections perpendicular to this axis are circles. Spheres, right circular cylinders, and right circular cones are surfaces of revolution.

To find the equation of a surface generated by rotating a curve in one of the coordinate planes about a coordinate axis in that plane, first select the two variables not measured along that axis. Then, substitute, in the curve equation, the square root of the sum of their squares for that one of these two variables that occurs in the curve equation.

• Find the equation of the surface generated by rotating the parabola $y^2 = 8x$ about the x axis.
The equation is obtained by substituting $\sqrt{y^2 + z^2}$ for y, since y and z are not measured along the x axis and y also appears in the given equation. Hence, the surface is given by

$$y^2 + z^2 = 8x$$

6-67. Quadric Surfaces. These are defined by equations of the second degree in x, y, and z.

By suitable selection of a coordinate system, the equation of a quadric surface can be written in the form

$$f(x,y,z) = \alpha x^2 + \beta y^2 + \gamma z^2 + (dx + ey + gz) + h = 0$$

where α, β, γ, d, e, g, and h are any real numbers.

α, β, and γ are independent of the choice of coordinate system. They are called characteristic roots.

If none of the three characteristic roots is zero and all have the same sign, the quadric surfaces are ellipsoids or point ellipsoids. There are no straight lines on such surfaces (Art. 6-68).

If the three roots differ in sign, the surface may be a cone (Art. 6-48), or a hyperboloid of one or two sheets (Arts. 6-69 and 6-70).

If one root is zero and the other two have the same sign, the surface may be an elliptic paraboloid (Art. 6-71), an elliptic cylinder (Art. 6-65), or a point.

If one root is zero and the other two differ in sign, the surface may be a hyperbolic paraboloid (Art. 6-72), a hyperbolic cylinder (Art. 6-65), or two intersecting planes.

If only one root is not zero, the surface may be a parabolic cylinder (Art. 6-65), two parallel planes, or two coinciding planes.

Every plane section of a quadric surface is a conic section (Art. 6-49). Equations of the form $\pm x^2/a^2 \pm y^2/b^2 \pm z^2/c^2 = 1$ are central quadrics. Equations of the form $x^2/a^2 \pm y^2/b^2 = 2cz$ are noncentral quadrics.

6-68. Ellipsoid Equations. The equation of an ellipsoid with center at the origin and semiaxes a, b, and c is

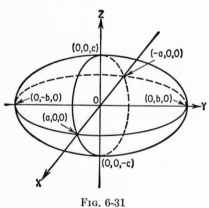

FIG. 6-31

$$\frac{x^2}{a^2} + \frac{y^2}{b^2} + \frac{z^2}{c^2} = 1$$

It is symmetric with respect to each coordinate plane, each axis, and the origin (Fig. 6-31). All signs in its equation are positive (see also Art. 6-67). All plane sections of this quadric surface are ellipses.

If two of the semiaxes a, b, and c are equal, the ellipsoid is a surface of revolution about the axis corresponding to the third variable. If all three axes are equal, the surface is a sphere (Art. 6-47).

An ellipse revolved about its minor axis is called an oblate spheroid; one revolved about its major axis is a prolate spheroid.

6-69. Equations of Hyperboloid of One Sheet. The equation of a hyperboloid of one sheet with origin at the center is

$$\frac{x^2}{a^2} + \frac{y^2}{b^2} - \frac{z^2}{c^2} = 1$$

It is symmetric with respect to each coordinate plane, each axis, and the

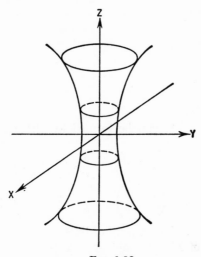

FIG. 6-32

origin (Fig. 6-32). There is only one negative sign in the equation (see also Art. 6-67).

The surface lies along the z axis but does not intersect it. If the negative sign were associated with x, the surface would lie along the x axis, and if associated with y, along the y axis.

If $a \neq b$ for a hyperboloid lying along the z axis, plane sections, $z = k$, parallel to the xy plane are ellipses, with axes increasing with distance from the xy plane. If $a = b$, the surface is a hyperboloid of revolution with the z axis as axis of revolution.

Plane sections parallel to the yz and xz planes are hyperbolas, with axes decreasing with distance from those planes. The hyperbolas degenerate into two intersecting lines when the sections are taken at $x = \pm a$ or $y = \pm b$.

A hyperboloid of one sheet is a ruled surface; it can be generated by a

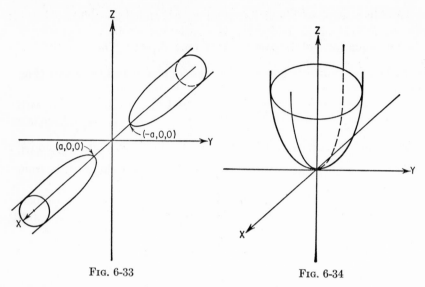

Fig. 6-33   Fig. 6-34

moving straight line. It can be formed by either of the two families of lines:

$$\frac{x}{a} + \frac{z}{c} = k\left(1 + \frac{y}{b}\right) \qquad \frac{x}{a} - \frac{z}{c} = \frac{1}{k}\left(1 - \frac{y}{b}\right)$$

$$\frac{x}{a} + \frac{z}{c} = k\left(1 - \frac{y}{b}\right) \qquad \frac{x}{a} - \frac{z}{c} = \frac{1}{k}\left(1 + \frac{y}{b}\right)$$

where k is a constant. The family of lines is called a regulus.

6-70. Equations of Hyperboloid of Two Sheets. The equation of a hyperboloid of two sheets with center at the origin is

$$\frac{x^2}{a^2} - \frac{y^2}{b^2} - \frac{z^2}{c^2} = 1$$

It represents two symmetrical surfaces lying along the x axis with vertices at $x = \pm a$ (Fig. 6-33). Each is symmetric with respect to the xy and xz planes and the x axis.

The equation has two negative signs. In general, the hyperboloids lie along the axis of the variable with the positive sign.

If hyperboloids lying along the x axis are cut by plane sections parallel to the yz plane, $x = k$, the sections are ellipses if $b \neq c$, with axes increasing with distance from the yz plane. If $b = c$, the surfaces are hyperboloids of revolution.

Sections cut by planes parallel to the xy and xz planes are hyperbolas with axes increasing with distance from those planes.

6-71. Elliptic Paraboloid Equations. The equation of an elliptic paraboloid with vertex at the origin and lying along the positive side of the z axis is

$$\frac{x^2}{a^2} + \frac{y^2}{b^2} = z$$

The surface is symmetric about the yz and xz planes and the z axis (Fig. 6-34).

Plane sections parallel to the xy plane, $z = k$, are ellipses if $a \neq b$, with axes increasing with distance from that plane. If $a = b$, the surface is a paraboloid of revolution.

Plane sections parallel to the xz and yz planes are parabolas with vertices receding from the xy plane as the distances of the sections from those planes increase.

6-72. Hyperbolic Paraboloid Equations. The equation of a hyperbolic paraboloid intersecting the coordinate axes only at the origin and lying along the z axis is

$$\frac{x^2}{a^2} - \frac{y^2}{b^2} = z$$

The surface is symmetric with respect to the xz and yz planes and the z axis. It has approximately the shape of a saddle (Fig. 6-35).

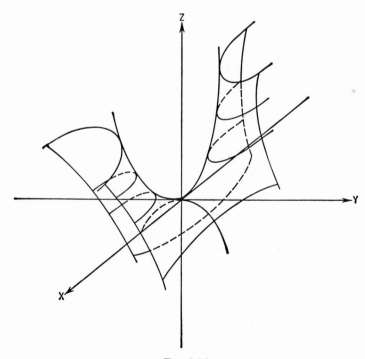

FIG. 6-35

Plane sections parallel to the xy plane, $z = k$, are hyperbolas. If k is positive, the transverse axes of the hyperbolas are parallel to the x axis; if k is negative, the transverse axes are parallel to the y axis. The traces on the xy plane are two straight lines $y = \pm (b/a)x$ through the origin. The farther the plane section is taken from the xy plane, the larger are the axes of the hyperbolas.

Sections parallel to the xz plane, $y = k$, are parabolas. They are concave upward, with vertices receding in the negative direction from the xy plane as k increases. The trace on the xz plane is the parabola $z = x^2/a^2$.

Sections parallel to the yz plane, $x = k$, also are parabolas, but these are concave downward, with vertices receding in the positive direction from the xy plane as k increases. The trace on the yz plane is the parabola $z = -y^2/b^2$.

A hyperbolic paraboloid is a ruled surface; it can be generated by a moving straight line. It can be formed by either of the two families of lines (reguluses):

$$\frac{x}{a} + \frac{y}{b} = k \qquad \frac{x}{a} - \frac{y}{b} = \frac{z}{k}$$

$$\frac{x}{a} + \frac{y}{b} = kz \qquad \frac{x}{a} - \frac{y}{b} = \frac{1}{k}$$

The first family is parallel to the plane $bx = ay$; the second is parallel to the plane $bx = -ay$. Hence, the surface can be generated by a line moving in contact with any two lines of the first regulus and parallel to $bx = -ay$, or in contact with two lines of the second regulus and parallel to $bx = ay$.

6-73. Reference Curves. (See pages 187 and 188.)

BIBLIOGRAPHY

Albert, Adrian: "Solid Analytic Geometry," McGraw-Hill Book Company, Inc., New York, 1949.

Middlemiss, R. R.: "Analytic Geometry," 2d ed., McGraw-Hill Book Company, Inc., New York, 1955.

Miller, Frederic H.: "Analytic Geometry and Calculus: A Unified Treatment," John Wiley & Sons, Inc., New York, 1949.

Catenary

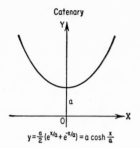

$$y = \frac{a}{2}(e^{x/a} + e^{x/a}) = a \cosh \frac{x}{a}$$

Cardioid

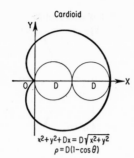

$$x^2 + y^2 + Dx = D\sqrt{x^2 + y^2}$$
$$\rho = D(1 - \cos\theta)$$

Folium of Descartes

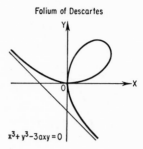

$$x^3 + y^3 - 3axy = 0$$

Limaçon

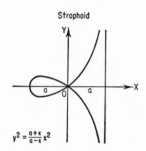

$$\rho = a\cos\theta + b; \quad b < a$$

Probability curve

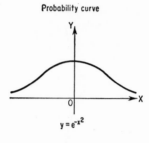

$$y = e^{-x^2}$$

Serpentine

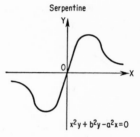

$$x^2 y + b^2 y - a^2 x = 0$$

Strophoid

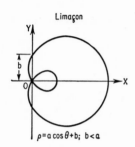

$$y^2 = \frac{a+x}{a-x} x^2$$

Spiral of Archimedes

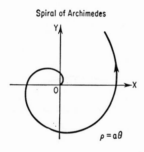

$$\rho = a\theta$$

Logarithmic or equiangular spiral

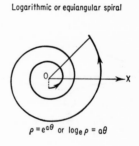

$$\rho = e^{a\theta} \text{ or } \log_e \rho = a\theta$$

Hyperbolic or reciprocal spiral

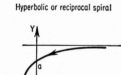

$$\rho\theta = a$$

Lituus

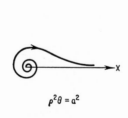

$$\rho^2 \theta = a^2$$

Roses

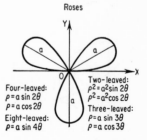

Four-leaved:
$\rho = a \sin 2\theta$
$\rho = a \cos 2\theta$
Eight-leaved:
$\rho = a \sin 4\theta$

Two-leaved:
$\rho^2 = a^2 \sin 2\theta$
$\rho^2 = a^2 \cos 2\theta$
Three-leaved:
$\rho = a \sin 3\theta$
$\rho = a \cos 3\theta$

Cubical parabola

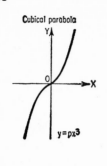

$$y = px^3$$

Semicubical parabola

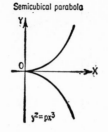

$$y^2 = px^3$$

Parabola

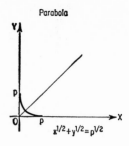

$$x^{1/2} + y^{1/2} = p^{1/2}$$

Witch of Agnesi

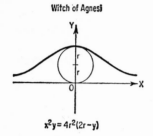

$$x^2 y = 4r^2(2r - y)$$

Cissoid of Diocles

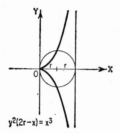

$$y^2(2r - x) = x^3$$

Lemniscate of Bernoulli

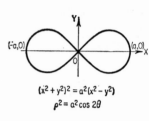

$$(x^2 + y^2)^2 = a^2(x^2 - y^2)$$
$$\rho^2 = a^2 \cos 2\theta$$

Conchoid of Nicomedes

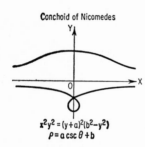

$$x^2 y^2 = (y + a)^2(b^2 - y^2)$$
$$\rho = a \csc \theta + b$$

Cycloid

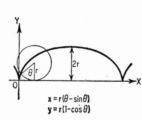

$$x = r(\theta - \sin\theta)$$
$$y = r(1 - \cos\theta)$$

Trochoid

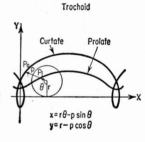

$$x = r\theta - p \sin\theta$$
$$y = r - p \cos\theta$$

Hypocycloid

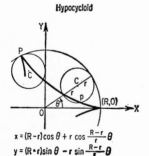

$$x = (R - r)\cos\theta + r \cos \frac{R - r}{r}\theta$$
$$y = (R + r)\sin\theta - r \sin \frac{R - r}{r}\theta$$

Hypocycloid of four cusps

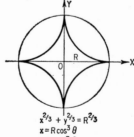

$$x^{2/3} + y^{2/3} = R^{2/3}$$
$$x = R\cos^3\theta$$
$$y = R\sin^3\theta$$

Epicycloid

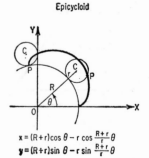

$$x = (R + r)\cos\theta - r \cos \frac{R + r}{r}\theta$$
$$y = (R + r)\sin\theta - r \sin \frac{R + r}{r}\theta$$

Chapter 7

DIFFERENTIAL CALCULUS

7-1. Limits. If, as a variable x increases or decreases, the absolute value of $a - x$ becomes less than any conceivable positive number, no matter how small, then a is the limit of x ($\lim x = a$ or $x \to a$).

$\lim [f(x)]_{x \to a} = b$ signifies that as x approaches a as a limit, $f(x)$ approaches b as a limit.

$$\lim (x + y) = \lim x + \lim y$$
$$\lim xy = \lim x \lim y$$
$$\lim \frac{x}{y} = \frac{\lim x}{\lim y} \qquad \text{if } \lim y \neq 0$$

• Find $\lim (1 - x)/(1 - x^2)$ if $x \to 1$.

$$\frac{1 - x}{1 - x^2} = \frac{1 - x}{(1 - x)(1 + x)} = \frac{1}{1 + x}$$
$$\lim \left[\frac{1 - x}{1 - x^2} \right]_{x \to 1} = \left[\lim \frac{1}{1 + x} \right]_{x \to 1} = \tfrac{1}{2}$$

• Find $\lim [(6x^3 - 2x^2)/(2x^3 - 3x + 2)]_{x \to \infty}$.
Divide numerator and denominator by x^3.

$$\lim \left[\frac{6x^3 - 2x^2}{2x^3 - 3x + 2} \right]_{x \to \infty} = \lim \left[\frac{6 - 2/x}{2 - (3/x^2) + 2/x^3} \right]_{x \to \infty} = \tfrac{6}{2} = 3$$

since $2/x$, $3/x^2$, and $2/x^3$ approach zero as $x \to \infty$.

• Evaluate $e = \lim [(1 + x)^{1/x}]_{x \to 0}$.
Expand the function by the binomial theorem.

$$(1 + x)^{1/x} = 1 + 1 + \frac{1}{2!} (1 - x) + \frac{1}{3!} (1 - x)(1 - 2x)$$
$$+ \frac{1}{4!} (1 - x)(1 - 2x)(1 - 3x) + \cdots$$
$$\lim [(1 + x)^{1/x}]_{x \to 0} = 2 + \frac{1}{2!} + \frac{1}{3!} + \frac{1}{4!} + \cdots = 2.7183 = e$$

• Evaluate $\lim [\sin x/x]_{x \to 0}$ if x is measured in radians.
In a unit circle, draw a chord AB and tangents PA and PB from an external point P to its extremities. Let x be half the central angle subtended by AB (Fig. 7-1). Then arc $AB = 2x$, since the radius is unity and x is measured in

189

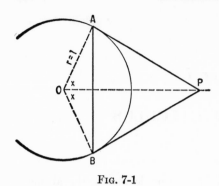

Fig. 7-1

radians. Also, $AB = 2 \sin x$ and $PA + PB = 2 \tan x$.

$$AB < \text{arc } AB < PA + PB$$
$$2 \sin x < \quad 2x \quad < 2 \tan x$$

Divide the inequality by $2 \sin x$ and invert.

$$1 > \frac{\sin x}{x} > \cos x$$

As $x \to 0$, $\cos x \to 1$. Hence, $\sin x/x$ also must have 1 as its limit.

7-2. Infinity and Infinitesimal. A function that increases without bound under the law that governs its variation is said to become infinite.

$$\lim \left[\frac{1}{x} \right]_{x \to 0} = \infty$$

Note, however, that $1/0$ has no meaning.

A variable that approaches zero as a limit becomes an infinitesimal.

The sum or product of infinitesimals is an infinitesimal. The product of a constant and an infinitesimal or division of an infinitesimal by a constant yields an infinitesimal.

7-3. Derivative of a Function of One Variable. If the independent variable is given a small increment, the derivative of the function with respect to that variable is the ratio of the increment of the function to the increment of the variable as the latter approaches zero as a limit. If $y = f(x)$, then the derivative may be denoted by $\frac{dy}{dx}$, $\frac{d}{dx} f(x)$, $D_x y$. $f'(x)$, y', or $\dot{y}$.

$$\frac{dy}{dx} = \lim \left[\frac{\Delta y}{\Delta x} \right]_{\Delta x \to 0} = \lim \left[\frac{f(x + \Delta x) - f(x)}{\Delta x} \right]_{\Delta x \to 0}$$

• What is the derivative with respect to x of $y = x^2$?

$$\frac{dy}{dx} = \frac{(x + \Delta x)^2 - x^2}{\Delta x} \bigg]_{\Delta x \to 0} = (2x + \Delta x)]_{\Delta x \to 0} = 2x$$

The process of finding a derivative is called differentiation.

If the derivative of a function is positive for a given value of the variable, the function will increase with increasing values of the variable; if the derivative is negative, the function will decrease.

A derivative represents a rate of change of a function. Hence, the derivative at any point of a curve is equal to the slope of the curve (or its tangent) at that point. Also, the velocity of a moving body at any instant is equal to the derivative of distance with respect to time.

• If a dropped weight falls in accordance with the law $s = 16t^2$, what is its speed at the end of 4 seconds?

$$\frac{ds}{dt}\bigg]_{t=4} = \frac{d}{dt}[16t^2]_{t=4} = 16 \times 2t\bigg]_{t=4} = 128$$

7-4. Derivative of a Function of a Function. If y is a function of a variable u, and u is a function of another variable x, then

$$\frac{dy}{dx} = \frac{dy}{du}\frac{du}{dx}$$

• Differentiate $y = \sqrt[3]{x^2 + 2}$.
Let $u = x^2 + 2$; hence, $y = u^{1/3}$.

$$\frac{dy}{du} = \frac{d}{du} u^{1/3} = \tfrac{1}{3}u^{1/3-1} = \tfrac{1}{3}u^{-2/3} = \frac{1}{3(x^2 + 2)^{2/3}}$$

$$\frac{du}{dx} = \frac{d}{dx}(x^2 + 2) = 2x$$

$$\frac{dy}{dx} = \frac{dy}{du}\frac{du}{dx} = \frac{2x}{3(x^2 + 2)^{2/3}}$$

7-5. Derivative of an Inverse Function. If $y = f_1(x)$ is solved for x, the solution $x = f_2(y)$ is called its inverse. For example, $x = \sin^{-1} y$ is the inverse of $y = \sin x$; $x = \pm \sqrt{y}$ is the inverse of $y = x^2$.

$$\frac{dy}{dx} = \frac{1}{\dfrac{dx}{dy}} \qquad \text{or} \qquad f'_2(y) = \frac{1}{f'_1(x)}$$

• Differentiate $y = \arcsin x$, given $\dfrac{d}{dy} \sin y = \cos y$.

Since $x = \sin y$ and $\dfrac{dx}{dy} = \cos y$,

$$\frac{dy}{dx} = \frac{1}{\dfrac{dx}{dy}} = \frac{1}{\cos y} = \frac{1}{\sqrt{1 - \sin^2 y}} = \frac{1}{\sqrt{1 - x^2}}$$

7-6. Differentiation of Parametric Functions. If a function is given in parametric form, $x = f_1(t)$, $y = f_2(t)$, then

$$\frac{dy}{dx} = \frac{\dfrac{dy}{dt}}{\dfrac{dx}{dt}}$$

• What is the slope of the cycloid $x = a(\theta - \sin\theta)$, $y = a(1 - \cos\theta)$, where θ is a parameter? $\left(\text{The slope equals } \dfrac{dy}{dx}.\right)$

$$\frac{dx}{d\theta} = a(1 - \cos\theta)$$

$$\frac{dy}{d\theta} = a\sin\theta$$

$$\frac{dy}{dx} = \frac{\dfrac{dy}{d\theta}}{\dfrac{dx}{d\theta}} = \frac{a\sin\theta}{a(1 - \cos\theta)} = \frac{\sin\theta}{1 - \cos\theta}$$

7-7. Differentiation of Implicit Functions. Given $f(x,y) = 0$, to find $\dfrac{dy}{dx}$ differentiate the function, treating y as a function of x and solve for $\dfrac{dy}{dx}$. $\left(\text{NOTE: } \dfrac{dx}{dx} = 1\right)$

• What is the slope at any point of the ellipse $4x^2 + 9y^2 = 36$?

The slope is equal to the derivative $y' = \dfrac{dy}{dx}$. Differentiate.

$$4(2x) + 9(2y)y' = 0$$

$$y' = -\frac{4x}{9y}$$

As an alternate method, use

$$\frac{dy}{dx} = -\frac{\dfrac{\partial f}{\partial x}}{\dfrac{\partial f}{\partial y}}$$

(For partial derivatives, see Art. 7-10.)

• Solve the previous problem by the alternate method.

$$\frac{\partial f}{\partial x} = \frac{\partial}{\partial x}(4x^2 + 9y^2 - 36) = 8x$$

$$\frac{\partial f}{\partial y} = \frac{\partial}{\partial y}(4x^2 + 9y^2 - 36) = 18y$$

$$\frac{dy}{dx} = -\frac{\dfrac{\partial f}{\partial x}}{\dfrac{\partial f}{\partial y}} = -\frac{8x}{18y} = -\frac{4x}{9y}$$

7-8. Derivatives of Second and Higher Order. The derivative of the first derivative of a function is called the second derivative. If $y = f(x)$, the second derivative with respect to x may be denoted by $\dfrac{d^2y}{dx^2}$, $D_x{}^2y$, $\dfrac{dy'}{dx}$, $f''(x)$, y'', or $\ddot{y}$.

Similarly, the derivative with respect to x of the second derivative is called the third derivative and may be denoted by $\dfrac{d^3y}{dx^3}$, D_x^3y, $f^{(3)}(x)$, or $y^{(3)}$.

- What is the mth derivative of x^n?

$$y' = nx^{n-1} \qquad y'' = n(n-1)x^{n-2} \qquad y^{(3)} = n(n-1)(n-2)x^{n-3}$$

From examination of this sequence of derivatives, it can be concluded that

$$y^{(m)} = n(n-1)(n-2) \cdots (n-m+1)x^{n-m}$$

- If the distance of a body from a fixed point is given by $s = 1/(t+1)$, where t is the time in minutes, what are its velocity and acceleration after 2 min?

$$\text{Velocity} = v = \frac{ds}{dt} = \frac{d}{dt}(t+1)^{-1} = -(t+1)^{-2} = -\frac{1}{(t+1)^2}$$

After 2 min
$$v = -\frac{1}{(2+1)^2} = -\frac{1}{9}$$

$$\text{Acceleration} = a = \frac{dv}{dt} = \frac{d^2s}{dt^2} = \frac{d}{dt} - (t+1)^{-2} = 2(t+1)^{-3} = \frac{2}{(t+1)^3}$$

After 2 min
$$a = \frac{2}{(2+1)^3} = \frac{2}{27}$$

If a function is given in parametric form $x = f_1(t)$, $y = f_2(t)$, then

$$\frac{dy}{dx} = \frac{dy}{dt}\frac{dt}{dx} = \frac{\dfrac{dy}{dt}}{\dfrac{dx}{dt}}$$

$$\frac{d^2y}{dx^2} = \frac{dy'}{dt}\frac{dt}{dx} = \frac{\dfrac{dx}{dt}\dfrac{d^2y}{dt^2} - \dfrac{d^2x}{dt^2}\dfrac{dy}{dt}}{\left(\dfrac{dx}{dt}\right)^3}$$

7-9. Differentials. On a curve AB defined by $y = f(x)$, let $P(x,y)$ and $Q(x + \Delta x,\ y + \Delta y)$ be nearby points. As Q is moved toward P along the curve, Δx and Δy approach zero, and the limiting position of the line PQ is the tangent at P. The slope of the tangent is $\dfrac{dy}{dx}$, the limit of the ratio $\Delta y/\Delta x$ as $\Delta x \to 0$. That is, dy is the increment of the ordinate of the tangent at P for an increment dx at P, whereas Δy is the increment of the ordinate to AB (Fig. 7-2).

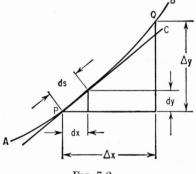

Fig. 7-2

The increments dx and dy are called differentials. The differential of a function $y = f(x)$ equals its derivative multiplied by the differential of the independent variable.

$$dy = f'(x)\, dx$$

• Differentiate $x^2 + y^2 = r^2$.

Apply the definition of differential to both x^2 and y^2, noting that r is a constant and therefore its differential is zero.

$$2x\, dx + 2y\, dy = 0$$
$$\frac{dy}{dx} = -\frac{x}{y}$$

In Fig. 7-2, the length of the straight line PQ is given by

$$(\Delta s)^2 = (\Delta x)^2 + (\Delta y)^2$$

Similarly, the differential length of curve is given by

$$ds^2 = dx^2 + dy^2$$
$$ds = \sqrt{1 + y'^2}\, dx = \sqrt{1 + x'^2}\, dy$$

In polar coordinates,

$$ds = \sqrt{d\rho^2 + \rho^2\, d\theta^2} = \sqrt{\rho^2 + \left(\frac{d\rho}{d\theta}\right)^2}\, d\theta$$

7-10. Partial Derivative. The derivative of a function with respect to one variable, all other variables being treated as constants.

The partial derivative of $z = f(x,y)$ with respect to x is denoted by

$$\frac{\partial z}{\partial x} \qquad z_x \qquad f_x(x,y)$$

and the partial derivative of z with respect to y is denoted by

$$\frac{\partial z}{\partial y} \qquad z_y \qquad f_y(x,y)$$

The partial derivative of z with respect to x is defined by

$$\frac{\partial z}{\partial x} = \lim \left[\frac{f(x+\Delta x,y) - f(x,y)}{\Delta x} \right]_{\Delta x \to 0}$$

It equals the slope of the curve of intersection of the surface represented by $z = f(x,y)$ and a plane $y = $ constant. Similarly, $\dfrac{\partial z}{\partial y}$ equals the slope in a plane $x = $ constant.

• What is the slope of $xyz = a^3$ in the planes $z = k$.

Differentiate the equation with respect to x, holding z constant, and solve for the derivative?

$$\frac{\partial}{\partial x} xyz = yz + xz \frac{\partial y}{\partial x} = 0$$

$$\frac{\partial y}{\partial x} = -\frac{y}{x}$$

If the function to be differentiated is in the form $f(x,y,z) = 0$, it sometimes is easier to use

$$\frac{\partial z}{\partial x} = -\frac{\dfrac{\partial f}{\partial x}}{\dfrac{\partial f}{\partial z}} \qquad \frac{\partial z}{\partial y} = -\frac{\dfrac{\partial f}{\partial y}}{\dfrac{\partial f}{\partial z}}$$

• Solve the previous problem with the above formulas.

$f(x,y,z) = xyz - a^3 = 0$.

$$\frac{\partial f}{\partial x} = yz \qquad \frac{\partial f}{\partial y} = xz$$

$$\frac{\partial y}{\partial x} = -\frac{yz}{xz} = -\frac{y}{x}$$

If x and y in $z = f(x,y)$ are changed to $x = f_1(u,v)$ and $y = f_2(u,v)$, then the partial derivatives of z with respect to the new variables are given by

$$\frac{\partial z}{\partial u} = \frac{\partial z}{\partial x}\frac{\partial x}{\partial u} + \frac{\partial z}{\partial y}\frac{\partial y}{\partial u}$$

$$\frac{\partial z}{\partial v} = \frac{\partial z}{\partial x}\frac{\partial x}{\partial v} + \frac{\partial z}{\partial y}\frac{\partial y}{\partial v}$$

7-11. Partial Derivatives of Second or Higher Order. In general, the partial derivatives of $z = f(x,y)$ also are functions of x and y and can, in turn, be differentiated.

$$\frac{\partial}{\partial x}\frac{\partial z}{\partial x} = \frac{\partial^2 z}{\partial x^2} = \frac{\partial^2 f}{\partial x^2} = f_{xx} = z_{xx}$$

$$\frac{\partial}{\partial y}\frac{\partial z}{\partial y} = \frac{\partial^2 z}{\partial y^2} = \frac{\partial^2 f}{\partial y^2} = f_{yy} = z_{yy}$$

$$\frac{\partial}{\partial x}\frac{\partial z}{\partial y} = \frac{\partial}{\partial y}\frac{\partial z}{\partial x} = \frac{\partial^2 z}{\partial x\,\partial y} = \frac{\partial^2 f}{\partial x\,\partial y} = f_{xy} = z_{xy}$$

In obtaining the second derivative with respect to x and y, the order of differentiation is immaterial if the derivatives are continuous.

• Show that $\dfrac{\partial^2 z}{\partial y\, \partial x} = \dfrac{\partial^2 z}{\partial x\, \partial y}$ if $z = (x + y)/(x - y)$.

$$\frac{\partial z}{\partial x} = \frac{(x - y) - (x + y)}{(x - y)^2} = -\frac{2y}{(x - y)^2} = -2y(x - y)^{-2}$$

$$\frac{\partial^2 z}{\partial x\, \partial y} = \frac{\partial}{\partial y}[-2y(x - y)^{-2}] = -2(x - y)^{-2} - 4y(x - y)^{-3} = -\frac{2(x + y)}{(x - y)^3}$$

$$\frac{\partial z}{\partial y} = \frac{(x - y) + (x + y)}{(x - y)^2} = \frac{2x}{(x - y)^2} = 2x(x - y)^{-2}$$

$$\frac{\partial^2 z}{\partial y\, \partial x} = \frac{\partial}{\partial x}[2x(x - y)^{-2}] = 2(x - y)^{-2} - 4x(x - y)^{-3} = -\frac{2(x + y)}{(x - y)^3}$$

$$\frac{\partial^2 z}{\partial x\, \partial y} = \frac{\partial^2 z}{\partial y\, \partial x} = -\frac{2(x + y)}{(x - y)^3}$$

7-12. Directional Derivative. If $z = f(x,y)$, the slope of the curve of intersection of the surface with a plane parallel to the z axis and at an angle θ with the xz plane is given by

$$\frac{\partial z}{\partial l} = \frac{\partial z}{\partial x} \cos \theta + \frac{\partial z}{\partial y} \sin \theta$$

• If the temperature at any point of a plate lying in the xy plane is given by $t = e^{-x^2-y^2}t_0$, what is the rate of change of temperature at a point $(1,1)$ in the direction making an angle of $60°$ with the x axis?
Find the directional derivative of t with $\theta = 60°$.

$$\left.\frac{\partial t}{\partial x}\right]_{1,1} = [-2xe^{-x^2-y^2}t_0]_{1,1} = -2e^{-2}t_0$$

$$\left.\frac{\partial t}{\partial y}\right]_{1,1} = [-2ye^{-x^2-y^2}t_0]_{1,1} = -2e^{-2}t_0$$

$$\left.\frac{\partial t}{\partial l}\right]_{1,1} = -2e^{-2}t_0 \cos 60° - 2e^{-2}t_0 \sin 60° = -e^{-2}t_0(1 + \sqrt{3})$$

7-13. Total Differential. If $u = f(x,y,z, \ldots ,t)$ and x is given an increment dx, y is given an increment dy, etc., then the increment in u is the total differential of u.

$$du = \frac{\partial u}{\partial x}\, dx + \frac{\partial u}{\partial y}\, dy + \frac{\partial u}{\partial z}\, dz + \cdots + \frac{\partial u}{\partial t}\, dt$$

• Find the total differential of u if $u = xyz$.

$$du = yz\, dx + xz\, dy + xy\, dz$$

7-14. Total Derivative. If $u = f(x,y,z)$ and $x = f_1(t)$, $y = f_2(t)$, and $z = f_3(t)$, the total derivative of u with respect to t is

$$\frac{du}{dt} = \frac{\partial u}{\partial x}\frac{dx}{dt} + \frac{\partial u}{\partial y}\frac{dy}{dt} + \frac{\partial u}{\partial z}\frac{dz}{dt}$$

If $u = f(x,y,z)$ and y and z are both functions of x, then

$$\frac{du}{dx} = \frac{\partial u}{\partial x} + \frac{\partial u}{\partial y}\frac{dy}{dx} + \frac{\partial u}{\partial z}\frac{dz}{dx}$$

Note that to obtain $\frac{\partial u}{\partial x}$ only x is assumed to vary, but to determine $\frac{du}{dx}$ all the variables must be assumed to change with an increment dx in x.

• Find $\frac{dz}{dt}$ for $z = x^2 + 4y^2$ when $x = \sin t$ and $y = \cos t$.

$$\frac{dz}{dt} = \frac{\partial z}{\partial x}\frac{dx}{dt} + \frac{\partial z}{\partial y}\frac{dy}{dt} = 2x \cos t - 8y \sin t$$

7-15. Basic Differentiation Formulas for Algebraic Functions

$$\frac{dc}{dx} = 0 \qquad c = \text{constant}$$

$$\frac{dx}{dx} = 1$$

$$\frac{d}{dx}(u + v - w) = \frac{du}{dx} + \frac{dv}{dx} - \frac{dw}{dx}$$

$$\frac{d}{dx}cv = c\frac{dv}{dx} \qquad c = \text{constant}$$

$$\frac{d}{dx}uv = u\frac{dv}{dx} + v\frac{du}{dx}$$

$$\frac{d}{dx}(u_1 u_2 \cdots u_n) = (u_2 u_3 \cdots u_n)\frac{du_1}{dx}$$

$$+ (u_1 u_3 \cdots u_n)\frac{du_2}{dx} + \cdots + (u_1 u_2 \cdots u_{n-1})\frac{du_n}{dx}$$

$$\frac{dx^n}{dx} = nx^{n-1} \qquad n = \text{constant}$$

$$\frac{du^n}{dx} = nu^{n-1}\frac{du}{dx} \qquad u = f(x)$$

$$\frac{d}{dx}\frac{u}{c} = \frac{1}{c}\frac{du}{dx} \qquad c = \text{constant}$$

$$\frac{d}{dx}\frac{u}{v} = \frac{v\frac{du}{dx} - u\frac{dv}{dx}}{v^2}$$

• Differentiate $y = x^3 - 3x^2 + 4 + 1/x^2 + \sqrt[3]{x}$.

$$\frac{dy}{dx} = \frac{d}{dx}x^3 - 3\frac{d}{dx}x^2 + \frac{d}{dx}4 + \frac{d}{dx}x^{-2} + \frac{d}{dx}x^{1/3}$$

$$= 3x^{3-1} - 3 \times 2x^{2-1} + 0 - 2x^{-2-1} + \tfrac{1}{3}x^{1/3-1}$$

$$= 3x^2 - 6x - \frac{2}{x^3} + \frac{1}{3x^{2/3}}$$

• Differentiate $s = (t^2 - 3)^4$.

$$\frac{ds}{dt} = 4(t^2 - 3)^{4-1} \frac{d}{dt}(t^2 - 3) = 4(t^2 - 3)^3(2t^{2-1} + 0) = 8t(t^2 - 3)^3$$

• Differentiate $y = x^2/(x - 1)$.

$$\frac{dy}{dx} = \frac{(x - 1)\frac{d}{dx}x^2 - x^2\frac{d}{dx}(x - 1)}{(x - 1)^2} = \frac{2x(x - 1) - x^2}{(x - 1)^2} = \frac{x^2 - 2x}{(x - 1)^2}$$

This also could have been differentiated as the product $x^2(x - 1)^{-1}$.

$$\frac{dy}{dx} = x^2 \frac{d}{dx}(x - 1)^{-1} + (x - 1)^{-1}\frac{d}{dx}x^2 = -x^2(x - 1)^{-2} + 2x(x - 1)^{-1}$$
$$= \frac{-x^2 + 2x(x - 1)}{(x - 1)^2} = \frac{x^2 - 2x}{(x - 1)^2}$$

7-16. Basic Differentiation Formulas for Transcendental Functions

$$\frac{d}{dx}\log_e u = \frac{1}{u}\frac{du}{dx}$$

$$\frac{d}{dx}\log_{10} u = \frac{\log_{10} e}{u}\frac{du}{dx}$$

$$\frac{d}{dx}a^u = a^u \frac{du}{dx}\log_e a \qquad a = \text{constant}$$

$$\frac{d}{dx}e^u = e^u \frac{du}{dx}$$

$$\frac{d}{dx}u^v = vu^{v-1}\frac{du}{dx} + u^v \frac{dv}{dx}\log_e u$$

$$\frac{d}{dx}\sin u = \cos u \frac{du}{dx}$$

$$\frac{d}{dx}\cos u = -\sin u \frac{du}{dx}$$

$$\frac{d}{dx}\tan u = \sec^2 u \frac{du}{dx}$$

$$\frac{d}{dx}\cot u = -\csc^2 u \frac{du}{dx}$$

$$\frac{d}{dx}\sec u = \sec u \tan u \frac{du}{dx}$$

$$\frac{d}{dx}\csc u = -\csc u \cot u \frac{du}{dx}$$

$$\frac{d}{dx}\arcsin u = \frac{1}{\sqrt{1 - u^2}}\frac{du}{dx}$$

$$\frac{d}{dx} \arccos u = - \frac{1}{\sqrt{1 - u^2}} \frac{du}{dx}$$

$$\frac{d}{dx} \arctan u = \frac{1}{1 + u^2} \frac{du}{dx}$$

$$\frac{d}{dx} \operatorname{arccot} u = - \frac{1}{1 + u^2} \frac{du}{dx}$$

$$\frac{d}{dx} \operatorname{arcsec} u = \frac{1}{u \sqrt{u^2 - 1}} \frac{du}{dx}$$

$$\frac{d}{dx} \operatorname{arccsc} u = - \frac{1}{u \sqrt{u^2 - 1}} \frac{du}{dx}$$

$$\frac{d}{dx} \sinh u = \cosh u \frac{du}{dx}$$

$$\frac{d}{dx} \cosh u = \sinh u \frac{du}{dx}$$

$$\frac{d}{dx} \tanh u = \operatorname{sech}^2 u \frac{du}{dx}$$

$$\frac{d}{dx} \coth u = - \operatorname{csch}^2 u \frac{du}{dx}$$

$$\frac{d}{dx} \operatorname{sech} u = - \operatorname{sech} u \tanh u \frac{du}{dx}$$

$$\frac{d}{dx} \operatorname{csch} u = \operatorname{csch} u \coth u \frac{du}{dx}$$

$$\frac{d}{dx} \operatorname{argsinh} u = \frac{1}{\sqrt{u^2 + 1}} \frac{du}{dx}$$

$$\frac{d}{dx} \operatorname{argcosh} u = \pm \frac{1}{\sqrt{u^2 - 1}} \frac{du}{dx} \qquad u > 1$$

$$\frac{d}{dx} \operatorname{argtanh} u = \frac{1}{1 - u^2} \frac{du}{dx} \qquad -1 < u < 1$$

$$\frac{d}{dx} \operatorname{argcoth} u = \frac{1}{1 - u^2} \frac{du}{dx} \qquad |u| > 1$$

$$\frac{d}{dx} \operatorname{argsech} u = \pm \frac{1}{u \sqrt{1 - u^2}} \frac{du}{dx} \qquad 0 < u < 1$$

$$\frac{d}{dx} \operatorname{argcsch} u = - \frac{1}{u \sqrt{1 + u^2}} \frac{du}{dx} \qquad u > 0$$

$$= \frac{1}{u \sqrt{1 + u^2}} \frac{du}{dx} \qquad u < 0$$

• Differentiate $y = (x + 3)^4 (x - 1)^3$.

This may be done by taking logarithms of both sides of the equation.

$$\log_e y = 4 \log_e (x + 3) + 3 \log_e (x - 1)$$
$$\frac{d}{dx} \log_e y = \frac{1}{y} \frac{dy}{dx} = \frac{4}{x + 3} + \frac{3}{x - 1}$$

Substitute $(x + 3)^4(x - 1)^3$ for y and solve for $\dfrac{dy}{dx}$.

$$\frac{dy}{dx} = 4(x + 3)^3(x - 1)^3 + 3(x + 3)^4(x - 1)^2 = (x + 3)^3(x - 1)^2(7x + 5)$$

• Differentiate $y = e^x \sin x$.

$$\log_e y = \log_e e^x + \log_e \sin x = x + \log_e \sin x$$
$$\frac{d}{dx} \log_e y = \frac{1}{y} \frac{dy}{dx} = 1 + \frac{\cos x}{\sin x}$$

Substitute $e^x \sin x$ for y and solve for $\dfrac{dy}{dx}$.

$$\frac{dy}{dx} = e^x \sin x + e^x \cos x = e^x(\sin x + \cos x)$$

7-17. Maximum and Minimum Values of a Function. To find a maximum or minimum value of a function of one variable, set the first derivative of the function equal to zero. Solve the equation to find the critical values of the variable.

Test the derivative first with a value a trifle less than each of the critical values, then with a value a trifle larger than each. If the sign of the derivative is first plus, then minus, the function is at a maximum, but if the sign is first minus, then plus, the function is at a minimum.

Another test is to substitute the critical values in the second derivative. If this derivative is negative, the function is at a maximum; if the second derivative is positive, the function is at a minimum.

The second test fails when the second derivative also is zero. In that case, test successive derivatives with the critical value until a derivative is found that is not zero. If that derivative is an even derivative, the function is at a maximum if the critical value makes the derivative negative and at a minimum if the derivative is positive.

If the first nonvanishing derivative is an odd derivative, the function is at an inflection point.

• What is the maximum value of $y = x^3 - 3x^2$?
Differentiate and set $y' = 0$.

$$y' = 3x^2 - 6x = 0$$

The critical values are $x = 0,2$. Substitute in the second derivative.

$$y''\Big]_0 = [6x - 6]_0 = -6 \qquad y''\Big]_2 = [6x - 6]_2 = 12 - 6 = 6$$

The function has a maximum at $x = 0$, where y'' is negative.

• What is the area of the largest rectangular field that can be enclosed with 900 lin ft of fencing?

Let x be one side of the field, y the other, and A the area. Then $2x + 2y = 900$ and $A = xy$. Elimination of y yields

$$A = x\left(\frac{900 - 2x}{2}\right) = 450x - x^2$$

If A is a maximum, $\dfrac{dA}{dx} = 0$.

$$\frac{dA}{dx} = 450 - 2x = 0$$
$$x = 225 \text{ ft}$$

Substitution in the original equations gives $y = x = 225$ ft,

$$A = 50{,}625 \text{ ft}^2.$$

To find a maximum or minimum value of $z = f(x,y)$, set

$$\frac{\partial z}{\partial x} = 0 \qquad \frac{\partial z}{\partial y} = 0$$

and solve for the critical values of x and y.

To test for a maximum or minimum, calculate for these values:

$$Q = \left(\frac{\partial^2 z}{\partial x\, \partial y}\right)^2 - \frac{\partial^2 z}{\partial x^2}\frac{\partial^2 z}{\partial y^2}$$

If Q and $\dfrac{\partial^2 z}{\partial x^2}\left(\text{or } \dfrac{\partial^2 z}{\partial y^2}\right)$ are negative, z will be at a maximum.

If $Q < 0$ and $\dfrac{\partial^2 z}{\partial x^2}\left(\text{or } \dfrac{\partial^2 z}{\partial y^2}\right)$ is positive, z will be at a minimum.

If $Q > 0$, z will be neither at a maximum nor minimum.
If $Q = 0$, the test fails.

• Find maximum and minimum values of $z = x^2 + xy + y^2 - 6x - 3y$.

$$\frac{\partial z}{\partial x} = 2x + y - 6 = 0 \qquad \frac{\partial z}{\partial y} = x + 2y - 3 = 0$$

Simultaneous solution of these equations yields $x = 3$, $y = 0$. Hence, $z = -9$. To test for a maximum or minimum, compute

$$\frac{\partial^2 z}{\partial x\, \partial y}\bigg]_{3,0} = 1 \qquad \frac{\partial^2 z}{\partial x^2}\bigg]_{3,0} = 2 \qquad \frac{\partial^2 z}{\partial y^2}\bigg]_{3,0} = 2$$
$$Q = (1)^2 - 2 \times 2 = 1 - 4 = -3$$

Since $Q < 0$ and the second derivatives are positive, $(3,0,-9)$ is a minimum.

To find maximum and minimum values of u when $u = f(x,y,z)$ and the variables are related by $\varphi(x,y,z) = 0$, solve the following equations simultaneously for x, y, z, and k:

$$\varphi(x,y,z) = 0 \qquad \frac{\partial f}{\partial x} + k\frac{\partial \varphi}{\partial x} = 0$$

$$\frac{\partial f}{\partial y} + k\frac{\partial \varphi}{\partial y} = 0 \qquad \frac{\partial f}{\partial z} + k\frac{\partial \varphi}{\partial z} = 0$$

• Find the volume of the largest rectangular parallelepiped that can be inscribed in the ellipsoid $x^2/a^2 + y^2/b^2 + z^2/c^2 = 1$.

Assume the parallelepiped to be centered on the center of the ellipsoid and let its sides be $2x$, $2y$, and $2z$. Hence, the volume $V = 8xyz$, which is to be a maximum. The variable must also satisfy $\varphi = x^2/a^2 + y^2/b^2 + z^2/c^2 - 1 = 0$.

$$\frac{\partial V}{\partial x} = 8yz \qquad \frac{\partial V}{\partial y} = 8xz \qquad \frac{\partial V}{\partial z} = 8xy$$

$$\frac{\partial \varphi}{\partial x} = \frac{2x}{a^2} \qquad \frac{\partial \varphi}{\partial y} = \frac{2y}{b^2} \qquad \frac{\partial \varphi}{\partial z} = \frac{2z}{c^2}$$

Next, solve simultaneously

$$\frac{x^2}{a^2} + \frac{y^2}{b^2} + \frac{z^2}{c^2} = 1$$

$$\frac{\partial f}{\partial x} + k\frac{\partial \varphi}{\partial x} = 8yz + \frac{2x}{a^2} k = 0$$

$$\frac{\partial f}{\partial y} + k\frac{\partial \varphi}{\partial y} = 8xz + \frac{2y}{b^2} k = 0$$

$$\frac{\partial f}{\partial z} + k\frac{\partial \varphi}{\partial z} = 8xy + \frac{2z}{c^2} k = 0$$

Let k be a negative number. Divide the second equation by the third and fourth and solve for y^2 and z^2. Substitute these values in the first equation and solve for x^2. The result is

$$x = \frac{a\sqrt{3}}{3} \qquad y = \frac{b\sqrt{3}}{3} \qquad z = \frac{c\sqrt{3}}{3}$$

$$V = 8xyz = \frac{8\sqrt{3}\,abc}{9}$$

7-18. Rolle's Theorem. If $f(x)$ becomes zero when $x = a$ and $x = b$, and if $f(x)$ and $f'(x)$ are continuous for all values of x in the interval, then there is at least one value of x between a and b for which $f'(x) = 0$.

• At what values are y and y' zero for $y = x^3 - 3x$?
Set $y = 0$.

$$x^3 - 3x = x(x^2 - 3) = 0$$
$$x = 0,\ \pm\sqrt{3}$$

Hence, y' must be zero for a value of x between 0 and $\sqrt{3}$ and between 0 and $-\sqrt{3}$, since $y' = 3x^2 - 3$ is continuous in these intervals.
Check: When $y' = 0$, $x = \pm 1$.

7-19. Theorem of the Mean. If $f(x)$ is continuous in the interval $a \leq x \leq b$ and has a derivative for every value of x ($a < x < b$), then there is a value x_1 of x between a and b for which

$$f(b) - f(a) = (b - a)f'(x_1)$$

where $f'(x_1)$ is the value of the derivative of $f(x)$ at x_1.

Cauchy's formula (extended theorem of the mean) states that

$$\frac{f(b) - f(a)}{g(b) - g(a)} = \frac{f'(x_1)}{g'(x_1)} \qquad a < x_1 < b$$

where g is any other function that is continuous and has derivatives in the interval $a \leq x \leq b$.

7-20. L'Hôpital's Rule for Indeterminate Forms. If $f(x)/g(x)$ approaches $0/0$ or ∞/∞ as $x \to a$ and if $f(x)$ and $g(x)$ have derivatives of all orders up to n, then the limit of $f(x)/g(x)$ equals the first of the following limits that is not indeterminate (if a limit exists):

$$\lim \left[\frac{f'(x)}{g'(x)} \right]_{x \to a}; \lim \left[\frac{f''(x)}{g''(x)} \right]_{x \to a}; \ldots ; \lim \left[\frac{f^{(n)}(x)}{g^{(n)}(x)} \right]_{x \to a}$$

If the first of these limits becomes infinite, then $f(x)/g(x)$ also becomes infinite.

• Evaluate $\lim [(x^2 - 4)/(x^2 + x - 6)]_{x \to 2}$.

If $x = 2$ is substituted in this function, it takes the form $0/0$. Apply l'Hôpital's rule; differentiate numerator and denominator.

$$\lim \left[\frac{x^2 - 4}{x^2 + x - 6} \right]_{x \to 2} = \left[\frac{2x}{2x + 1} \right]_{x \to 2} = \frac{4}{5}$$

• Evaluate $\lim [(x + \log x)/x \log x]_{x \to \infty}$.

As x increases without bound, the function approaches ∞/∞. Differentiate numerator and denominator.

$$\lim \left[\frac{x + \log x}{x \log x} \right]_{x \to \infty} = \left[\frac{1 + 1/x}{\log x + 1} \right]_{x \to \infty} = 0$$

As $x \to \infty$, the numerator approaches 1 but the denominator approaches ∞, so the limit is zero.

7-21. Evaluation of Indeterminate Form $0 \cdot \infty$. If as $x \to a, f(x) g(x) \to 0 \cdot \infty$, put the product in the form

$$\frac{f(x)}{1/g(x)} \qquad \text{or} \qquad \frac{g(x)}{1/f(x)}$$

so that it approaches $0/0$ or ∞/∞. Then find the limit (if one exists) by l'Hôpital's rule (Art. 7-20).

• Determine $\lim [x \log (1 + 1/x)]_{x \to \infty}$.

If $x = \infty$ is substituted in this function, it takes the form $\infty \cdot 0$. Rewrite it as

$$\frac{\log (1 + 1/x)}{1/x}$$

which approaches $0/0$ as $x \to \infty$. To evaluate this function, differentiate numerator and denominator.

$$\lim \left\{ \frac{-1/[(1 + 1/x)x^2]}{-1/x^2} \right\}_{x \to \infty} = \left[\frac{1}{1 + 1/x} \right]_{x \to \infty} = 1$$

7-22. Evaluation of Indeterminate Form $\infty - \infty$. If, as $x \to a$, $f(x) - g(x) \to \infty - \infty$, transform the function into one of the forms $0/0$ or ∞/∞. Then find the limit by l'Hôpital's rule (Art. 7-20).

• Determine $\lim (\csc \theta - \cot \theta)$ as $\theta \to 0$ from the positive side.

If $\theta = 0$ is substituted in this function, it takes the form $\infty - \infty$. Rewrite it as

$$\csc \theta - \cot \theta = \frac{1}{\sin \theta} - \frac{\cos \theta}{\sin \theta} = \frac{1 - \cos \theta}{\sin \theta}$$

The function on the right approaches $0/0$ as $\theta \to 0$. Differentiate numerator and denominator and let $\theta \to 0^+$.

$$\lim [\csc \theta - \cot \theta]_{\theta \to 0^+} = \left[\frac{\sin \theta}{\cos \theta} \right]_{\theta \to 0^+} = 0$$

7-23. Evaluation of Indeterminate Forms 0^0, 1^∞, ∞^0. If, as $x \to a$, $y = f(x)^{g(x)} \to 0^0$, 1^∞, or ∞^0, take the logarithm of both sides of the equation.

$$\log_e y = g(x) \log_e f(x)$$

This approaches $0 \cdot \infty$ as $x \to a$. Rewrite it as

$$\log_e y = \frac{\log_e f(x)}{1/g(x)}$$

and apply l'Hôpital's rule (Art. 7-20) to find the limit of $\log_e y$. If this limit is z, then $y = e^z$.

• Determine $\lim [(\sin \theta)^{\sec \theta}]_{\theta \to \pi/2}$.

If $\theta = \pi/2$ is substituted in this function, it takes the form 1^∞.

Let $\varphi = (\sin \theta)^{\sec \theta}$. Then, $\log \varphi = \sec \theta \log \sin \theta$. This takes the form $\infty \cdot 0$ when $\theta \to \pi/2$. Rewrite it as

$$\log \varphi = \frac{\log \sin \theta}{1/\sec \theta} = \frac{\log \sin \theta}{\cos \theta}$$

which takes the form $0/0$ when θ approaches $\pi/2$. Differentiate numerator and denominator and pass to the limit.

$$\lim [\log \varphi]_{\theta \to \pi/2} = \left[\frac{\cos \theta}{-\sin^2 \theta} \right]_{\theta \to \pi/2} = 0$$

$$\lim [(\sin \theta)^{\sec \theta}]_{\theta \to \pi/2} = e^0 = 1$$

BIBLIOGRAPHY

Bacon, H. M.: "Differential and Integral Calculus," McGraw-Hill Book Company, Inc., New York, 1955.

Dull, R. W., and Richard Dull: "Mathematics for Engineers," McGraw-Hill Book Company, Inc., New York, 1951.

Rice, H. S., and R. M. Knight: "Technical Mathematics with Calculus," McGraw-Hill Book Company, Inc., New York, 1957.

Smith, E. S., M. Salkover, and H. K. Justice: "Calculus," John Wiley & Sons, Inc., New York, 1958.

Chapter 8

INTEGRAL CALCULUS

8-1. Indefinite Integral. Integration is the process of finding a function when given its differential. The operation is indicated by a symbol resembling an elongated S.

$$\int f(x)\ dx = F(x) + C$$

The left-hand side of the above equation is called an indefinite integral. The sign in front of $f(x)$ is the integration sign; $f(x)$ is called the integrand. The expression is read "the integral of $f(x)\ dx$."

On the right-hand side of the equation, $F(x)$ is any function whose derivative is $f(x)$ and C is a constant. (Since the derivative of a constant is zero, all functions differing only by a constant have the same derivative.) $F(x)$ plus any constant is called an integral of $f(x)$. For methods of integration, see Arts. 8-2 to 8-6.

• What is the equation of a curve whose slope at any point is always twice the abscissa of the point and which passes through (3,4)?

The slope is equal to $\dfrac{dy}{dx} = 2x$. Hence, $dy = 2x\ dx$. Integration of both sides of this equation yields

$$\int dy = y = \int 2x\ dx = x^2 + C$$

As a check, note that the differential of $x^2 + C = 2x\ dx$. C can be found by substituting (3,4) for (x,y).

$$4 = (3)^2 + C$$
$$C = 4 - 9 = -5$$

Therefore, the required curve is $y = x^2 - 5$.

• If a body is dropped from a height of 1,000 ft and has a constant acceleration g of 32 ft per sec², what is its velocity after 3 sec? How high above the ground will it be?

If the downward direction is taken as negative, the acceleration g equals $-\dfrac{dv}{dt}$, where v is the velocity of the body and t is time. Therefore,

$$\int dv = v = -\int g\ dt = -gt + C_1$$

At the start, both t and v are zero. Hence, $C_1 = 0$, so

$$v = -gt$$

At the end of 3 sec, $v = -32 \times 3 = -96$ fps.

Now, velocity equals $\dfrac{ds}{dt}$, the rate of change of distance with respect to time. Thus,

$$\int ds = s = \int v \, dt = -\int gt \, dt = -\tfrac{1}{2}gt^2 + C_2$$

At the start, $t = 0$ and $s = 1{,}000$ ft; therefore, $C_2 = 1{,}000$. Then

$$s = 1{,}000 - \tfrac{1}{2}gt^2$$

and at the end of 3 sec, the height of the body will be

$$s = 1{,}000 - \tfrac{1}{2} \times 32(3)^2 = 856 \text{ ft}$$

8-2. Basic Properties of Integrals. By the definition of integration (Art. 8-1), for any function u of the variable x, whose differential is du,

$$\int du = u + C \qquad \int \frac{du}{dx} dx = u + C \qquad \frac{d}{dx} \int u \, dx = u + C$$

where C is a constant.

If k is a constant,

$$\int k \, du = k \int du$$

Thus, a constant factor can be moved outside the integral sign.

If u, v, and w are functions of a single variable,

$$\int (du + dv - dw) = \int du + \int dv - \int dw$$

The integral of the algebraic sum of a finite number of functions is the algebraic sum of the integrals of the functions.

• Find $\int (3x^2 - 4x + 5) \, dx$.

$$\int (3x^2 - 4x + 5) \, dx = \int 3x^2 \, dx - \int 4x \, dx + \int 5 \, dx = 3\int x^2 \, dx - 4\int x \, dx + 5\int dx$$
$$= 3(\tfrac{1}{3}x^3) - 4(\tfrac{1}{2}x^2) + 5x + C = x^3 - 2x^2 + 5x + C$$

8-3. Standard Elementary Integrals

$$\int dx = x + C$$

$$\int u^n \, du = \frac{u^{n+1}}{n+1} + C \qquad n \neq -1$$

$$\int \frac{du}{u} = \log_e u + C = \log_e cu$$

$$\int e^u \, du = e^u + C$$

$$\int a^u \, du = \frac{a^u}{\log_e a} + C$$

$$\int \sin u \, du = -\cos u + C$$

$$\int \cos u \, du = \sin u + C$$

$$\int \tan u \, du = \log_e \sec u + C$$

$$\int \cot u \, du = \log_e \sin u + C$$

$$\int \sec u \, du = \log_e (\sec u + \tan u)C$$

$$\int \csc u \, du = \log_e (\csc u - \cot u)C$$

$$\int \sec^2 u \, du = \tan u + C$$

$$\int \csc^2 u \, du = -\cot u + C$$

$$\int \sec u \tan u \, du = \sec u + C$$

$$\int \csc u \cot u \, du = -\csc u + C$$

$$\int \frac{du}{u^2 + a^2} = \frac{1}{a} \arctan \frac{u}{a} + C$$

$$\int \frac{du}{u^2 - a^2} = \frac{1}{2a} \log_e \frac{a - u}{a + u} C$$

$$\int \frac{du}{a^2 - u^2} = \frac{1}{2a} \log_e \frac{a + u}{a - u} C$$

$$\int \frac{du}{\sqrt{u^2 \pm a^2}} = \log_e (u + \sqrt{u^2 \pm a^2})C$$

$$\int \frac{du}{\sqrt{a^2 - u^2}} = \arcsin \frac{u}{a} + C$$

$$\int \sqrt{a^2 - u^2} \, du = \frac{u}{2} \sqrt{a^2 - u^2} + \frac{a^2}{2} \arcsin \frac{u}{a} + C$$

$$\int \sqrt{u^2 \pm a^2} \, du = \frac{u}{2} \sqrt{u^2 \pm a^2} \pm \frac{a^2}{2} \log_e (u + \sqrt{u^2 \pm a^2})C$$

$$\int \sinh u \, du = \cosh u + C$$

$$\int \cosh u \, du = \sinh u + C$$

$$\int \tanh u \, du = \log_e \cosh u + C$$

$$\int \coth u \, du = \log_e |\sinh u| + C$$

8-4. Integration Methods. The first step in integration is to simplify the integral, if necessary. If the integrand is a sum of functions, it may be easier to integrate if it is written as the sum of the integrals of the functions (Art. 8-2). If the integrand is a rational fraction with numerator of degree equal to or greater than that of the denominator, carry out the indicated division until the remainder is of lower degree than the denominator. If necessary, break the remainder into partial fractions for further simplification. Sometimes, substitution of another variable for a portion of the integrand may facilitate integration.

Compare the original or simplified integrand with the standard elementary forms (Art. 8-3) or with solutions in a more elaborate table of integrals (see Bibliography at the end of this chapter). If the solution is not now apparent by inspection, try algebraic or trigonometric manipulation to put the integral in a recognizable form for solution with a table, or try substitution of another variable.

• Find $\int \sin \theta \cos \theta \, d\theta$.
Let $u = \sin \theta$; then $du = \cos \theta \, d\theta$. Substitute in the integral.

$$\int \sin \theta \cos \theta \, d\theta = \int u \, du = \tfrac{1}{2} u^2 + C = \tfrac{1}{2} \sin^2 \theta + C$$

• Find $\int (x^2 + 2)^9 x \, dx$.
This could be integrated by expanding $(x^2 + 2)^9$ and integrating the result term by term. Simpler, however, is the substitution $u = x^2 + 2$, with $du = 2x \, dx$, $x \, dx = \tfrac{1}{2} du$.

$$\int (x^2 + 2)^9 x \, dx = \int u^9 (\tfrac{1}{2} du) = \tfrac{1}{2} \int u^9 \, du = \frac{1}{2} \frac{u^{10}}{10} + C = \frac{(x^2 + 2)^{10}}{20} + C$$

• Find $\int dx/(x^2 + 6x + 13)$.
Rewrite the denominator as $(x + 3)^2 + (2)^2$. Let $u = x + 3$, $du = dx$.

$$\int \frac{dx}{x^2 + 6x + 13} = \int \frac{du}{u^2 + (2)^2} = \frac{1}{2} \arctan \frac{u}{2} + C = \frac{1}{2} \arctan \frac{x + 3}{2} + C$$

In general, if an integral contains

$$\begin{aligned}
f\left(\sqrt{a^2 - u^2}\right) & \quad \text{try} \quad u = a \sin \theta \\
f\left(\sqrt{a^2 + u^2}\right) & \quad \text{try} \quad u = a \tan \theta \\
f\left(\sqrt{u^2 - a^2}\right) & \quad \text{try} \quad u = a \sec \theta \\
f(x^{p/q}) & \quad \text{try} \quad x = z^n
\end{aligned}$$

where n is the least common denominator of the fractional exponents of x.

$$f[x, (a + bx)^{p/q}] \quad \text{try} \quad a + bx = z^n$$

where n is the least common denominator of the fractional exponents of $a + bx$.

For integrals of the form

$$x^m(a + bx^n)^{p/q}$$

where $q > 0$, $n \neq 0$, m, n, p, q are integers

$$\text{try} \qquad z^q = a + bx^n \text{ if } \frac{m + 1}{n} \text{ is zero or an integer}$$

$$z^q x^n = a + bx^n \text{ if } \frac{m + 1}{n} + \frac{p}{q} \text{ is zero or an integer}$$

Integrals of the binomial differential form $\int x^m(a + bx^n)^p \, dx$ may be integrated with the aid of reduction formulas. These contain an integral, but of lower degree in x. Successive reduction may yield the required integral.

$$\int x^m(a + bx^n)^p \, dx = \frac{x^{m-n+1}(a + bx^n)^{p+1}}{(np + m + 1)b}$$
$$- \frac{(m - n + 1)a}{(np + m + 1)b} \int x^{m-n}(a + bx^n)^p \, dx$$

This reduces m by n, but fails when $np + m + 1 = 0$.

$$\int x^m(a + bx^n)^p \, dx = \frac{x^{m+1}(a + bx^n)^p}{np + m + 1}$$
$$+ \frac{anp}{np + m + 1} \int x^m(a + bx^n)^{p-1} \, dx$$

This decreases p by 1, but fails when $np + m + 1 = 0$.

$$\int x^m(a + bx^n)^p \, dx = \frac{x^{m+1}(a + bx^n)^{p+1}}{(m + 1)a}$$
$$- \frac{(np + n + m + 1)b}{(m + 1)a} \int x^{m+n}(a + bx^n)^p \, dx$$

This increases m by n and is useful when m is negative, but fails when $m = -1$.

$$\int x^m(a + bx^n)^p \, dx = -\frac{x^{m+1}(a + bx^n)^{p+1}}{n(p + 1)a}$$
$$+ \frac{np + n + m + 1}{n(p + 1)a} \int x^m(a + bx^n)^{p+1} \, dx$$

This increases p by 1 and is useful when p is negative, but fails when $p = -1$.

If an integral contains

$$f(x, \sqrt{p + qx + x^2}) \qquad \text{try} \qquad x = \frac{z^2 - p}{2z + q}$$

so that $\sqrt{p + qx + x^2} = (z^2 + qz + p)/(2z + q) = z - x$.

$$f(x, \sqrt{p + qx - x^2}) \qquad \text{try} \qquad x = \frac{az^2 + b}{z^2 + 1}$$

where $x - a$ and $b - x$ are real factors of $p + qx - x^2$, so that

$$\sqrt{p + qx - x^2} = (x - a)z = \frac{z(b - a)}{z^2 + 1}$$

(For additional integration methods, see Arts. 8-5 and 8-6.)

8-5. Integration by Parts. If u and v are functions of a single variable

$$\int u \, dv = uv - \int v \, du$$

Hence, $u \, dv$ can be integrated if $v \, du$ can be integrated.

• Find $\int x \sin x \, dx$.
Let $u = x$, $dv = \sin x \, dx$. Then

$$du = dx \qquad v = \int \sin x \, dx = - \cos x$$
$$\int x \sin x \, dx = -x \cos x - \int - \cos x \, dx = -x \cos x + \sin x + C$$

• Find $y = \int e^x \sin x \, dx$.
Let $u = e^x$, $dv = \sin x \, dx$. Then

$$du = e^x \, dx \qquad v = \int \sin x \, dx = - \cos x$$
$$y = -e^x \cos x + \int e^x \cos x \, dx \tag{8-1}$$

Since the result contains an integral of the same form as the given integral, try $u = \sin x$, $dv = e^x \, dx$. Then

$$du = \cos x \, dx, \, v = \int e^x \, dx = e^x$$
$$y = e^x \sin x - \int e^x \cos x \, dx \tag{8-2}$$

This, too, contains an integral of the same form as the given integral. However, note that both Eqs. (8-1) and (8-2) contain the same integral. Eliminate it between the two equations and solve for y.

$$2y = e^x \sin x - e^x \cos x$$
$$y = \tfrac{1}{2}e^x(\sin x - \cos x) \tag{8-3}$$

A solution could have been obtained from Eq. (8-1) by integrating $e^x \cos x \, dx$ by parts. The result contains the given integral y.

$$y = -e^x \cos x + e^x \sin x - \int e^x \sin x \, dx = e^x(\sin x - \cos x) - y$$

Solution of this equation for y yields Eq. (8-3).

8-6. Integration of Trigonometric Functions. In addition to the methods described in Arts. 8-4 and 8-5, the following may be used to reduce a trigonometric integrand to a standard form (Art. 8-3):

If an integral contains $f(\sin \theta, \cos \theta)$, try substituting $\theta = 2 \arctan x$, so that $x = \tan (\theta/2)$, $\sin \theta = 2x/(1 + x^2)$, $\cos \theta = (1 - x^2)/(1 + x^2)$ and $d\theta = 2dx/(1 + x^2)$.

• Find $\int d\theta/(1 + \sin \theta + \cos \theta)$.
Let $x = \tan (\theta/2)$.

$$\int \frac{d\theta}{1 + \sin \theta + \cos \theta} = \int \frac{2dx}{1 + x^2} \frac{1}{1 + 2x/(1 + x^2) + (1 - x^2)/(1 + x^2)}$$

$$= 2 \int \frac{dx}{1 + x^2 + 2x + 1 - x^2} = \int \frac{dx}{1 + x}$$

$$= \log_e (1 + x) + C = \log_e \left(1 + \tan \frac{\theta}{2}\right) + C$$

If an integral has the form $\int \sin^{2m+1} \theta \cos^n \theta \, d\theta$, where m is any integer or zero, rewrite the integrand in the form

$$\sin^{2m} \theta \cos^n \theta \sin \theta = (1 - \cos^2 \theta)^m \cos^n \theta \sin \theta$$

After the multiplication is carried out, the integrand becomes a sum of powers of $\cos \theta$ multiplied by $\sin \theta$ and can be integrated term by term.

If an integral has the form $\int \sin^p \theta \cos^{2r+1} \theta \, d\theta$, where r is any integer or zero, rewrite the integrand in the form

$$\sin^p \theta \cos^{2r} \theta \cos \theta = \sin^p \theta (1 - \sin^2 \theta)^r \cos \theta$$

After the multiplication is carried out, the integrand becomes a sum of powers of $\sin \theta$ multiplied by $\cos \theta$ and can be integrated term by term.

• Find $\int \sin^3 \theta \, d\theta$.
Here $m = 1$ and $n = 0$, so

$$\int \sin^3 \theta \, d\theta = \int \sin^2 \theta \sin \theta \, d\theta = \int (1 - \cos^2 \theta) \sin \theta \, d\theta$$
$$= \int \sin \theta \, d\theta - \int \cos^2 \theta \sin \theta \, d\theta = - \cos \theta + \tfrac{1}{3} \cos^3 \theta + C$$

• Find $\int \sin^2 \theta \cos^3 \theta d\theta$.
Here $p = 2$ and $r = 1$, so

$$\int \sin^2 \theta \cos^3 \theta \, d\theta = \int \sin^2 \theta \cos^2 \theta \cos \theta \, d\theta = \int \sin^2 \theta (1 - \sin^2 \theta) \cos \theta \, d\theta$$
$$= \int \sin^2 \theta \cos \theta \, d\theta - \int \sin^4 \theta \cos \theta \, d\theta$$
$$= \tfrac{1}{3} \sin^3 \theta - \tfrac{1}{5} \sin^5 \theta + C$$

If an integral has the form $\int \sin^{2m} \theta \cos^{2n} \theta \, d\theta$, where m and n are integers or zero, use the identities

$$\sin^2 \theta = \tfrac{1}{2}(1 - \cos 2\theta) \qquad \cos^2 \theta = \tfrac{1}{2}(1 + \cos 2\theta) \qquad \sin \theta \cos \theta = \frac{\sin 2\theta}{2}$$

• Find $\int \sin^2 \theta \cos^4 \theta \, d\theta$.

The integration can be simplified by first transforming the integrand from a function of θ to a function of 2θ, then to a function of 4θ.

$$\int \sin^2 \theta \cos^4 \theta \, d\theta = \int (\sin \theta \cos \theta)^2 \cos^2 \theta \, d\theta = \int \left(\frac{1}{4} \sin^2 2\theta\right) \left(\frac{1 + \cos 2\theta}{2}\right) d\theta$$

$$= \frac{1}{8} \int \sin^2 2\theta \, d\theta + \frac{1}{8} \int \sin^2 2\theta \cos 2\theta \, d\theta$$

$$= \frac{1}{8} \int \frac{1 - \cos 4\theta}{2} \, d\theta + \frac{1}{8} \frac{1}{3} \frac{1}{2} \sin^3 2\theta$$

$$= \frac{\theta}{16} - \frac{1}{64} \sin 4\theta + \frac{1}{48} \sin^3 2\theta + C$$

If an integral has the form $\int \tan^m \theta \sec^n \theta \, d\theta$ or $\int \cot^m \theta \csc^n \theta \, d\theta$, try the identities $\sec^2 \theta = 1 + \tan^2 \theta$ or $\csc^2 \theta = 1 + \cot^2 \theta$.

Integrals of the form $\int \sin^m \theta \cos^n \theta \, d\theta$ also may be integrated with the aid of reduction formulas. These contain an integral, but of lower degree in $\sin \theta$ or $\cos \theta$. Successive reduction may yield the required integral.

$$\int \sin^m \theta \cos^n \theta \, d\theta = \frac{\sin^{m+1} \theta \cos^{n-1} \theta}{m + n} + \frac{n - 1}{m + n} \int \sin^m \theta \cos^{n-2} \theta \, d\theta$$

This reduces n by 2 but fails when $m + n = 0$.

$$\int \sin^m \theta \cos^n \theta \, d\theta = -\frac{\sin^{m-1} \theta \cos^{n+1} \theta}{m + n} + \frac{m - 1}{m + n} \int \sin^{m-2} \theta \cos^n \theta \, d\theta$$

This reduces m by 2 but fails when $m + n = 0$.

$$\int \sin^m \theta \cos^n \theta \, d\theta = -\frac{\sin^{m+1} \theta \cos^{n+1} \theta}{n + 1}$$
$$+ \frac{m + n + 2}{n + 1} \int \sin^m \theta \cos^{n+2} \theta \, d\theta$$

This increases n by 2 and is useful when n is negative, but fails when $n = -1$.

$$\int \sin^m \theta \cos^n \theta \, d\theta = \frac{\sin^{m+1} \theta \cos^{n+1} \theta}{m + 1} + \frac{m + n + 2}{m + 1} \int \sin^{m+2} \theta \cos^n \theta \, d\theta$$

This increases m by 2 and is useful when m is negative, but fails when $m = -1$.

• Find $\int \sin^3 \theta \cos^{1/2} \theta \, d\theta$.

Here $m = 3$ and $n = \frac{1}{2}$. Use the second reduction formula to reduce m by 2.

$$\int \sin^3 \theta \cos^{1/2} \theta \, d\theta = -\frac{\sin^2 \theta \cos^{3/2} \theta}{3 + \frac{1}{2}} + \frac{2}{3 + \frac{1}{2}} \int \sin \theta \cos^{1/2} \theta \, d\theta$$

$$= -\frac{2}{7} \sin^2 \theta \cos^{3/2} \theta - \frac{8}{21} \cos^{3/2} \theta + C$$

8-7. Definite Integral.   If $\int f(x)\ dx = F(x) + C$, then

$$F(b) - F(a) = \int_a^b f(x)\ dx$$

The expression on the right is called a definite integral. It is read "definite integral from a to b." a is called the lower limit; b is the upper limit. It is evaluated by integrating, substituting the limits for the variable, and subtracting the result obtained with the lower limit from that obtained with the upper limit.

$$\int_a^b f(x)\ dx = F(x)\Big]_a^b = F(b) - F(a)$$

Note that a definite integral is a function of the limits and is independent of the variable of integration. Also, the constant of integration is eliminated in the subtraction.

If another variable is substituted in integrating, the limits change too, in accordance with the equation of substitution.

• Find $\int_1^4 dx/\sqrt{x}$.

Let $x = z^2$, $dx = 2z\ dz$. To change the limits, note that when $x = 1$, $z = 1$, and when $x = 4$, $z = 2$.

$$\int_1^4 \frac{dx}{\sqrt{x}} = \int_1^2 \frac{2z\ dz}{z} = 2z\Big]_1^2 = 2 \times 2 - 2 \times 1 = 2$$

Reversal of the limits of a definite integral changes its sign.

$$\int_a^b f(x)\ dx = - \int_b^a f(x)\ dx$$

If $a < b < c$, then

$$\int_a^c f(x)\ dx = \int_a^b f(x)\ dx + \int_b^c f(x)\ dx$$

A definite integral is the limit of a sum. If $f(x)$ is continuous in the interval $a \le x \le b$, consider the interval divided into n subintervals $\Delta x_1, \Delta x_2, \ldots, \Delta x_n$, such that the largest Δx approaches zero when n approaches infinity. Also, let $x_1, x_2, \ldots, x_n$ be points in each subinterval. Then

$$\lim [f(x_1)\ \Delta x_1 + f(x_2)\ \Delta x_2 + \cdots + f(x_n)\ \Delta x_n]_{n \to \infty} = \int_a^b f(x)\ dx$$

• If a fluid weighs w lb per ft³, how much pressure does it exert on a 1-ft width of wall 10 ft high, assuming the unit pressure in pounds per square foot varies directly with depth?

Consider a horizontal strip of wall dx high at a distance x from the top of the wall. If y is the unit pressure at that level, the total pressure on the strip is

$y \, dx$. Summation of the pressures on such strips for the full height of the wall yields the total pressure.

$$P = \int_0^{10} y \, dx$$

The unit pressure y equals the weight of a column of fluid x ft high and 1 ft square in cross section; that is, $y = wx$. Hence,

$$P = \int_0^{10} wx \, dx = \tfrac{1}{2} \, wx^2 \Big]_0^{10} = \tfrac{1}{2}w[100 - 0] = 50w$$

8-8. Improper Integrals. If either or both of the limits of a definite integral are infinitely large, or if the integrand becomes infinite in the interval of integration, the definite integral is called an improper integral.

Integrals with infinite limits may be evaluated by considering their limits as variables approaching infinity as a limit, if the integral approaches a finite value as a limit.

$$\int_2^\infty \frac{dx}{x^2} = \lim \left[\int_2^b \frac{dx}{x^2} \right]_{b \to \infty} = \lim \left[-\frac{1}{b} + \frac{1}{2} \right]_{b \to \infty} = \frac{1}{2}$$

If the integrand becomes infinite at a limit of integration a or b, the integral may be evaluated by considering a or b as a variable approaching a limit, if the integral approaches a finite value as a limit.

$$\int_0^4 \frac{dx}{\sqrt{x}} = \lim \left[\int_a^4 \frac{dx}{\sqrt{x}} \right]_{a \to 0} = \lim \left[2\sqrt{4} - 2\sqrt{a} \right]_{a \to 0} = 4$$

If $a < b < c$, so that

$$\int_a^c f(x) \, dx = \int_a^b f(x) \, dx + \int_b^c f(x) \, dx$$

and $f(x)$ becomes infinitely large at $x = b$, the integral on the left-hand side of this equation may be evaluated if both integrals on the right-hand side approach finite values as b is approached as a limit.

$$\int_{-1}^1 \frac{dx}{x^{2/3}} = \lim \left[\int_a^1 \frac{dx}{x^{2/3}} \right]_{a \to 0+} + \lim \left[\int_{-1}^b \frac{dx}{x^{2/3}} \right]_{b \to 0-}$$
$$= [3 - 3a^{1/3}]_{a \to 0+} + [3b^{1/3} + 3]_{b \to 0-} = 3 + 3 = 6$$

8-9. Multiple Integrals. The symbols

$$\int_a^b dx \int_m^n f(x,y) \, dy = \int_a^b \int_m^n f(x,y) \, dy \, dx$$

where m and n may be functions of x, represent an iterated or repeated integral, that is, an integral of an integral. The symbols indicate that the integration is to be carried out first with respect to y, holding x constant, then with respect to x.

Let $f(x,y)$ be continuous throughout the region S of the xy plane, and consider S subdivided into n subregions ΔS_1, ΔS_2, . . . , ΔS_n. Let x_1,y_1; x_2,y_2; . . . ; x_n,y_n be pairs of points, one pair in each subregion. Then

$$\lim [f(x_1,y_1)\,\Delta S_1 + f(x_2,y_2)\,\Delta S_2 + \cdots + f(x_n,y_n)\,\Delta S_n]_{n\to\infty} = \iint\limits_{S} f(x,y)\,dS$$

The expression on the right-hand side of this equation, to be integrated over the entire region S, is called a double integral. It can be evaluated as an iterated integral.

$$\iint\limits_{S} f(x,y)\,dS = \int_a^b \int_m^n f(x,y)\,dy\,dx = \int_c^d \int_p^q f(x,y)\,dx\,dy$$

where m and n may be functions of x, and p and q may be functions of y.

• Find the double integral of x^2y^3 over the rectangle A whose vertices are $(0,0)$, $(3,0)$, $(3,2)$, and $(0,2)$.

$$\iint\limits_{A} x^2y^3\,dA = \int_0^3 \int_0^2 x^2y^3\,dy\,dx = \int_0^3 \left[\tfrac{1}{4}x^2y^4\right]_0^2 dx = \int_0^3 4x^2\,dx$$

$$= \left[\tfrac{4}{3}x^3\right]_0^3 = 36$$

• Find the double integral of xy^2 over the quarter circle $x^2 + y^2 = a^2$ lying in the first quadrant.

$$\iint\limits_{S} x^2y\,dS = \int_0^a \int_0^{\sqrt{a^2-x^2}} x^2y\,dy\,dx = \int_0^a \left[\tfrac{1}{2}x^2y^2\right]_0^{\sqrt{a^2-x^2}} dx$$

$$= \int_0^a \tfrac{1}{2}(a^2 - x^2)x^2\,dx = \left[\tfrac{1}{6}a^2x^3 - \tfrac{1}{10}x^5\right]_0^a = \tfrac{1}{15}a^5$$

8-10. Line Integral. If $M(x,y)$ and $N(x,y)$ are continuous functions that are single-valued at every point of a plane curve $C(x,y)$ between two points $A(x_1,y_1)$ and $B(x_2,y_2)$, then

$$I = \int_C [M(x,y)\,dx + N(x,y)\,dy]$$

integrated over C between A and B, is called a line integral.

• If a body moves along the parabola $y = x^2$, acted upon by a force F whose component parallel to the x axis is $X = k$, a constant, and whose component parallel to the y axis is $Y = 2kx$, how much work is done when the body moves from $(0,0)$ to $(2,4)$?

$$W = \int_C F\,ds = \int_C (X\,dx + Y\,dy) = \int_{0,0}^{2,4} (k\,dx + 2kx\,dy)$$

$$= k\int_0^2 dx + 2k\int_0^4 y^{1/2}\,dy = kx\Big]_0^2 + \tfrac{4}{3}ky^{3/2}\Big]_0^4 = 2k + 3\tfrac{2}{3}k = 3\tfrac{8}{3}k$$

If $\dfrac{\partial M}{\partial y} = \dfrac{\partial N}{\partial x}$, the line integral around any closed curve is zero. Also, the value of the integral between any two points is independent of the path between them and hence is a function of the end points only.

If $P(x,y,z)$, $Q(x,y,z)$, and $R(x,y,z)$ are single-valued continuous functions along a space curve C, then the line integral is

$$I = \int_C (P\,dx + Q\,dy + R\,dz)$$

If $\dfrac{\partial P}{\partial y} = \dfrac{\partial Q}{\partial x}$, $\dfrac{\partial Q}{\partial z} = \dfrac{\partial R}{\partial y}$, and $\dfrac{\partial R}{\partial x} = \dfrac{\partial P}{\partial z}$, the line integral around any closed curve in space is zero. Also, the value of the integral between any two points is independent of the path between them.

8-11. Green's Theorems. In a plane, if $M(x,y)$, $N(x,y)$, $\dfrac{\partial M}{\partial y}$, and $\dfrac{\partial N}{\partial x}$ are continuous single-valued functions over a closed region R and its boundary curve C, then

$$\iint_R \left(\frac{\partial M}{\partial y} - \frac{\partial N}{\partial x} \right) dx\,dy = - \int_C (M\,dx + N\,dy)$$

The double integral is taken over the given region (Art. 8-9) and the line integral is taken counterclockwise along C (Art. 8-10).

• Find the value of $\int_C (x^2 y\,dx + y\,dy)$ along the closed curve C formed by $x = y^2$ and $x = y$, between $(0,0)$ and $(1,1)$.
Here $M = x^2 y$, $\dfrac{\partial M}{\partial y} = x^2$; $N = y$, $\dfrac{\partial N}{\partial x} = 0$.

$$\int_C (x^2 y\,dx + y\,dy) = - \int_0^1 \int_{y^2}^{y} x^2\,dx\,dy = - \int_0^1 \left[\tfrac{1}{3}x^3 \right]_{y^2}^{y} dy$$

$$= -\tfrac{1}{3} \int_0^1 (y^3 - y^6)\,dy = -\tfrac{1}{3} \left[\tfrac{1}{4}y^4 - \tfrac{1}{7}y^7 \right]_0^1 = -\tfrac{1}{28}$$

In space, if $P(x,y,z)$, $Q(x,y,z)$, $R(x,y,z)$, $\dfrac{\partial P}{\partial x}$, $\dfrac{\partial Q}{\partial y}$, and $\dfrac{\partial R}{\partial z}$ are continuous, single-valued functions in a region V bounded by a closed surface S, then

$$\int_S (P \cos \alpha + Q \cos \beta + R \cos \gamma)\,dS = \int_V \left(\frac{\partial P}{\partial x} + \frac{\partial Q}{\partial y} + \frac{\partial R}{\partial z} \right) dV$$

where $\cos \alpha$, $\cos \beta$, and $\cos \gamma$ are the direction cosines of the exterior normal to S. The surface integral can be written in the form

$$\iint_{S'} (P\,dy\,dz + Q\,dz\,dx + R\,dx\,dy) = \iiint_V \left(\frac{\partial P}{\partial x} + \frac{\partial Q}{\partial y} + \frac{\partial R}{\partial z} \right) dx\,dy\,dz$$

This formula is also known as the divergence theorem. (See also Art. 14-31.)

8-12. Stokes' Theorem. If $P(x,y,z)$, $Q(x,y,z)$, and $R(x,y,z)$ and their partial derivatives with respect to x, y, and z are continuous functions in a region containing a surface S bounded by a closed curve C, then

$$\int_C (P\,dx + Q\,dy + R\,dz) = \iint_S \left[\left(\frac{\partial R}{\partial y} - \frac{\partial Q}{\partial z}\right) \cos \alpha \right.$$

$$+ \left(\frac{\partial P}{\partial z} - \frac{\partial R}{\partial x}\right) \cos \beta + \left.\left(\frac{\partial Q}{\partial x} - \frac{\partial P}{\partial y}\right) \cos \gamma \right] dS$$

$$= \iint_S \left[\left(\frac{\partial Q}{\partial x} - \frac{\partial P}{\partial y}\right) dx\,dy \right.$$

$$+ \left(\frac{\partial R}{\partial y} - \frac{\partial Q}{\partial z}\right) dy\,dz + \left.\left(\frac{\partial P}{\partial z} - \frac{\partial R}{\partial x}\right) dz\,dx \right]$$

where $\cos \alpha$, $\cos \beta$, and $\cos \gamma$ are the direction cosines of the exterior normal to the surface S. (See also Art. 14-31.)

8-13. Differentiation under the Integral Sign. If the integrand is a function of a parameter as well as the variable of integration

$$F(\alpha) = \int_a^b f(x,\alpha)\,dx$$

it may be differentiated with respect to the parameter before integration if the partial derivative of the integrand with respect to the parameter exists and is continuous.

$$\frac{dF}{d\alpha} = \frac{d}{d\alpha} \int_a^b f(x,\alpha)\,dx = \int_a^b \frac{\partial f}{\partial \alpha}\,dx$$

• Evaluate $y = \int_{-\infty}^0 [(e^{\alpha x} - e^x)/x]\,dx$, where $\alpha \geq 1$.
Differentiate with respect to α.

$$\frac{dy}{d\alpha} = \int_{-\infty}^0 e^{\alpha x}\,dx = \frac{1}{\alpha} e^{\alpha x} \Big]_{-\infty}^0 = \frac{1}{\alpha} - 0$$

$$dy = \frac{d\alpha}{\alpha}$$

Integration of both sides yields $y = \log_e \alpha + C$. To evaluate C, return to the original integral and let $\alpha = 1$.

$$y(1) = \int_{-\infty}^0 \frac{e^x - e^x}{x}\,dx = 0 = \log_e 1 + C$$

Since $\log_e 1 = 0$, $C = 0$. Hence, the required integral is $y = \log_e \alpha$.

If the limits of a definite integral, as well as the integrand, are functions of a parameter α,

$$\frac{dF}{d\alpha} = \int_{a(\alpha)}^{b(\alpha)} \frac{\partial f(x,\alpha)}{\partial \alpha}\, dx + f(b,\alpha)\frac{db}{d\alpha} - f(a,\alpha)\frac{da}{d\alpha}$$

• Find $\dfrac{dy}{d\alpha}$ if $y = \displaystyle\int_{\pi/2\alpha}^{\pi/\alpha} [(\sin \alpha x)/x]\, dx$, where $\alpha \neq 0$.

Differentiate under the integral sign.

$$\frac{dy}{d\alpha} = \int_{\pi/2\alpha}^{\pi/\alpha} \cos \alpha x\, dx + \frac{\sin \pi}{\pi/\alpha}\left(-\frac{\pi}{\alpha^2}\right) - \frac{\sin \pi/2}{\pi/2\alpha}\left(-\frac{\pi}{2\alpha^2}\right)$$

$$= \frac{1}{\alpha}\sin \alpha x \Big]_{\pi/2\alpha}^{\pi/\alpha} + 0 + \frac{1}{\alpha} = 0 - \frac{1}{\alpha} + \frac{1}{\alpha} = 0$$

8-14. Jacobian. If u_1, u_2, . . . , u_n are continuous functions of the variables x_1, x_2, . . . , x_n, the jacobian or functional determinant is given by

$$J\left(\frac{u_1,u_2, \ldots ,u_n}{x_1,x_2, \ldots ,x_n}\right) = \begin{vmatrix} \dfrac{\partial u_1}{\partial x_1} & \dfrac{\partial u_2}{\partial x_1} & \cdots & \dfrac{\partial u_n}{\partial x_1} \\[2mm] \dfrac{\partial u_1}{\partial x_2} & \dfrac{\partial u_2}{\partial x_2} & \cdots & \dfrac{\partial u_n}{\partial x_2} \\[2mm] \cdots & \cdots & \cdots & \cdots \\[2mm] \dfrac{\partial u_1}{\partial x_n} & \dfrac{\partial u_2}{\partial x_n} & \cdots & \dfrac{\partial u_n}{\partial x_n} \end{vmatrix}$$

If $x = f_1(u,v)$ and $y = f_2(u,v)$, the differential of area is given by

$$dA = \left| J\left(\frac{x,y}{u,v}\right) \right| du\, dv$$

The absolute value of the jacobian is to be used in the formula.

• Transform $dA = dx\, dy$ from rectangular coordinates to polar.
The transformation equations are $x = \rho \cos \theta$ and $y = \rho \sin \theta$. Hence,

$$\frac{\partial x}{\partial \rho} = \cos \theta,\ \frac{\partial x}{\partial \theta} = -\rho \sin \theta,\ \frac{\partial y}{\partial \rho} = \sin \theta,\ \text{and}\ \frac{\partial y}{\partial \theta} = \rho \cos \theta.$$

$$J\left(\frac{x,y}{\rho,\theta}\right) = \begin{vmatrix} \cos \theta & \sin \theta \\ -\rho \sin \theta & \rho \cos \theta \end{vmatrix} = \rho \cos^2 \theta + \rho \sin^2 \theta = \rho$$

$$dA = |J|\, d\rho\, d\theta = \rho\, d\rho\, d\theta$$

If $x_1 = f_1(u,v,w)$, $y = f_2(u,v,w)$, and $z = f_3(u,v,w)$, the differential of volume is given by

$$dV = \left| J\left(\frac{x,y,z}{u,v,w}\right) \right| du\, dv\, dw$$

The absolute value of the jacobian is to be used in the formula.

• Transform $dV = dx\ dy\ dz$ from rectangular coordinates to spherical.
The transformation equations are $x = \rho \sin \theta \cos \varphi$, $y = \rho \sin \theta \sin \varphi$, and $z = \rho \cos \theta$. Hence,

$$\frac{\partial x}{\partial \rho} = \sin \theta \cos \varphi \qquad \frac{\partial x}{\partial \varphi} = -\rho \sin \theta \sin \varphi \qquad \frac{\partial x}{\partial \theta} = \rho \cos \theta \cos \varphi$$

$$\frac{\partial y}{\partial \rho} = \sin \theta \sin \varphi \qquad \frac{\partial y}{\partial \varphi} = \rho \sin \theta \cos \varphi \qquad \frac{\partial y}{\partial \theta} = \rho \cos \theta \sin \varphi$$

$$\frac{\partial z}{\partial \rho} = \cos \theta \qquad \frac{\partial z}{\partial \varphi} = 0 \qquad \frac{\partial z}{\partial \theta} = -\rho \sin \theta$$

$$J\left(\frac{x,y,z}{\rho,\varphi,\theta}\right) = \begin{vmatrix} \sin \theta \cos \varphi & \sin \theta \sin \varphi & \cos \theta \\ -\rho \sin \theta \sin \varphi & \rho \sin \theta \cos \varphi & 0 \\ \rho \cos \theta \cos \varphi & \rho \cos \theta \sin \varphi & -\rho \sin \theta \end{vmatrix} = -\rho^2 \sin \theta$$

$$dV = |J|\ d\rho\ d\varphi\ d\theta = \rho^2 \sin \theta\ d\rho\ d\varphi\ d\theta$$

If a jacobian $J\left(\dfrac{u,v,w,\ \cdots}{x,y,z,\ \cdots}\right)$ is zero, the variables x, y, z, . . . are related by some function of u, v, w, . . . , $\varphi(u,v,w,\ \ldots) = 0$

A set of equations

$$u_1 = f_1(x_1,x_2,\ \ldots\ ,x_n)$$
$$u_2 = f_2(x_1,x_2,\ \ldots\ ,x_n)$$
$$\cdot\ \cdot\ \cdot\ \cdot\ \cdot\ \cdot\ \cdot\ \cdot\ \cdot\ \cdot\ \cdot\ \cdot$$
$$u_n = f_n(x_1,x_2,\ \ldots\ ,x_n)$$

defined for a region R can be solved uniquely for x_1, x_2, . . . , x_n in terms of u_1, u_2, . . . , u_n if

$$J\left(\frac{u_1,u_2,\ \ldots\ ,u_n}{x_1,x_2,\ \ldots\ ,x_n}\right) \neq 0$$

for all values of x_1, x_2, . . . , x_n in R.

8-15. Areas of Plane Figures by Integration. If $y = f(x)$, the area bounded by $f(x)$, the x axis, and the ordinates $x = a$ and $x = b$ is given by

$$A = \int_a^b y\ dx$$

The area above the x axis is taken as positive, that below the x axis as negative.

• Find the area under the curve $y = \sin x$ between $x = 0$ and $x = \pi$.

$$A = \int_0^\pi y\ dx = \int_0^\pi \sin x\ dx = -\cos x\Big]_0^\pi = 1 + 1 = 2$$

• Find the area of the ellipse $x = a \cos \theta$, $y = b \sin \theta$.
Compute the area in the first quadrant, then multiply by 4 to take advantage

of symmetry. When $x = 0$, $\theta = \pi/2$; when $x = a$, $\theta = 0$. Differentiation of the equation for x yields $dx = -a \sin \theta \, d\theta$.

$$A = 4 \int_0^a y \, dx = -4 \int_{\pi/2}^0 ab \sin^2 \theta \, d\theta = 4ab \int_0^{\pi/2} \tfrac{1}{2}(1 - \cos 2\theta) \, d\theta$$

$$= 2ab \left[\theta - \tfrac{1}{2} \sin 2\theta \right]_0^{\pi/2} = \pi ab$$

In polar coordinates, the area under a curve is given by

$$A = \tfrac{1}{2} \int_\alpha^\beta \rho^2 \, d\theta$$

• Find the area of the circle $\rho = a \sin \theta$.

Compute the area of the semicircle in the first quadrant; then multiply by 2 to take advantage of symmetry.

$$A = \tfrac{2}{2} \int_0^{\pi/2} \rho^2 \, d\theta = \int_0^{\pi/2} a^2 \sin^2 \theta \, d\theta = \frac{a^2}{2} \int_0^{\pi/2} (1 - \cos 2\theta) \, d\theta$$

$$= \frac{a^2}{2} \left[\theta - \tfrac{1}{2} \sin 2\theta \right]_0^{\pi/2} = \frac{a^2 \pi}{4}$$

In rectangular coordinates, the area under a curve is also given by the double integral

$$A = \iint_A dA = \iint dx \, dy$$

• Find the finite area between the parabola $y^2 = x + 1$ and the line $x + y = 1$.

The points of intersection are $(3, -2)$ and $(0, 1)$. First integrate with respect to x, holding y constant, to obtain the area of the horizontal strip, dy high, between the parabola and the line. Then integrate with respect to y to add the strips between $(3, -2)$ and $(0, 1)$.

$$A = \int_{-2}^1 \int_{y^2-1}^{1-y} dx \, dy = \int_{-2}^1 (1 - y - y^2 + 1) \, dy = \left[2y - \tfrac{1}{2}y^2 - \tfrac{1}{3}y^3 \right]_{-2}^1 = 4\tfrac{1}{2}$$

In polar coordinates, the double integral takes the form

$$A = \iint \rho \, d\rho \, d\theta$$

• Find the area inside the circle $\rho = 3 \cos \theta$ but outside the circle $\rho = \cos \theta$.

Compute the area above the axis and multiply by 2 to take advantage of symmetry.

$$A = 2 \int_0^{\pi/2} \int_{\cos \theta}^{3 \cos \theta} \rho \, d\rho \, d\theta = 2 \int_0^{\pi/2} \left[\tfrac{1}{2}\rho^2 \right]_{\cos \theta}^{3 \cos \theta} d\theta = \int_0^{\pi/2} 8 \cos^2 \theta \, d\theta$$

$$= 8 \int_0^{\pi/2} \tfrac{1}{2}(1 + \cos 2\theta) \, d\theta = 4 \left[\theta + \tfrac{1}{2} \sin 2\theta \right]_0^{\pi/2} = 2\pi$$

For change of coordinate systems with double and triple integrals see Art. 8-14.

The area bounded by a closed curve C also can be obtained from the line integral

$$A = \tfrac{1}{2} \int_C (-y \, dx + x \, dy)$$

The integration along C is to be carried out with the area always on the left (counterclockwise integration) (see also Art. 8-10).

• Find the area between $x^2 = 4y$ and $y^2 = 4x$.

The curves intersect at (0,0) and (4,4). The integration should be carried out first along $x^2 = 4y$ from (0,0) to (4,4), then along $y^2 = 4x$ from (4,4) to (0,0).

$$A = \frac{1}{2} \int_{0,0}^{4,4} (-y\,dx + x\,dy) + \frac{1}{2} \int_{4,4}^{0,0} (-y\,dx + x\,dy)$$

$$= \frac{1}{2} \int_0^4 \left(-\frac{x^2}{4}\,dx + x\,\frac{x}{2}\,dx\right) + \frac{1}{2} \int_4^0 \left(-y\,\frac{y}{2}\,dy + \frac{y^2}{4}\,dy\right)$$

$$= \frac{x^3}{24}\bigg]_0^4 - \frac{y^3}{24}\bigg]_4^0 = 16\frac{2}{3}$$

8-16. Length of a Curve by Integration. In rectangular coordinates the length of a curve is given by

$$s = \int_a^b (1 + y'^2)^{\frac{1}{2}}\,dx = \int_c^d (1 + x'^2)^{\frac{1}{2}}\,dy$$

• Find the length of the catenary $y = a \cosh (x/a)$ from $x = 0$ to $x = 1$.

$$y' = \sinh \frac{x}{a} \qquad (1 + y'^2) = 1 + \sinh^2 \frac{x}{a} = \cosh^2 \frac{x}{a}$$

$$s = \int_0^1 \left(\cosh^2 \frac{x}{a}\right)^{\frac{1}{2}}\,dx = \int_0^1 \cosh \frac{x}{a}\,dx = a \sinh \frac{x}{a}\bigg]_0^1 = a \sinh \frac{1}{a}$$

In polar coordinates the length of curve is given by

$$s = \int_\alpha^\beta \left[\rho^2 + \left(\frac{d\rho}{d\theta}\right)^2\right]^{\frac{1}{2}} d\theta = \int_{\rho_1}^{\rho_2} \left[\rho^2 \left(\frac{d\theta}{d\rho}\right)^2 + 1\right]^{\frac{1}{2}} d\rho$$

• Find the perimeter of the cardioid $\rho = a(1 + \cos \theta)$.

Compute the length above the axis; then multiply by 2 to take advantage of symmetry.

$$\frac{d\rho}{d\theta} = -a \sin \theta$$

$$s = 2 \int_0^\pi [a^2(1 + \cos \theta)^2 + a^2 \sin^2 \theta]^{\frac{1}{2}}\,d\theta$$

$$= 2a \int_0^\pi (1 + 2\cos \theta + \cos^2 \theta + \sin^2 \theta)^{\frac{1}{2}}\,d\theta$$

$$= 2a \int_0^\pi (2 + 2\cos \theta)^{\frac{1}{2}}\,d\theta = 2a \int_0^\pi 2 \cos \frac{\theta}{2}\,d\theta = 4a\left[2 \sin \frac{\theta}{2}\right]_0^\pi = 8a$$

8-17. Areas of Surfaces by Integration. If a plane curve is rotated about an axis in its plane, the area of the surface of revolution is given by

$$S = 2\pi \int_a^b y\,ds$$

where $y =$ distance from axis to curve

$ds =$ differential of length of arc (Art. 7-9)

• Find the surface area generated by rotating the hypocycloid $x^{2/3} + y^{2/3} = a^{2/3}$ about the x axis.

Compute the area from $x = 0$ to $x = a$ and multiply by 2 to take advantage of symmetry.

$$y = (a^{2/3} - x^{2/3})^{3/2} \qquad y' = -\frac{y^{1/3}}{x^{1/3}} \qquad 1 + y'^2 = 1 + \frac{y^{2/3}}{x^{2/3}} = \frac{a^{2/3}}{x^{2/3}}$$

$$ds = (1 + y'^2)^{1/2}\,dx = \frac{a^{1/3}}{x^{1/3}}\,dx$$

$$S = 4\pi \int_0^a (a^{2/3} - x^{2/3})^{3/2} \left(\frac{a^{1/3}}{x^{1/3}}\right) dx$$

To integrate, let $z^2 = a^{2/3} - x^{2/3}$. Then $2z\,dz = -\frac{2}{3}x^{-1/3}\,dx$; $dx = -3zx^{1/3}\,dz$. When $x = 0$, $z = a^{1/3}$; when $x = a$, $z = 0$.

$$S = 4\pi a^{1/3} \int_{a^{1/3}}^0 z^3 x^{-1/3}(-3zx^{1/3}\,dz) = 12\pi a^{1/3} \int_0^{a^{1/3}} z^4\,dz = 12\pi a^{1/3} \left[\tfrac{1}{5} z^5\right]_0^{a^{1/3}}$$

$$= \frac{12\pi a^2}{5}$$

The area of any curved surface in rectangular coordinates is given by

$$A = \iint_S \left[1 + \left(\frac{\partial z}{\partial x}\right)^2 + \left(\frac{\partial z}{\partial y}\right)^2\right]^{1/2} dy\,dx$$

$$= \iint_S \left[1 + \left(\frac{\partial y}{\partial x}\right)^2 + \left(\frac{\partial y}{\partial z}\right)^2\right]^{1/2} dz\,dx$$

$$= \iint_S \left[1 + \left(\frac{\partial x}{\partial y}\right)^2 + \left(\frac{\partial x}{\partial z}\right)^2\right]^{1/2} dz\,dy$$

• Find the area of the surface of the sphere $x^2 + y^2 + z^2 = r^2$.

Compute the area in the first octant; then multiply by 8 to take advantage of symmetry.

$$\frac{\partial z}{\partial x} = -\frac{x}{z} \qquad \frac{\partial z}{\partial y} = -\frac{y}{z}$$

$$1 + \left(\frac{\partial z}{\partial x}\right)^2 + \left(\frac{\partial z}{\partial y}\right)^2 = 1 + \frac{x^2}{z^2} + \frac{y^2}{z^2} = \frac{z^2 + x^2 + y^2}{z^2} = \frac{r^2}{r^2 - x^2 - y^2}$$

Integrate first with respect to y, holding x constant, to find the area of a strip dx wide and extending from the xz plane to the xy plane. Then integrate with respect to x, to add the strips from $x = 0$ to $x = r$.

$$A = 8 \int_0^r \int_0^{\sqrt{r^2 - x^2}} \left[\frac{r^2}{r^2 - x^2 - y^2}\right]^{1/2} dy\,dx = 8r \int_0^r \left[\arcsin \frac{y}{\sqrt{r^2 - x^2}}\right]_0^{\sqrt{r^2 - x^2}} dx$$

$$= 8r \int_0^r \frac{\pi}{2}\,dx = 4\pi r \left[x\right]_0^r = 4\pi r^2$$

(For change of coordinate systems with double integrals, see Art. 8-14.)

8-18. Volumes by Integration. If a plane curve is rotated about an axis in its plane, the volume of the solid generated is

$$V = \pi \int_a^b y^2 \, dx$$

where y = distance from axis to curve
  dx = differential length along axis
The integration is equivalent to adding circular cross sections of thickness dx.

• Find the volume of the ellipsoid generated by revolving $x^2/a^2 + y^2/b^2 = 1$ about the x axis.

Compute the area on one side of the y axis and double it to take advantage of symmetry.

$$V = 2\pi \int_0^a y^2 \, dx = 2\pi \int_0^a \frac{b^2}{a^2} (a^2 - x^2) \, dx = \frac{2\pi b^2}{a^2} \left[a^2 x - \tfrac{1}{3} x^3 \right]_0^a = \frac{4\pi ab^2}{3}$$

Volume of revolution also is given by

$$V = 2\pi \int_{r_1}^r rh \, dr$$

where r = distance from axis to curve
  h = length of element of volume normal to r
The integration is equivalent to adding cylinders of thickness dr.

• Find the volume of the torus obtained by rotating the area bounded by the circle $x^2 + y^2 = a^2$ about the line $x = b$, where $b > a$.
Take elements of width dx parallel to $x = b$ for rotation about $x = b$. $h = 2y$; $r = b - x$; $dr = -dx$.

$$V = 2\pi \int_{-a}^a - (b - x)2y \, dx = -4\pi \int_{-a}^a (b - x) \sqrt{a^2 - x^2} \, dx = 2\pi^2 a^2 b$$

In polar coordinates, volume of revolution is given by

$$V = 2\pi \int_\alpha^\beta \int_c^d r^2 \sin \theta \, dr \, d\theta$$

(See also Art. 8-19.)

If the areas of parallel cross sections of a solid can be expressed conveniently as a function of distance h along the axis perpendicular to these sections, the volume can be computed from

$$V = \int_{h_1}^{h_2} A \, dh$$

where A is the cross-sectional area as a function of h.

• Find the volume of the paraboloid $x^2 + 4y^2 = z$ between the planes $z = 0$ and $z = 1$.

Consider the elliptical cross sections $x^2 + 4y^2 = z_i$. Each has as half its major axis $\sqrt{z_i}$ and as half its minor axis $\frac{1}{2} \sqrt{z_i}$, and the area of each is $\pi \sqrt{z_i} (\frac{1}{2} \sqrt{z_i}) = (\pi/2)z_i$. Let each section have thickness dz; integrate.

$$V = \int_0^1 \frac{\pi}{2} z \, dz = \frac{\pi}{2} \Big[\frac{1}{2}z^2 \Big]_0^1 = \frac{\pi}{4}$$

In general, the volume of a solid in rectangular coordinates is given by

$$V = \int_a^b \int_m^n f(x,y) \, dy \, dx = \int_c^d \int_p^q f(x,y) \, dx \, dy$$

where m and n may be functions of x, and p and q may be functions of y. The integration is equivalent to first adding elements $z = f(x,y)$ high, dy (or dx) wide, and dx (or dy) thick, to obtain the volume of a strip parallel to the yz (or xz) plane and dx (or dy) thick. Then the strips are added by integrating.

• Find the volume in the first octant bounded by the cylinder $x^2 + y^2 = a^2$, the coordinate planes, and the plane $z = x + y$.

Consider an element $z = x + y$ high, dy wide, and dx thick. First sum the elements with x held constant, from $y = 0$ to the boundary of the cylinder $y = \sqrt{a^2 - x^2}$. Then sum these strips, each dx thick, from $x = 0$ to the boundary of the cylinder, $x = a$.

$$V = \int_0^a \int_0^{\sqrt{a^2-x^2}} (x + y) \, dy \, dx = \int_0^a \Big[xy + \frac{1}{2}y^2 \Big]_0^{\sqrt{a^2-x^2}} dx$$

$$= \int_0^a [x \sqrt{a^2 - x^2} + \frac{1}{2}(a^2 - x^2)] \, dx = \frac{2}{3}a^3$$

In cylindrical coordinates, volume is given by

$$V = \int_\alpha^\beta \int_a^b f(r,\theta) r \, dr \, d\theta = \int_c^d \int_\gamma^\delta f(r,\theta) r \, d\theta \, dr$$

where a and b may be functions of r, and γ and δ may be functions of θ.

Volume also can be calculated by triple integration.

$$V = \iiint_V dV = \int_a^b \int_m^n \int_u^v dz \, dy \, dx$$

where u and v may be functions of y, and m and n may be functions of x. The integration is equivalent to first adding elementary parallelepipeds dz high, dy wide, and dx thick, to find the volume of elements extending from u to v, dy wide and dx thick. Next, these elements are added to obtain the volumes of strips parallel to the yz plane and dx thick. Finally, the strips are added.

• Find the volume of the conoid $a^2y^2 = (z - a)^2(r^2 - x^2)$ between the planes $z = 0$ and $z = a$.

Compute the volume in the first octant and multiply by 4 to take advantage of symmetry. First integrate from $z_1 = 0$ to $z_2 = a - ay/\sqrt{r^2 - x^2}$, then from $y_1 = 0$ to $y_2 = \sqrt{r^2 - x^2}$, and finally from $x_1 = 0$ to $x = r$.

$$V = 4 \int_0^r \int_{y_1}^{y_2} \int_{z_1}^{z_2} dz\, dy\, dx = 4 \int_0^r \int_{y_1}^{y_2} \Big[z \Big]_{z_1}^{z_2} dy\, dz$$

$$= 4 \int_0^r \int_{y_1}^{y_2} \Big(a - \frac{ay}{\sqrt{r^2 - x^2}} \Big)\, dy\, dx$$

$$= 4a \int_0^r \Big[y - \frac{y^2}{2\sqrt{r^2 - x^2}} \Big]_{y_1}^{y_2} dx = 4a \int_0^r (\sqrt{r^2 - x^2} - \tfrac{1}{2}\sqrt{r^2 - x^2})\, dx$$

$$= 2a \int_0^r \sqrt{r^2 - x^2}\, dx = \frac{\pi a r^2}{2}$$

(For change of coordinate systems with double and triple integrals, see Art. 8-14.)

8-19. Pappus's Theorems

1. If a plane arc is revolved about an axis in its plane and not cutting it, the area of the surface generated equals the product of the length of the arc and the circumference of the circle generated by the centroid of the arc.

$$A = 2\pi r s$$

• Find the surface area of the torus formed by rotating the circle $x^2 + y^2 = a^2$ about the line $x = b$, where $b > a$.

The length of the curve is $s = 2\pi a$. By symmetry, its centroid is at the origin $(0,0)$. Hence, the radius of the circle generated by the centroid is $r = b$. The surface area is

$$A = 2\pi b(2\pi a) = 4\pi^2 ab$$

2. If a plane area is revolved about an axis lying in its plane and not crossing it, the volume of the solid generated equals the product of the area and the circumference of the circle generated by the centroid of the area.

$$V = 2\pi r A$$

• Find the volume of the torus of cross section πa^2 and diameter of centerline $2b$ (see previous example).

$$V = 2\pi b(\pi a^2) = 2\pi^2 a^2 b$$

BIBLIOGRAPHY

Bacon, H. M.: "Differential and Integral Calculus," 2d ed., McGraw-Hill Book Company, Inc., New York, 1955.

Buck, R. C.: "Advanced Calculus," McGraw-Hill Book Company, Inc., New York, 1956.

Burington, R. S.: "Handbook of Mathematical Tables and Formulas," 3d ed., McGraw-Hill Book Company, Inc., New York, 1948.

Dwight, H. B.: "Tables of Integrals and Other Mathematical Data," 2d ed., The Macmillan Company, New York, 1947.

Peirce, B. O., and R. M. Foster: "A Short Table of Integrals," Ginn & Company, Boston, 1956.

Rice, H. S., and R. M. Knight: "Technical Mathematics with Calculus," McGraw-Hill Book Company, Inc., New York, 1957.

Smith, E. S., M. Salkover, and H. K. Justice: "Calculus," 2d ed., John Wiley & Sons, Inc., New York, 1958.

Chapter 9

DIFFERENTIAL EQUATIONS

9-1. Initial-value Problems. A differential equation has a solution (when one exists) containing arbitrary constants or functions. These may be determined for specific problems by conditions imposed by geometric, physical, or other considerations. When conditions are imposed only at one point, the problem is called an initial-value problem.

• If $y = c_1 \sin x + c_2 \cos x$ results from integration of a differential equation, find the solution for the conditions $y = 0$, $y' = 1$ when $x = 0$.

The condition $y = 0$ when $x = 0$ requires that $c_2 = 0$. The condition $y'(0) = 1$ requires that $c_1 = 1$. Hence, the solution is

$$y = \sin x$$

9-2. Boundary-value Problems. When solutions of a differential equation are required to satisfy conditions at more than one point, the problem is called a boundary-value problem. If solutions exist, they need not be unique.

The conditions imposed may be met by selecting suitable values of the arbitrary constants in the general solution.

• If $y = c_1 e^x + c_2 e^{-x}$ results from integration of a differential equation, find the solution if $y = 0$ when $x = 0$ and $y = e^2 - 1$ when $x = 1$.

The condition $y(0) = 0$ requires that $c_2 = -c_1$. The condition $y(1) = e^2 - 1$ requires that

$$y = c_1 e + c_2 e^{-1} = c_1 \left(e - \frac{1}{e} \right) = c_1 \left(\frac{e^2 - 1}{e} \right) = e^2 - 1$$

Hence, $c_1 = e$, $c_2 = -e$, and the solution is

$$y = ee^x - ee^{-x} = e^{x+1} - e^{1-x}$$

9-3. Characteristic-value (Eigenvalue) Problems. These are generally concerned with boundary-value problems of differential equations containing parameters.

Consider, for example, a differential equation of the form

$$Ly - \lambda y = 0$$

228

where L is a differential operator and λ a parameter. The general solution may have the form

$$y(x,\lambda) = c_1 y_1(x,\lambda) + c_2 y_2(x,\lambda) + \cdots + c_n y_n(x,\lambda)$$

where the c's are arbitrary constants. Physical or geometric considerations may impose n conditions leading to n equations, for determination of the c's, of the form

$$f(x_0,\lambda,c) = 0$$

This system of equations will have a nontrivial solution only if the determinant of the coefficients of the c's is zero.

The determinant is a function of λ. The equation resulting from setting the determinant equal to zero is called a characteristic equation, and its solutions $\lambda = \lambda_i$ are called characteristic values or eigenvalues. The solution corresponding to each eigenvalue λ_i is called a characteristic function or eigenfunction.

• If $y = c_1 \sin kx + c_2 \cos kx$ is the result of integrating a differential equation, find the solution if $y = 0$ at $x = 0$ and at $x = L$ and the maximum value of y is a.

The condition $y(0) = 0$ requires that $c_2 = 0$. By inspection, if $y_{max} = a$, then $c_1 = a$, since the maximum value of $\sin kx$ is 1. The condition $y(L) = 0$ requires that

$$y = a \sin kL = 0$$

Thus, $\sin kL = 0$, and $k = n\pi/L$, where n is an integer. The solution is

$$y = a \sin \frac{n\pi}{L} x$$

9-4. Ordinary Differential Equations. These are equations involving one independent variable and derivatives with respect to it.

The order of the highest derivative is the order of the equation. Thus, $y'' + y'^3 = x^4$ is a second-order ordinary differential equation, because the highest-order derivative of y present is the second.

A solution is any relationship between the variables that reduces the equation to an identity (see also Arts. 9-1 to 9-3). An equation of order n has a solution containing n arbitrary constants, called a general solution. One with values assigned to the constants is a particular solution.

Some differential equations may have singular solutions—solutions that cannot be obtained from the general solution by specifying the arbitrary constants. For example, if the general solution has an envelope, its equation is a singular solution. [An envelope of a family of curves consists of one or more curves tangent at each point to a curve of the family. The equation of the envelope is obtained by eliminating c from the equation of the family $\varphi(x,y,c)$ and $\dfrac{\partial \varphi}{\partial c}$.]

There is no general method for finding the solutions of all types of differential equations. However, methods of solution have been developed for certain classes of equations. These are indicated in the following articles and Chaps. 10 and 11 for those classes of equations likely to be encountered in physical problems; for other types, refer to one of the books listed in the bibliography at the end of this chapter.

9-5. First-order Differential Equations. These may be written in the form $y' = \dfrac{dy}{dx} = f(x,y)$ or $M(x,y)\,dx + N(x,y)\,dy = 0$.

If the variables are separable, the equation can be written

$$P(x)\,dx + Q(y)\,dy = 0$$

Then the solution is

$$\int P(x)\,dx + \int Q(y)\,dy = c$$

where c is an arbitrary constant.

• Solve $y' = xe^{2y}$.
Rewrite in the form

$$e^{-2y}\,dy = x\,dx$$

Integration of both sides of the equation yields

$$-\tfrac{1}{2}e^{-2y} = \tfrac{1}{2}x^2 + c'$$
$$e^{-2y} + x^2 = c$$

If the variables are not separable, see if $\dfrac{\partial M}{\partial y} = \dfrac{\partial N}{\partial x}.$ If so, the equation has the solution $F(x,y) = c$, where c is a constant and

$$F = \int (M\,dx + N\,dy)$$

for

$$M\,dx + N\,dy = \frac{\partial F}{\partial x}\,dx + \frac{\partial F}{\partial y}\,dy = dF$$

(see Art. 7-13). The equation is then an exact differential equation. For example, the solution of $x\,dy + y\,dx = 0$ is $xy = c$, since

$$x\,dy + y\,dx = d(xy)$$

• Solve $y' = (2x + y)/(3 - x)$.
Rewrite as

$$(2x + y)\,dx + (x - 3)\,dy = 0$$

Then determine whether the equation is exact by taking partial derivatives.

$$\frac{\partial}{\partial y}\,(2x + y) = 1 \qquad \frac{\partial}{\partial x}\,(x - 3) = 1$$

Since the partial derivatives are equal, the equation is exact.

$$M = \frac{\partial F}{\partial x} = 2x + y$$

$$F = \int(2x + y)\,dx = x^2 + xy + f(y)$$

$$N = x - 3 = \frac{\partial F}{\partial y} = \frac{\partial}{\partial y}(x^2 + xy) + f'(y) = x + f'(y)$$

$$f'(y) = -3$$

$$f(y) = -3y + c'$$

Hence, the solution is $F = x^2 + xy - 3y = c$.

Sometimes a first-order equation may be made exact by multiplying it by a function of the variables. Such a function is called an integrating factor. For example, x^2, y^2, $x^2 + y^2$, and $x^2 - y^2$ can be used as integrating factors for $x\,dy - y\,dx$.

$$\frac{x\,dy - y\,dx}{x^2} = d\left(\frac{y}{x}\right)$$

$$\frac{x\,dy - y\,dx}{y^2} = d\left(-\frac{x}{y}\right)$$

$$\frac{x\,dy - y\,dx}{x^2 + y^2} = d\left(\tan^{-1}\frac{y}{x}\right)$$

$$\frac{x\,dy - y\,dx}{x^2 - y^2} = d\left(\frac{1}{2}\log_e\frac{x + y}{x - y}\right)$$

Similarly, x^2 and y^2 are integrating factors for $2xy\,dy - y^2\,dx$ and $2xy\,dx - x^2\,dy$, respectively.

$$\frac{2xy\,dy - y^2\,dx}{x^2} = d\left(\frac{y^2}{x}\right)$$

$$\frac{2xy\,dx - x^2\,dy}{y^2} = d\left(\frac{x^2}{y}\right)$$

If a first-order equation $y' = f(x,y)$ can be written in the form $y' = g(y/x)$ or $y' = h(x/y)$, try the substitution $y = vx$ or $x = vy$.

• Solve $x^2y\,dx = (x^3 - y^3)\,dy$.

After division of both sides by $x^2y\,dy$, the equation can be written

$$\frac{dx}{dy} = \frac{x}{y} - \frac{y^2}{x^2}$$

Let $x = vy$, $\dfrac{dx}{dy} = y\dfrac{dv}{dy} + v$. Substitution in the equation gives

$$y\frac{dv}{dy} + v = \frac{vy}{y} - \frac{y^2}{v^2y^2} = v - \frac{1}{v^2}$$

The variables now are separable.

$$-v^2\,dv = \frac{dy}{y}$$

Integration yields

$$-\tfrac{1}{3}v^3 = \log_e y + c_1$$

Substitution of $v = x/y$ gives the solution,

$$\log_e c_2 y = -\frac{x^3}{3y^3}$$

If a first-order equation is linear (of the first degree in the dependent variable and its derivative), it can be written in the form

$$y' + Py = Q$$

where P and Q are functions of x. Then e^z, where $z = \int P\,dx$, is an integrating factor. The solution can be determined from

$$y = e^{-z}\int Qe^z\,dx + ce^{-z}$$

where c is a constant.

• Solve $y' - y = e^x$.
Here, $Q = e^x$, $P = -1$, and $z = \int P\,dx = -\int dx = -x$.

$$y = e^x \int e^x e^{-x}\,dx + ce^x = e^x \int dx + ce^x = xe^x + ce^x = e^x(x + c)$$

The Clairaut equation, which has the form

$$y = xy' + f(y') \text{ or } F(y - xy', y') = 0$$

has the general solution $F(y - cx, c) = 0$.

• Solve $xy'^2 - yy' + 3 = 0$.
This can be written

$$-y'(y - xy') + 3 = 0$$

or

$$y - xy' - \frac{3}{y'} = 0$$

This is a Clairaut equation, and the solution is

$$y - cx - \frac{3}{c} = 0$$

The Bernoulli equation

$$y' + P(x)y = Q(x)y^n$$

can be reduced to a linear equation by setting $y = z^{1/(1-n)}$.

• Solve $y' + y = xy^3$.
Let $y = z^{1/(1-3)} = z^{-\frac{1}{2}}$, $y' = -\tfrac{1}{2}z^{-\frac{3}{2}}z'$. Substitution in the equation gives

$$-\tfrac{1}{2}z^{-\frac{3}{2}}z' + z^{-\frac{1}{2}} = xz^{-\frac{3}{2}}$$

Multiply the equation by $-2z^{3/2}$.

$$z' - 2z = -2x$$

In this linear equation, $Q = -2x$, $P = -2$, and $\int P\, dx = -2x$.   The solution is

$$z = y^{-2} = e^{2x}\int - 2xe^{-2x}\, dx + ce^{2x} = e^{2x}(xe^{-2x} + \tfrac{1}{2}e^{-2x}) + ce^{2x}$$
$$y^2(x + \tfrac{1}{2} + ce^{2x}) = 1$$

When a first-order equation is encountered in a form for which a solution cannot be found by any of the preceding methods, see if substitution of a new variable can transform the equation into one that can be solved.

• Solve $y' = (x - y - 1)/(x + y + 3)$.
Let $x = u + a$, $y = v + b$.   Substitution in the equation gives

$$\frac{dv}{du} = \frac{u - v + (a - b - 1)}{u + v + (a + b + 3)}$$

Select a and b so that $a - b - 1 = 0$ and $a + b + 3 = 0$; that is, set $a = -1$ and $b = -2$.   The resulting equation can be written in the form $\dfrac{dv}{du} = f(u/v)$ and thus can be solved by the substitution $v = zu$, $\dfrac{dv}{du} = u\dfrac{dz}{du} + z$.   The solution is

$$(x + 1)^2 - 2(x + 1)(y + 2) - (y + 2)^2 = c$$

9-6. Solution by Successive Approximations.   When other methods do not work, Picard's method of solving $y' = f(x,y)$ by successive approximations may succeed.

Assume that a condition imposed on the solution is that $y = b$ when $x = a$.   Then the differential equation has the solution

$$y = b + \int_a^x f(x,y)\, dx$$

As a first approximation, take y in the integral equal to b (or any other reasonable value).   Compute the new value of y and substitute that in the integral.   Repeat as many times as desired.

• Solve $y' = 1 + y^2$ when $y(0) = 0$.
For a first approximation take $y_1 = 0$.

$$y_2 = 0 + \int_0^x (1 + y^2)\, dx = \int_0^x dx = x$$

As a second approximation, substitute x for y in the integral.

$$y_3 = \int_0^x (1 + x^2)\, dx = x + \tfrac{1}{3}x^3$$

Continue.

$$y_4 = \int_0^x [1 + (x + \tfrac{1}{6}x^3)^2]\, dx = \int_0^x (1 + x^2 + \tfrac{2}{3}x^4 + \tfrac{1}{9}x^6)\, dx$$

$$= x + \tfrac{1}{3}x^3 + \tfrac{2}{15}x^5 + \tfrac{1}{63}x^7$$

$$y_5 = x + \tfrac{1}{3}x^3 + \tfrac{2}{15}x^5 + \tfrac{17}{315}x^7 + \tfrac{38}{2,835}x^9 + \tfrac{134}{51,927}x^{11} + \cdots$$

(This solution will be a converging series only when $|x| < \pi/2$.)

9-7. Differential Operators. The symbol D may be used to indicate that the function following it is to be differentiated with respect to the independent variable. Thus, $Dy = \dfrac{dy}{dx}$ and $D(y - 3)^2 = 2(y - 3)y'$.

More generally, D^n indicates that the function following it is to be differentiated n times. Thus, $D^2y = \dfrac{d^2y}{dx^2}$.

$$(D^3 + 3D^2 - 4D + 1)y \equiv \frac{d^3y}{dx^3} + 3\frac{d^2y}{dx^2} - 4\frac{dy}{dx} + y$$

The symbol L may be used to represent $f(D)$. Thus, Ly can represent $(D^3 + 3D^2 - 4D + 1)y$ in the expression above.

A polynomial in D with constant coefficients can be manipulated in addition, subtraction, and multiplication as if D were an algebraic quantity. For example,

$$(D^2 + D - 2)\sin x \equiv (D - 1)(D + 2)\sin x \equiv (D - 1)(\cos x + 2\sin x)$$
$$\equiv -\sin x + 2\cos x - \cos x - 2\sin x$$
$$\equiv \cos x - 3\sin x$$

Similarly, if

$$(D^2 - 2D + 1)y = 0$$

then

$$(D - 1)^2 y = 0$$

If r is an integer, m is any constant, and y is a function of x,

$$D^r(e^{mx}y) \equiv e^{mx}(D + m)^r y$$

Thus, $D(e^{mx}y) \equiv e^{mx}(D + m)y$, $D^2(e^{mx}y) \equiv e^{mx}(D + m)^2 y$, etc. Hence,

$$L(D)e^{mx}y \equiv e^{mx}L(D + m)y$$

(See also Art. 9-9.)

9-8. Wronskian. A determinant in which the elements of each column consist of a function or its derivatives with respect to the independent variable.

The necessary and sufficient condition that a given set of functions of one variable, such as the solutions $y_1, y_2, \ldots, y_n$ of an nth order linear differential equation, be linearly independent is that their Wronskian should not be zero. (Functions are linearly dependent if a set of constants not all zero can be found such that the sum of the products of the constants and the functions is zero.)

$$W = \begin{vmatrix} y_1 & y_2 & \cdots & y_n \\ y_1' & y_2' & \cdots & y_n' \\ y_1'' & y_2'' & \cdots & y_n'' \\ \cdot & \cdot & \cdot & \cdot & \cdot & \cdot & \cdot & \cdot \\ y_1^{(n-1)} & y_2^{(n-1)} & \cdots & y_n^{(n-1)} \end{vmatrix} \neq 0$$

• Show that e^x and e^{-x} are linearly independent.
The Wronskian for these functions is

$$\begin{vmatrix} e^x & e^{-x} \\ e^x & -e^{-x} \end{vmatrix} = -1 - 1 = -2$$

Since the Wronskian is not zero for any value of x, the functions are independent. Hence, no constants a and b can be found by which e^x and e^{-x} can be multiplied to make the sum $ae^x + be^{-x}$ equal zero.

9-9. Linear Differential Equations. These are equations of the first degree in the dependent variable and its derivatives. Thus, a linear differential equation of the nth order can be written in the form

$$a_0(x) \frac{d^n y}{dx^n} + a_1(x) \frac{d^{n-1} y}{dx^{n-1}} + \cdots + a_{n-1}(x) \frac{dy}{dx} + a_n(x)y = f(x)$$

A formula for a general solution of a linear first-order equation is given in Art. 9-5. No such formula is available when $n > 1$. Methods of solution of linear equations with constant coefficients are given in Art. 9-10.

If $f(x)$ in the general linear equation is zero, the resulting equation is said to be homogeneous. Its solution is called the complementary function. The solution when $f(x) \neq 0$ is called a particular integral of the equation.

If a linear differential equation has y_c as a complementary function and y_p as a particular integral, its solution, containing n arbitrary constants, is $y = y_c + y_p$.

If $y = y_1$ is any solution of the homogeneous equation, then $y = cy_1$ also is a solution. If $y = y_2, y_3, \ldots, y_n$ are also solutions, and all are linearly independent, then the linear combination

$$y_c = c_1 y_1 + c_2 y_2 + \cdots + c_n y_n$$

is the general solution (see Art. 9-8).

To solve a linear differential equation with variable coefficients for which a method of solution is not immediately obvious, look for a substitution that will convert it to one with constant coefficients.

For example, the Cauchy or Euler equation

$$(b_0 x^n D^n + b_1 x^{n-1} D^{n-1} + \cdots + b_{n-1} x D + b_n)y = f(x)$$

where the b's are constants and $D = \dfrac{d}{dx}$, can be transformed into a linear equation with constant coefficients by the substitution $x = e^z$.

Substitutions sometimes may be used to reduce the order of an equation. One that may work for a second-order equation is $y = y_1 v$, where v is a new dependent variable and y_1 is a function of x that simplifies the resulting differential equation. If a particular integral can be found for the corresponding homogeneous differential equation

$$y'' + Py' + Qy = 0$$

let y_1 be that integral. Sometimes $y = e^z v$, where $z = -\frac{1}{2}\int P\, dx$, may convert the given equation to a linear equation with constant coefficients or to a Cauchy equation (see also Arts. 9-10 to 9-15 and Chaps. 10 and 11).

9-10. Linear Equations with Constant Coefficients. These can be written in the form

$$Ly = (a_0 D^n + a_1 D^{n-1} + \cdots + a_{n-1} D + a_n)y = f(x)$$

where $\qquad L = $ linear operator representing $f(D)$
$a_0, a_1, \ldots, a_n = $ constants
$$D = \frac{d}{dx}$$

(See Art. 9-7.)

To solve such an equation, first find the complementary function y_c, which is the general solution of the homogeneous equation $Ly = 0$ and contains n arbitrary constants (Art. 9-9). Then obtain the particular integral y_p which satisfies $Ly = f(x)$. The complete solution will be $y_c + y_p$.

To obtain the complementary function, solve the auxiliary polynomial equation $L(m) = 0$. If none of the roots $m_1, m_2, \ldots, m_n$ are repeated,

$$y_c = c_1 e^{m_1 x} + c_2 e^{m_2 x} + \cdots + c_n e^{m_n x}$$

If any root m_r is repeated r times, the corresponding term in the complementary function is not just $ce^{m_r x}$ but $(c_1 + c_2 x + c_3 x^2 + \cdots + c_r x^{r-1})e^{m_r x}$.

If $L(m)$ has a complex root $\alpha + \beta i$, it also has a root $\alpha - \beta i$. The corresponding terms for these roots in the complementary function may be written

$$e^{\alpha x}(A \cos \beta x + B \sin \beta x) \quad \text{or} \quad ce^{\alpha x} \sin (\beta x + \varphi) \quad \text{or} \quad ce^{\alpha x} \cos (\beta x + \theta)$$

If such a complex set of roots is repeated r times, the corresponding terms in the complementary function are

$$e^{\alpha x}[(A_1 + A_2 x + \cdots + A_r x^{r-1}) \cos \beta x$$
$$+ (B_1 + B_2 x + \cdots + B_r x^{r-1}) \sin \beta x]$$

• Solve $(D^2 + 4D + 3)y = 0$.
The auxiliary equation

$$m^2 + 4m + 3 = (m + 3)(m + 1) = 0$$

has roots $m = -3, -1$. Hence, the complementary function is

$$y_c = c_1 e^{-3x} + c_2 e^{-x}$$

• Solve $(D^3 + 4D)y = 0$.
The auxiliary equation

$$m^3 + 4m = m(m^2 + 4) = 0$$

has roots $m = 0, \pm 2i$. Hence, the complementary function is

$$y_c = c_1 e^0 + c_2 e^{2i} + c_3 e^{-2i} = c_1 + c_2 \cos 2x + c_3 \sin 2x$$

• Solve $(D^4 - D^2)y = 0$.
The auxiliary equation

$$m^4 - m^2 = m^2(m^2 - 1) = 0$$

has roots $m = \pm 1$ and the twice-repeated root $m = 0$. Hence,

$$y_c = c_1 e^x + c_2 e^{-x} + e^0(c_3 + c_4 x) = A \cosh x + B \sinh x + c_3 + c_4 x$$

Several methods may be used in search of the particular integral y_p of $Ly = f(x)$. If $f(x)$ consists of the sum of two or more terms, a different method may be used to find the particular integral corresponding to each term; the complete particular integral is the sum of those integrals.

The method of undetermined coefficients usually is the simplest method when $f(x)$ is any one or a sum of the following terms:

$$cx^p, \ ce^{qx}, \ cx^p e^{qx}, \ c \sin \beta x, \ c \cos \beta x, \ cx^p e^{\alpha x} \sin \beta x, \ cx^p e^{\alpha x} \cos \beta x$$

where $p =$ positive integer or zero
$q, \alpha, \beta =$ constants

Select, for the particular integral, $f(x)$ and all its derivatives that are different from each other. Arrange the terms in groups so that all terms derived from a single term of $f(x)$ appear in only one group. If any group contains a term that is in the complementary function y_c, multiply all terms in the group by the lowest integral power of x that will make every term different from any term in y_c. The particular integral then consists of the sum of the products of each term and a coefficient to be determined by the following procedure:

Substitute the particular integral in Ly and equate the resulting expression to $f(x)$. Equate the coefficients of terms on the left-hand side of the equation to the corresponding terms on the right and solve for the coefficients.

• Solve $(D^2 + 1)y = 2e^x + x^3 - x$.

The complementary function is

$$y_c = c_1 \sin x + c_2 \cos x$$

For the particular integral try

$$y_p = Ae^x + Bx^3 + Cx^2 + Dx + E$$

None of these terms appear in y_c. Substitute y_p in the given equation, noting that $y_p' = Ae^x + 3Bx^2 + 2Cx + D$ and $y_p'' = Ae^x + 6Bx + 2C$.

$$(Ae^x + 6Bx + 2C) + (Ae^x + Bx^3 + Cx^2 + Dx + E) = 2e^x + x^3 - x$$

Equate coefficients of like terms.

$$
\begin{array}{ll}
2A = 2 & A = 1 \\
 & B = 1 \\
 & C = 0 \\
6B + D = -1 & D = -7 \\
 & E = 0
\end{array}
$$

Hence, the complete solution is

$$y = y_c + y_p = c_1 \sin x + c_2 \cos x + e^x + x^6 - 7x$$

• Solve $(D^4 + 2D^2 + 1)y = \cos x$.

The roots of the auxiliary equation are $\pm i$ repeated twice. Hence, the complementary function is

$$y_c = (c_1 + c_2 x) \sin x + (c_3 + c_4 x) \cos x$$

The particular integral would have been chosen as $A \cos x$, except that this term and its first derivative already appear in the complementary function, as do terms containing $x \cos x$ and $x \sin x$. Try instead

$$y_p = Ax^2 \cos x$$

Substitution of this term and its derivatives in the given equation yields

$$(-12A \cos x + 8Ax \sin x + Ax^2 \cos x)$$
$$+ 2(2A \cos x - 4Ax \sin x - Ax^2 \cos x) + (Ax^2 \cos x) = \cos x$$

Thus, $A = -\frac{1}{8}$, and the complete solution is

$$y = (c_1 + c_2 x) \sin x + (c_3 + c_4 x) \cos x - \frac{1}{8}x^2 \cos x$$

When $f(x)$ contains terms X that do not permit use of the method of undetermined coefficients, the particular integrals corresponding to X

sometimes may be computed from

$$y_p = e^{m_n x} \int e^{(m_{n-1} - m_n)x} \int \cdots \int e^{(m_1 - m_2)x} \int e^{-m_1 x} X (dx)^n$$

where $\qquad n = $ order of equation

$m_1, m_2, \ldots, m_n = $ roots of auxiliary equation

Select the roots in an order that will simplify the repeated integration.

• Solve $(D^2 + D)y = 1/x$.

The roots of the auxiliary equation are $m_1 = 0$, $m_2 = -1$, so the complementary function is

$$y_c = c_1 + c_2 e^{-x}$$

The particular integral is

$$y_p = e^{-x} \int e^x \int \frac{dx}{x}\, dx = e^{-x} \int e^x \log_e x\, dx$$

$$= e^{-x} \left(\log_e |x| + x + \frac{x^2}{2 \cdot 2!} + \cdots + \frac{x^n}{n \cdot n!} \right)$$

Lagrange's method of variation of parameters is another way of finding a particular integral for X. If $y_1, y_2, \ldots, y_n$ are terms of the complementary function, without arbitrary constants, seek a particular integral in the form

$$y_p = v_1 y_1 + v_2 y_2 + \cdots + v_n y_n$$

where $v_1, v_2, \ldots, v_n$ are unknown functions. Differentiate y_p successively $n - 1$ times. Set up $n - 1$ equations by equating to zero the sum of all terms in each derivative that involve the derivatives of v_1, $v_2, \ldots, v_n$. Obtain an nth equation by substituting the derivatives of y_p, less the terms equated to zero, in the differential equation. Then solve the n equations for $v_1, v_2, \ldots, v_n$.

• Solve $(D^2 + 1)y = \sec x$.

The complementary function is $y_c = c_1 \sin x + c_2 \cos x$. To find the particular integral take

$$y_p = v_1 \sin x + v_2 \cos x$$
$$Dy_p = v_1 \cos x - v_2 \sin x + v_1' \sin x + v_2' \cos x$$

Set $v_1' \sin x + v_2' \cos x = 0$. Then

$$D^2 y_p = -v_1 \sin x - v_2 \cos x + v_1' \cos x - v_2' \sin x$$

When this is substituted in the given differential equation, it simplifies to

$$v_1' \cos x - v_2' \sin x = \sec x$$

Simultaneous solution of this equation and

$$v_1' \sin x + v_2' \cos x = 0$$

yields $v_1' = 1$, $v_1 = x$, $v_2' = -\tan x$, and $v_2 = \log_e \cos x$.

Therefore, the particular integral is

$$y_p = x \sin x + \cos x \log_e \cos x$$

(See also Arts. 9-11 and 9-13 to 9-15 and Chap. 11.)

9-11. Reduction of Order. The order of a differential equation some-times can be reduced by the substitution $y' = p$.

If y is absent from an nth order equation, it has the form

$$f(x, y', y'', \ldots, y^{(n)}) = 0$$

Substitution of $y' = p$, $y'' = p'$, $\ldots$, $y^{(n)} = p^{(n-1)}$ transforms the equation to

$$f(x, p, p', \ldots, p^{(n-1)}) = 0$$

reducing the order by 1. After solution for p, integrate to obtain y.

• Solve $x^2 y'' = 1 - x$.

Let $y' = p$, $y'' = p'$. Then

$$x^2 p' = 1 - x$$

$$dp = \frac{1 - x}{x^2} \, dx$$

$$p = -\frac{1}{x} - \log_e x + c_1 = y'$$

$$y = -\log_e x - x \log_e x + c_1 x + c_2$$

If x is absent, let y' be the new dependent variable p. Take y as the independent variable.

$$y'' = \frac{dp}{dx} = \frac{dp}{dy}\frac{dy}{dx} = p\frac{dp}{dy} = pp'$$

$$y^{(3)} = p^2 p'' + pp'^2$$

$$y^{(4)} = p^3 p^{(3)} + 4p^2 p' p'' + pp'^3$$

After solution for p, integrate to find y.

• Solve $yy'' + y'^2 + 1 = 0$.

Let $y' = p$ and $y'' = pp'$, where $p' = \dfrac{dp}{dy}$. Then

$$ypp' + p^2 + 1 = 0$$

$$\frac{p}{p^2 + 1} \, dp = -\frac{dy}{y}$$

$$\tfrac{1}{2} \log_e (p^2 + 1) = -\log_e y + c_1 = -\log_e c_2 y$$

$$p^2 + 1 = \frac{c_3}{y^2}$$

$$p = \frac{dy}{dx} = \sqrt{\frac{c_3}{y^2} - 1}$$

The solution of this last equation is

$$c_3 - y^2 = (x + c)^2$$

9-12. Simultaneous Differential Equations. To solve a system of n linear equations involving one independent variable t and n dependent variables $x, y, z, \ldots$, treat the operator $D = \dfrac{d}{dt}$ as an algebraic quantity.

• Solve

$$(D + 1)x + (D + 3)y = t \tag{9-1}$$
$$x + (D + 1)y = e^t \tag{9-2}$$

Multiply Eq. (9-2) by $D + 1$ and subtract the result from Eq. (9-1).

$$(D + 3)y - (D + 1)^2y = t - (D + 1)e^t = t - 2e^t$$
$$(D^2 + D - 2)y = 2e^t - t$$
$$y = (c_1 + \tfrac{2}{3}t)e^t + c_2e^{-2t} + \frac{t}{2} + \frac{1}{4}$$

Now multiply Eq. (9-1) by $D + 1$ and Eq. (9-2) by $-(D + 3)$ and add.

$$(D + 1)^2x - (D + 3)x = (D + 1)t - (D + 3)e^t = 1 + t - 4e^t$$

The solution, with the arbitrary constants evaluated by substitution in Eq. (9-2), is

$$x = (\tfrac{1}{3} - 2c_1 - \tfrac{2}{3}t)e^t + c_2e^{-2t} - \frac{t}{2} - \frac{3}{4}$$

9-13. Solving Differential Equations with Power Series. Some differential equations, particularly homogeneous linear equations with variable coefficients, can be solved by assuming a solution in the form of an infinite series. First try a Maclaurin series:

$$y = a_0 + a_1x + a_2x^2 + \cdots + a_nx^n = \sum_0^\infty a_nx^n$$

Substitute in the differential equation and equate coefficients of like powers of x to evaluate the a's.

• Solve $y'' - xy = 0$.
Assume $y = \Sigma a_nx^n$, $y' = \Sigma na_nx^{n-1}$, and $y'' = \Sigma n(n - 1)a_nx^{n-2}$. Substitution in the given equation yields

$$\Sigma n(n - 1)a_nx^{n-2} - x\Sigma a_nx^n = 0$$

Equate coefficients of like powers of x to zero.

$$2a_2 = 0 \qquad a_0 = 3 \cdot 2a_3 \qquad a_1 = 4 \cdot 3a_4 \qquad 5 \cdot 4a_5 = a_2 = 0$$

Hence, $a_2 = a_5 = a_8 = \cdots = 0$; $a_3 = \dfrac{a_0}{2 \cdot 3}$, $a_6 = \dfrac{a_3}{5 \cdot 6} = \dfrac{a_0}{2 \cdot 3 \cdot 5 \cdot 6}$, $\cdots$;

$a_4 = \dfrac{a_1}{3 \cdot 4}$, $a_7 = \dfrac{a_4}{6 \cdot 7} = \dfrac{a_1}{3 \cdot 4 \cdot 6 \cdot 7}$, $\cdots$, and

$$y = a_0\left(1 + \frac{x^3}{2 \cdot 3} + \frac{x^6}{2 \cdot 3 \cdot 5 \cdot 6} + \cdots\right) + a_1\left(x + \frac{x^4}{3 \cdot 4} + \frac{x^7}{3 \cdot 4 \cdot 6 \cdot 7} + \cdots\right)$$

If a Maclaurin series does not yield the general solution, try $y = x^c \Sigma a_n x^n$, where c is a constant to be determined by substitution in the differential equation (method of Frobenius).

• Solve $4xy'' + 2y' - y = 0$.

Assume $y = x^c \Sigma a_n x^n = \Sigma a_n x^{n+c}$. This yields $y' = \Sigma (n + c) a_n x^{n+c-1}$, and $y'' = \Sigma (n + c)(n + c - 1) a_n x^{n+c-2}$. Substitute in the equation.

$$4x\Sigma (n + c)(n + c - 1) a_n x^{n+c-2} + 2\Sigma (n + c) a_n x^{n+c-1} - \Sigma a_n x^{n+c} = 0$$
$$2\Sigma (n + c)(2n + 2c - 1) a_n x^{n+c-1} - \Sigma a_n x^{n+c} = 0$$

Set the coefficients of x^{c-1} equal to zero to solve for c, noting that $n \geq 0$.

$$(0 + c)(0 + 2c - 1) = 0$$
$$c = 0, \tfrac{1}{2}$$

With $c = 0$, equate coefficients of like powers of x to zero. The result is

$$y = a_0 \left(1 + \frac{x}{2!} + \frac{x^2}{4!} + \frac{x^3}{6!} + \cdots \right) = a_0 \cosh \sqrt{x}$$

Next, with $c = \tfrac{1}{2}$, equate coefficients of like powers of x to zero. The result is

$$y = b_0 \left(x^{1/2} + \frac{x^{3/2}}{3!} + \frac{x^{5/2}}{5!} + \cdots \right) = b_0 \sinh \sqrt{x}$$

Therefore, the complete solution is $y = a \cosh \sqrt{x} + b \sinh \sqrt{x}$.

See also Chap. 11.

9-14. Laplace Transform. The function $F(p)$ given by

$$F(p) = \int_0^\infty f(x) e^{-px} \, dx$$

is called the Laplace transform of $f(x)$. (Sometimes the transform is defined as $1/p$ times the integral. The lower limit also may be taken as $-\infty$.) The operator $\mathcal{L}$ may be used to signify the transformation of f into F. Tables are available that give the transforms of various functions (see Bibliography at the end of this chapter).

Here are some important characteristics of Laplace transforms:

$$\mathcal{L}(f + g) = \mathcal{L}(f) + \mathcal{L}(g)$$
$$\mathcal{L}(cf) = c\mathcal{L}(f) \qquad c = \text{constant}$$
$$\mathcal{L}[f(x - c)] = e^{-pc}F(p) \qquad x > c, c = \text{positive constant}$$
$$\mathcal{L}[e^{cx}f(x)] = F(p - c)$$
$$\mathcal{L}[f(cx)] = \frac{1}{c} F\left(\frac{p}{c} \right)$$
$$\mathcal{L}\left[\frac{1}{c} f\left(\frac{x}{c} \right) \right] = F(cp)$$
$$\mathcal{L}[f'(x)] = pF(p) - f(0)$$
$$\mathcal{L}[f^{(n)}(x)] = p^n F(p) - p^{n-1}f(0) - p^{n-2}f'(0) - \cdots - f^{(n-1)}(0)$$

$$\mathcal{L}[xf(x)] = -F'(p)$$

$$\mathcal{L}[x^n f(x)] = (-1)^n F^{(n)}(p)$$

$$\mathcal{L}\left[\frac{1}{x} f(x)\right] = \int_p^\infty F(p)\, dp$$

$$\mathcal{L}\left[\int_a^x f(t)\, dt\right] = \frac{1}{p} F(p) + \frac{1}{p}\int_a^0 F(t)\, dt$$

$$\mathcal{L}(f)\mathcal{L}(g) = \mathcal{L}(h) \qquad h = \int_0^x f(\lambda) g(x-\lambda)\, d\lambda \qquad \text{convolution theorem}$$

Here are some common transforms:

$$\mathcal{L}(c) = \frac{c}{p} \qquad c = \text{constant}$$

$$\mathcal{L}(x) = \frac{1}{p^2}$$

$$\mathcal{L}(x^n) = \frac{n!}{p^{n+1}} \qquad n = \text{integer} > 0$$

$$\mathcal{L}(x^{-\frac{1}{2}}) = \sqrt{\frac{\pi}{p}}$$

$$\mathcal{L}(x^{\frac{1}{2}}) = \tfrac{1}{2}\sqrt{\pi}\, p^{-\frac{3}{2}}$$

$$\mathcal{L}(e^{cx}) = \frac{1}{p-c}$$

$$\mathcal{L}(x^n e^{cx}) = \frac{n!}{(p-c)^{n+1}} \qquad n = \text{integer} > 0$$

$$\mathcal{L}(\sin cx) = \frac{c}{p^2 + c^2}$$

$$\mathcal{L}(x \sin cx) = \frac{2cp}{(p^2 + c^2)^2}$$

$$\mathcal{L}(e^{ax} \sin bx) = \frac{b}{(p-a)^2 + b^2}$$

$$\mathcal{L}(\cos cx) = \frac{p}{p^2 + c^2}$$

$$\mathcal{L}(x \cos cx) = \frac{p^2 - a^2}{(p^2 + c^2)^2}$$

$$\mathcal{L}(e^{ax} \cos bx) = \frac{p-a}{(p-a)^2 + b^2}$$

$$\mathcal{L}(\sinh cx) = \frac{c}{p^2 - c^2}$$

$$\mathcal{L}(\cosh cx) = \frac{p}{p^2 - c^2}$$

Linear differential equations with constant coefficients can be solved with the formulas for the transforms of the derivatives of a function.

• Solve $y'' + y = e^{-x}$ when $y(0) = y'(0) = 0$.

Let $\mathcal{L}(y) = F(p)$ and take the transform of the equation term by term.

$$p^2 \mathcal{L}(y) + \mathcal{L}(y) = \frac{1}{p+1}$$

$$\mathcal{L}(y) = \frac{1}{(p^2 + 1)(p + 1)}$$

The solution is the function whose transform is $\mathcal{L}(y)$. To find it, first express $F(p)$ in partial fractions.

$$\mathcal{L}(y) = \frac{1}{2(p+1)} - \frac{p-1}{2(p^2+1)} = \frac{1}{2(p+1)} - \frac{p}{2(p^2+1)} + \frac{1}{2(p^2+1)}$$

Then look up the inverse transforms term by term, as given on page 243 (with $c = 1$) or in a more complete table.

$$y = \tfrac{1}{2}e^{-x} - \tfrac{1}{2} \cos x + \tfrac{1}{2} \sin x$$

9-15. Fourier Transforms. The function $\mathfrak{F}(n)$ given by

$$\mathfrak{F}(n) = \int_0^\pi f(x) \sin nx \, dx$$

where n is an integer greater than zero, is called the finite Fourier sine transform of $f(x)$. The inverse is

$$f(x) = \frac{2}{\pi} \sum_{n=1}^\infty \mathfrak{F}(n) \sin nx \qquad 0 < x < \pi$$

The Fourier sine transformation of $f(x)$ on the half axis is defined by

$$\mathfrak{F}(r) = \int_0^\infty f(x) \sin rx \, dx \qquad r \geq 0$$

and its inverse is given by

$$f(x) = \frac{2}{\pi} \int_0^\infty \mathfrak{F}(r) \sin rx \, dr \qquad x > 0$$

The finite Fourier cosine transformation of $f(x)$ is defined by

$$\mathfrak{F}(c) = \int_0^\pi f(x) \cos cx \, dx$$

where c is an integer greater than zero. The inverse is

$$f(x) = \frac{1}{\pi} \mathfrak{F}(0) + \frac{2}{\pi} \sum_{c=1}^\infty \mathfrak{F}(c) \cos cx \qquad 0 < x < \pi$$

The Fourier cosine transformation of $f(x)$ on the half axis is defined by

$$\mathfrak{F}(c) = \int_0^\infty f(x) \cos cx \, dx \qquad c \geq 0$$

and its inverse is given by

$$f(x) = \frac{2}{\pi} \int_0^\infty \mathfrak{F}(r) \cos rx \, dr \qquad x > 0$$

The Fourier exponential transformation of $f(x)$ is defined by

$$\mathfrak{F}(t) = \int_{-\infty}^\infty f(x) e^{itx} \, dx$$

The inverse is given by

$$f(x) = \frac{1}{2\pi} \int_{-\infty}^\infty \mathfrak{F}(t) e^{-itx} \, dt$$

Tables of transforms and operational properties simplify application of the transforms. The transforms may be used to solve differential equations in much the same way as are Laplace transforms (see R. V. Churchill, "Operational Mathematics" in Bibliography at the end of the chapter).

9-16. Linear Partial Differential Equations. These are differential equations in the first degree in the dependent variable and its various partial derivatives.

A typical linear equation of the second order in two independent variables x and t and the dependent variable z has the form

$$a_1 \frac{\partial^2 z}{\partial x^2} + a_2 \frac{\partial^2 z}{\partial x \, \partial t} + a_3 \frac{\partial^2 z}{\partial t^2} + a_4 \frac{\partial z}{\partial x} + a_5 \frac{\partial z}{\partial t} + a_6 z = f(x,t)$$

where the a's are functions of x and t. The equation also can be written in the form

$$a_1 z_{xx} + a_2 z_{xt} + a_3 z_{tt} + a_4 z_x + a_5 z_t + a_6 z = f(x,t)$$

If $f(x,t) = 0$, the equation is called homogeneous.

If $z_1, z_2, \ldots, z_n$ are n solutions of a linear homogeneous partial differential equation, then $c_1 z_1 + c_2 z_2 + \cdots + c_n z_n$ are also solutions, the c's being any constants.

Sometimes the solution of a homogeneous linear equation with independent variables x and t can be found in the form

$$z = XT$$

where X = function of x
T = function of t

If three independent variables x, y, and t are involved, try $z = XYT$, where Y is a function only of y.

• Solve $y_{tt} = a^2 y_{xx}$, the equation of a vibrating string, where $y = 0$ when $x = 0, L$ for all values of t, and $y \approx y_0 \sin (\pi x/L)$ and $y_t = 0$ when $t = 0$.

Assume $y(x,t) = XT$, $y_{tt} = XT''$, where $T'' = \dfrac{d^2T}{dt^2}$ and $y_{xx} = TX''$ where $X'' = \dfrac{d^2X}{dx^2}$. Then

$$XT'' = a^2TX''$$

Separate the variables.

$$\frac{T''}{a^2T} = \frac{X''}{X} = k$$

where k is a constant (since neither side of the equation changes when the variable on the other side varies). Hence,

$$T'' - a^2kT = 0 \qquad X'' - kX = 0$$

From physical conditions imposed, the solution must consist of periodic functions. These result when k is negative.

$$y = (c_1 \sin \sqrt{k}\, x + c_2 \cos \sqrt{k}\, x)(c_3 \sin at \sqrt{k} + c_4 \cos at \sqrt{k})$$

When $x = 0$, $y = 0$; hence, $c_2 = 0$. When $x = L$, $y = 0$; therefore, $\sqrt{k}\, L = n\pi$, where n is an integer. So far, then,

$$y = c_1 \sin \frac{n\pi x}{L} \left(c_3 \sin \frac{na\pi t}{L} + c_4 \cos \frac{na\pi t}{L} \right)$$

Now when $t = 0$, $y_t = 0$; hence, $c_3 = 0$. Also, when $t = 0$,

$$y_0 \sin \frac{\pi x}{L} = c_1 c_4 \sin \frac{n\pi x}{L}$$

from which $c_1 c_4 = y_0$ and $n = 1$. Therefore, the solution is

$$y = y_0 \sin \frac{\pi x}{L} \cos \frac{a\pi t}{L}$$

For other initial conditions, infinite series such as Fourier series, or Fourier transforms may be used to find the solution.

To find the general solution of the first-order linear equation

$$P_1 \frac{\partial z}{\partial x} + P_2 \frac{\partial z}{\partial y} + \cdots + P_n \frac{\partial z}{\partial t} = f$$

where z, P_1, P_2, $\ldots$, P_n and f are functions of x, y, $\ldots$, t, solve

$$\frac{dx}{P_1} = \frac{dy}{P_2} = \cdots = \frac{dt}{P_n} = \frac{dz}{f}$$

If the general solution of this system is $z_1 = c_1$, $z_2 = c_2$, $\ldots$, $z_n = c_n$, where the c's are constants, then the general solution of the partial differential equation is $\varphi(z_1, z_2, \ldots, z_n) = 0$, where φ is an arbitrary function of z_1, z_2, $\ldots$, z_n.

• Solve $yz_x - xz_y = x^2 - y^2$.

The solution of this equation can be obtained from

$$\frac{dx}{y} = -\frac{dy}{x} = \frac{dz}{x^2 - y^2}$$

The first pair of terms yields $x^2 + y^2 = c^2$. The first and third terms, after multiplication by $x^2 - y^2$, yield $x^2\,dx/y - y\,dx = dz$. Substitute $-y\,dy/x$ for dx in the first term of this equation and integrate to get $z + xy = k$. Therefore, the solution is

$$\varphi(x^2 + y^2, z + xy) = 0$$

If the orders of all the terms of a linear partial differential equation are equal, the solution parallels that of a linear ordinary differential equation with constant coefficients. The equation will have the form

$$(a_0D_x{}^n + a_1D_x{}^{n-1}D_y + \cdots + a_{n-1}D_xD_y{}^{n-1} + a_nD_y{}^n)z = 0$$

where D_x and D_y indicate partial differentiation with respect to x and y, respectively. Solve the auxiliary equation

$$a_0m^n + a_1m^{n-1} + \cdots + a_{n-1}m + a_n = 0$$

If the roots $m_1, m_2, \ldots, m_n$ are distinct, the general solution is

$$y = \varphi_1(y + m_1x) + \varphi_2(y + m_2x) + \cdots + \varphi_n(y + m_nx)$$

If m is an r-fold root, not only is $\varphi_1(y + mx)$ an integral but so also are $x\varphi_2(y + mx)$, $x^2\varphi_3(y + mx)$, $\ldots$, $x^{r-1}\varphi_r(y + mx)$ and $y\theta_2(y + mx)$, $y^2\theta_3(y + mx)$, $\ldots$, $y^{r-1}\theta_r(y + mx)$. When the right-hand side of the given equation is not zero, a particular solution can be found by methods analogous to those for linear ordinary differential equations.

• Solve $\dfrac{\partial^2 z}{\partial x^2} + \dfrac{\partial^2 z}{\partial x\,\partial y} - 2\dfrac{\partial^2 z}{\partial y^2} = (D_x{}^2 + D_xD_y - 2D_y{}^2)z = \sin(x + y)$.

The auxiliary equation is

$$m^2 + m - 2 = 0$$

and its roots are $m_1 = -2$ and $m_2 = 1$. Hence, the complementary function is

$$z = \varphi_1(y - 2x) + \varphi_2(y + x)$$

For a particular integral, try $z = cx \cos(x + y)$, since $f(x + y)$ is already a part of the complementary function. Substituting in the given equation and equating coefficients of $\sin(x + y)$ yield $c = -\frac{1}{3}$. The general solution then is

$$z = \varphi_1(y - 2x) + \varphi_2(y + x) - \tfrac{1}{3}x \cos(x + y)$$

BIBLIOGRAPHY

Agnew, P.: "Differential Equations," McGraw-Hill Book Company, Inc., New York, 1960.

Churchill, R. V.: "Operational Mathematics," 2d ed., McGraw-Hill Book Company, Inc., New York, 1958.

Coddington, E. A., and N. Levinson: "Theory of Ordinary Differential Equations," McGraw-Hill Book Company, Inc., New York, 1955.

Dwight, H. B.: "Tables of Integrals and Other Mathematical Data," The Macmillan Company, New York, 1947.

Jahnke, E., F. Emde, and F. Losch: "Tables of Higher Functions," 6th ed., McGraw-Hill Book Company, Inc., New York, 1960.

Kells, L. M.: "Elementary Differential Equations," 2d ed., McGraw-Hill Book Company, Inc., New York, 1960.

McLachlan, N. W.: "Modern Operational Calculus," The Macmillan Company, New York, 1948.

Peirce, B. O., and R. M. Foster: "A Short Table of Integrals," Ginn & Company, Boston, 1956.

Reddick, H. W., and D. E. Kibbey: "Differential Equations," 3d ed., John Wiley & Sons, Inc., New York, 1956.

Sneddon, I. N.: "Elements of Partial Differential Equations," McGraw-Hill Book Company, Inc., New York, 1957.

Van Der Pol, B., and H. Bremmer: "Operational Calculus Based on the Two-sided Laplace Integral," Cambridge University Press, New York, 1950.

Chapter 10

NONELEMENTARY SOLUTIONS OF
DIFFERENTIAL EQUATIONS

10-1. Nonelementary Functions. Elementary functions are finite combinations, obtained by addition, subtraction, multiplication, and division, of powers, roots, exponentials, logarithms, trigonometric and inverse trigonometric functions, and hyperbolic and inverse hyperbolic functions of a variable and constants.

x^2, e^x, $x^{1/3}$, log cos x, and $(1 + e^{-x})$ log sin x are examples of elementary functions.

Functions that satisfy differential equations involving only elementary functions and their derivatives, including integration of $y' = f(x)$, yet cannot be expressed as finite combinations of elementary functions, are called nonelementary functions. The following articles discuss a few such functions.

10-2. Elliptic Integrals. Three types of related integrals, one of which arises in the determination of the circumference of an ellipse, which cannot be evaluated in finite form. Tables are available that give the values of these integrals (see Bibliography at the end of the chapter).

The elliptic integral of the first kind is defined by

$$F(k,\varphi) = \int_0^\varphi \frac{d\theta}{\sqrt{1 - k^2 \sin^2 \theta}} \quad \text{or} \quad F(k,x) = \int_0^x \frac{dz}{\sqrt{(1 - z^2)(1 - k^2 z^2)}}$$

where $k = \sin \alpha = $ constant (modulus)
$\varphi = $ constant (amplitude)
$x = \sin \varphi$

When $\varphi = \pi/2$, the resulting integral

$$K = \int_0^{\pi/2} \frac{d\theta}{\sqrt{1 - k^2 \sin^2 \theta}} = \frac{\pi}{2}\left[1 + (1/2)^2 k^2 + \left(\frac{1 \cdot 3}{2 \cdot 4}\right)^2 k^4 + \left(\frac{1 \cdot 3 \cdot 5}{2 \cdot 4 \cdot 6}\right)^2 k^6 + \cdots \right]$$

is known as the complete integral of the first kind.

Values of elliptic integrals of the first kind need be tabulated only for $\varphi \leq \pi/2$, since

$$F(k,\varphi) = 2K - F(k, \pi - \varphi)$$
$$F(k, m\pi + \varphi) = 2mK + F(k,\varphi)$$

where m is an integer.

The elliptic integral of the second kind is defined by

$$E(k,\varphi) = \int_0^\varphi \sqrt{1 - k^2 \sin^2 \theta}\, d\theta \quad \text{or} \quad E(k,x) = \int_0^x \sqrt{\frac{1 - k^2 z^2}{1 - z^2}}\, dz$$

When $\varphi = \pi/2$, the integral becomes the complete integral of the second kind.

$$E = \int_0^{\pi/2} \sqrt{1 - k^2 \sin^2 \theta}\, d\theta = \frac{\pi}{2} \left[1 - (\tfrac{1}{2})^2 k^2 \right.$$
$$\left. - \left(\frac{1 \cdot 3}{2 \cdot 4}\right)^2 \frac{k^4}{3} - \left(\frac{1 \cdot 3 \cdot 5}{2 \cdot 4 \cdot 6}\right)^2 \frac{k^6}{5} - \cdots \right]$$

Values of elliptic integrals of the second kind need be tabulated only for $\varphi \leq \pi/2$, since

$$E(k,\varphi) = 2E - E(k, \pi - \varphi)$$
$$E(k, m\pi + \varphi) = 2mE + E(k,\varphi)$$

The elliptic integral of the third kind is defined by

$$\pi(n,k,\varphi) = \int_0^\varphi \frac{d\theta}{(1 + n \sin^2 \theta) \sqrt{1 - k^2 \sin^2 \theta}}$$

or

$$\pi(n,k,x) = \int_0^x \frac{dz}{(1 + nz^2) \sqrt{(1 - z^2)(1 - k^2 z^2)}}$$

where n is any constant.

Let $\varphi = $ am u (amplitude of u). Then the functions $\sin \varphi = x = $ sn u, $\cos \varphi = \sqrt{1 - x^2} = $ cn u, and $\Delta\varphi = $ dn $u = \sqrt{1 - k^2 x^2}$ denote elliptic functions, which resemble trigonometric functions.

10-3. Sine-, Cosine-, and Exponential-integral Functions. The sine-integral function is defined by

$$\text{Si}(x) = \int \frac{\sin x\, dx}{x} = x - \frac{x^3}{3 \cdot 3!} + \frac{x^5}{5 \cdot 5!} - \frac{x^7}{7 \cdot 7!} + \cdots$$

The cosine-integral function is defined by

$$\text{Ci}(x) = \int \frac{\cos x\, dx}{x} = \log |x| - \frac{x^2}{2 \cdot 2!} + \frac{x^4}{4 \cdot 4!} - \frac{x^6}{6 \cdot 6!} + \cdots$$

The exponential-integral function is defined by

$$\text{Ei}(x) = \int \frac{e^{-x}\,dx}{x} = \log|x| - \frac{x}{1!} + \frac{x^2}{2\cdot 2!} - \frac{x^3}{3\cdot 3!} + \cdots$$

10-4. Signum Function.   This is defined by

$$\begin{aligned}
\text{sgn } x &= 1 & x &> 0 \\
&= 0 & x &= 0 \\
&= -1 & x &< 0
\end{aligned}$$

It satisfies the Dirichlet discontinuous integral.

$$\text{sgn } x = \frac{1}{\pi}\int_{-\infty}^{\infty} \frac{\sin xt}{t}\,dt$$

A useful formula is

$$|x| = \int_0^x \text{sgn } t\,dt$$

10-5. Gamma Function.   Euler defined the gamma function as

$$\Gamma(n) = \int_0^{\infty} x^{n-1}e^{-x}\,dx \qquad n > 0$$

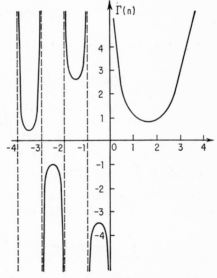

Integration by parts reveals that

$$\Gamma(n + 1) = n\Gamma(n)$$

and direct integration shows that

$$\Gamma(1) = 1$$

Therefore, when n is a positive integer

$$\begin{aligned}
&\Gamma(n + 1) \\
&= n(n - 1)(n - 2)\cdots 1 = n!
\end{aligned}$$

Hence, the gamma function is a generalization of the factorial concept.

The gamma function becomes infinite when n is zero or a nega-

Fig. 10-1

tive integer, but it can be evaluated for negative numbers other than integers (Fig. 10-1).

• Given $\Gamma(\tfrac{1}{2}) = \sqrt{\pi}$, find $\Gamma(-\tfrac{1}{2})$.

$$\Gamma(-\tfrac{1}{2}) = \frac{\Gamma(\tfrac{1}{2})}{-\tfrac{1}{2}} = -2\sqrt{\pi}$$

Tables are available giving values of the gamma function for the interval $1 < n < 2$. For other values of n, $\Gamma(n)$ can be computed by use of the recurrence formula $\Gamma(n + 1) = n\Gamma(n)$.

For $0 < n < 1$, $\Gamma(n)\Gamma(1 - n) = \pi/\sin n\pi$.

10-6. Gauss' Pi Function. This is defined by

$$\Pi(n) = \Gamma(n + 1)$$

Thus, $\Pi(n) = n\Pi(n - 1)$ (see also Art. 10-5).

When n is a positive integer

$$\Pi(n) = n!$$

10-7. Beta Function. This is defined by

$$\beta(m,n) = \int_0^1 x^{m-1}(1 - x)^{n-1}\,dx \qquad m > 0,\, n > 0$$

It is related to the gamma function (Art. 10-5) by

$$\beta(m,n) = \frac{\Gamma(m)\Gamma(n)}{\Gamma(m + n)}$$

The beta function may also occur in the forms

$$\beta(m,n) = 2\int_0^{\pi/2} \sin^{2m-1}\varphi \cos^{2n-1}\varphi\,d\varphi$$

$$= a^{1/(m+n-1)}\int_0^a y^{m-1}(a - y)^{n-1}\,dy$$

$$= \int_0^\infty \frac{z^{n-1}\,dz}{(1 + z)^{m+n}}$$

• Evaluate $\int_0^{\pi/2} \cos^r \theta\,d\theta$ for $r > -1$.

In the trigonometric form of $\beta(m,n)$, let $n = \frac{1}{2}(r + 1)$ and $m = \frac{1}{2}$ and use the gamma-function relationship, noting that $\Gamma(\frac{1}{2}) = \sqrt{\pi}$.

$$\int_0^{\pi/2} \cos^r \theta\,d\theta = \frac{\Gamma[(r + 1)/2]}{\Gamma(r/2 + 1)}\frac{\sqrt{\pi}}{2} = \int_0^{\pi/2} \sin^r \theta\,d\theta$$

10-8. Bessel Functions. Bessel's equation of order n,

$$x^2\frac{d^2y}{dx^2} + x\frac{dy}{dx} + (x^2 - n^2)y = 0$$

has the general solution

$$y = c_1 J_n(x) + c_2 J_{-n}(x) \qquad n \neq \text{integer}$$

But if n is an integer, the general solution is

$$y = c_1 J_n(x) + c_2 Y_n(x)$$

The expressions

$$J_n(x) = \sum_{m=0}^{\infty} \frac{(-1)^m}{\Pi(m)\,\Pi(m+n)} \left(\frac{x}{2}\right)^{2m+n}$$

$$J_{-n}(x) = \sum_{m=0}^{\infty} \frac{(-1)^m}{\Pi(m)\,\Pi(m-n)} \left(\frac{x}{2}\right)^{2m-n}$$

are called Bessel functions of the first kind of order n.

$$Y_n(x) = \lim \left[\frac{J_r(x)\cos r\pi - J_{-r}(x)}{\sin r\pi} \right]_{r \to n}$$

where n is an integer, is called a Bessel function of the second kind.
The functions defined by

$$H_n^{(1)}(x) = J_n(x) + iY_n(x)$$
$$H_n^{(2)}(x) = J_n(x) - iY_n(x) \qquad i = \sqrt{-1}$$

are called Bessel functions of the third kind or Hankel functions of order n.
The Bessel functions $J_1(x)$ and $J_0(x)$ resemble $\sin x$ and $\cos x$, respectively, with damped amplitudes (Fig. 10-2).

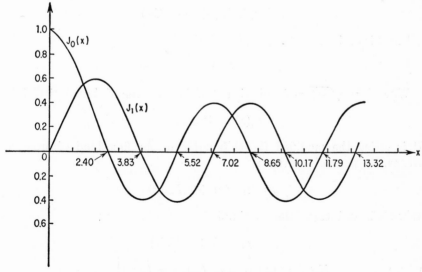

F<small>IG</small>. 10-2

Here are some important recurrence formulas for $J_n(x)$:

$$xJ'_n = nJ_n - xJ_{n+1} \qquad J'_n = \frac{d}{dx}J_n(x)$$

$$xJ'_n = -nJ_n + xJ_{n-1}$$

$$J'_n = \frac{J_{n-1} - J_{n+1}}{2}$$

$$J'_0 = -J_1$$

$$\frac{d}{dx}(x^{-n}J_n) = -x^{-n}J_{n+1}$$

$$\frac{d}{dx}(x^nJ_n) = x^nJ_{n-1}$$

$$J_n = \frac{x}{2n}(J_{n-1} + J_{n+1})$$

• If $J_{1/2}(x) = \sqrt{2/\pi x} \sin x$ and $J_{-1/2}(x) = \sqrt{2/\pi x} \cos x$, evaluate $J_{3/2}(x)$. From the recurrence formulas,

$$J_{3/2}(x) = \frac{1}{x}J_{1/2}(x) - J_{-1/2}(x) = \sqrt{\frac{2}{\pi x}}\left(\frac{\sin x}{x} - \cos x\right)$$

The general solution of Bessel's equation with parameter k,

$$\frac{d^2y}{dx^2} + \frac{1}{x}\frac{dy}{dx} + \left(k^2 - \frac{n^2}{x^2}\right)y = 0$$

where n is not an integer, is

$$y = c_1J_n(kx) + c_2J_{-n}(kx)$$

When n is an integer, the general solution is

$$y = c_1J_n(kx) + c_2Y_n(kx)$$

When $k = \sqrt{-1} = i$ and n is not an integer, one of the solutions is

$$I_n(x) = i^{-n}J_n(ix)$$

It is called the modified Bessel function of the first kind of order n. Hence, the general solution is

$$y = c_1I_n(x) + c_2I_{-n}(x)$$

When n is an integer, the solution is

$$y = c_1I_n(x) + c_2K_n(x)$$

where $\qquad K_n(x) = \lim\left\{\frac{\pi/2}{\sin r\pi}[I_{-r}(x) - I_r(x)]\right\}_{r\to n}$

When $n = 0$ and $k = \sqrt{-i} = i^{3/2}$, the solution is

$$y = c_1 J_0(i^{3/2}x) + c_2 K_0(i^{3/2}x)$$

The Bessel functions here are complex. New functions ber x, bei x, ker x, and kei x are defined by their real and complex parts.

$$J_0(i^{3/2}x) = \text{ber } x + i \text{ bei } x$$
$$K_0(i^{3/2}x) = \text{ker } x + i \text{ kei } x$$

A function $f(x)$ may be expanded in a series of Bessel functions in the interval $x = 0$ to $x = 1$.

$$f(x) = \sum_{m=1}^{\infty} C_m J_n(\alpha_m x)$$

where $\alpha_m =$ positive roots of $J_n(\alpha) = 0$

$$C_m = \frac{2}{J^2_{n+1}(\alpha_m)} \int_0^1 x J_n(\alpha_m x) f(x) \, dx$$

10-9. Legendre's Polynomials (Surface Zonal Harmonics). Legendre's differential equation

$$(1 - x^2) \frac{d^2y}{dx^2} - 2x \frac{dy}{dx} + n(n + 1)y = 0$$

has the general solution

$$y = c_1 P_n(x) + c_2 Q_n(x)$$

$P_n(x)$ is called Legendre's polynomial function of the first kind and $Q_n(x)$ is called Legendre's function of the second kind.

When n is a positive integer and $-1 \leq x \leq 1$,

$P_0(x) = 1$ $P_1(x) = x$
$P_2(x) = \frac{1}{2}(3x^2 - 1)$ $P_3(x) = \frac{1}{2}(5x^3 - 3x)$
$P_4(x) = \frac{1}{8}(35x^4 - 30x^2 + 3)$ $P_5(x) = \frac{1}{8}(63x^5 - 70x^3 + 15x)$

$$P_n(x) = \frac{1}{2^n n!} \frac{d^n}{dx^n} (x^2 - 1)^n = \sum_{r=0}^{N} (-1)^r \frac{(2n - 2r)!}{2^n r!(n - r)!(n - 2r)!} x^{n-2r}$$

where $N = n/2$ for even values of n and $N = (n - 1)/2$ for odd values of n.

When $-1 < x < 1$ and n is even,

$$Q_n(x) = b_1 \left[x - \frac{(n - 1)(n + 2)}{3!} x^3 \right.$$
$$\left. + \frac{(n - 1)(n - 3)(n + 2)(n + 4)}{5!} x^5 - \cdots \right]$$
$$b_1 = (-1)^{n/2} \frac{(2)(4) \cdots (n)}{(1)(3)(5) \cdots (n - 1)}$$

When n is odd,

$$Q_n(x) = b_0 \left[1 - \frac{n(n+1)}{2!} x^2 + \frac{n(n-2)(n+1)(n+3)}{4!} x^4 - \cdots \right]$$

$$b_0 = (-1)^{(n+1)/2} \frac{(2)(4) \cdots (n-1)}{(1)(3)(5) \cdots (n)}$$

When $|x| > 1$,

$$Q_n(x) = \sum_{r=0}^{\infty} \frac{2^n(n+r)!(n+2r)!}{r!(2n+2r+1)!} x^{-n-2r-1}$$

Both $P_n(x)$ and $Q_n(x)$ are special cases of the hypergeometric function (Art. 10-11).

The function φ defined by

$$\varphi = (1 - 2xZ + Z^2)^{-\frac{1}{2}} = \sum_{n=0}^{\infty} P_n(x)Z^n$$

is called the generating function for $P_n(x)$. Note that $P_n(x)$ represents the coefficients of Z in the binomial expansion of φ for $-1 \leq x \leq 1$, $-1 < Z < 1$.

A function $f(x)$ may be expanded in a series of Legendre polynomials in the interval $-1 \leq x \leq 1$.

$$f(x) = \sum_{n=0}^{\infty} a_n P_n(x)$$

$$a_n = \frac{2n+1}{2} \int_{-1}^{1} f(x)P_n(x) \, dx$$

The associated Legendre equation with an added term involving an integer m

$$(1 - x^2) \frac{d^2y}{dx^2} - 2x \frac{dy}{dx} + \left[n(n+1) - \frac{m^2}{(1-x^2)} \right] y = 0$$

is satisfied by the associated Legendre polynomial

$$P_n{}^m(x) = (1 - x^2)^{m/2} \frac{d^m}{dx^m} P_n(x)$$

If $m > n$, $P_n{}^m(x) = 0$.

10-10. Hermite Polynomials. The differential equation

$$\frac{d^2y}{dx^2} - x \frac{dy}{dx} + ny = 0$$

where n is a positive integer, is satisfied by the Hermite polynomials defined by

$$H_n(x) = (-1)^n e^z \frac{d^n e^{-z}}{dx^n}$$

where $z = x^2/2$.

Hermite functions are related by

$$\frac{dH_n(x)}{dx} = nH_{n-1}(x)$$

$$\frac{d^r H_n(x)}{dx^r} = n(n-1) \cdots (n-r+1)H_{n-r}(x)$$

The generating function is

$$\varphi = e^w = \sum_{n=0}^{\infty} H_n(x) \frac{z^n}{n!}$$

where $w = -z^2/2 + zx$.   Note that $H_n(x)$ represents the coefficients of $z^n/n!$ in the Taylor series expansion of φ.

10-11. Hypergeometric Function.   Gauss' hypergeometric equation

$$x(1-x)\frac{d^2y}{dx^2} + [(a+b+1)x - c]\frac{dy}{dx} + aby = 0$$

has the solution

$$y = c_1 \sum_{n=0}^{\infty} \frac{\Gamma(a+n)\Gamma(b+n)}{\Gamma(c+n)} \frac{x^n}{n!}$$

$$+ c_2 x^{1-c} \sum_{n=0}^{\infty} \frac{\Gamma(n+1+a-c)\Gamma(n+1+b-c)}{\Gamma(n+2-c)} \frac{x^n}{n!}$$

for $-1 < x < 1$, excluding $x = 0$.   But c cannot be an integer or zero, and a and b cannot be negative integers.

The hypergeometric series is derived from the first term of the solution.   It is defined by

$$F(a,b;c;x) = \frac{\Gamma(c)}{\Gamma(a)\Gamma(b)} \sum_{n=0}^{\infty} \frac{\Gamma(a+n)\Gamma(b+n)}{\Gamma(c+n)} \frac{x^n}{n!}$$

$$= 1 + \frac{ab}{c}\frac{x}{1!} + \frac{a(a+1)b(b+1)}{c(c+1)}\frac{x^2}{2!} + \cdots$$

$$+ \frac{a(a+1) \cdots (a+n-1)b(b+1) \cdots (b+n-1)}{c(c+1) \cdots (c+n-1)}\frac{x^n}{n!} + \cdots$$

where c is not zero or a negative integer. The function to which the series converges when $|x| < 1$ is called the hypergeometric function.

If a or b is zero or a negative integer, the function is a polynomial. Legendre's polynomial is a special case (Art. 10-8).

BIBLIOGRAPHY

Agnew, R. P.: "Differential Equations," McGraw-Hill Book Company, Inc., New York, 1960.

Dwight, H. B.: "Tables of Integrals and Other Mathematical Data," 2d ed., The Macmillan Company, New York, 1947.

Jahnke, E., F. Emde, and F. Losch: "Tables of Higher Functions," McGraw-Hill Book Company, Inc., New York, 1960.

Lass, Harry: "Elements of Pure and Applied Mathematics," McGraw-Hill Book Company, Inc., New York, 1957.

Peirce, B. O., and R. M. Foster: "A Short Table of Integrals," Ginn & Company, Boston, 1956.

Pipes, L. A.: "Applied Mathematics for Engineers and Physicists," 2d ed., McGraw-Hill Book Company, Inc., New York, 1958.

Sokolnikoff, I. S., and R. M. Redheffer: "Mathematics of Physics and Modern Engineering," McGraw-Hill Book Company, Inc., New York, 1958.

Wylie, C. R., Jr.: "Advanced Engineering Mathematics," McGraw-Hill Book Company, Inc., New York, 1960.

Chapter 11

NUMERICAL INTEGRATION OF
DIFFERENTIAL EQUATIONS

11-1. Trapezoidal Rule for Integration. To evaluate

$$y = \int_a^b f(x)\, dx$$

when a general solution cannot be found, divide the interval ab into n equal parts. The length of each part is then

$$h = \frac{b - a}{n}$$

Calculate

$$f_0 = f(a),\ f_1 = f(a + h),\ f_2 = f(a + 2h),\ \ldots,\ f_n = f(a + nh) = f(b)$$

A first approximation of the integral is

$$y_1 \approx h(\tfrac{1}{2}f_0 + f_1 + f_2 + \cdots + f_{n-1} + \tfrac{1}{2}f_n)$$

The accuracy of the result sometimes may be improved by repeating the calculation with another value of n, say m, and obtaining another approximation of y, say y_2. Then, more accurately,

$$y \approx \frac{m^2}{m^2 - n^2} y_2 - \frac{n^2}{m^2 - n^2} y_1$$

(See also Art. 11-2.)

• Evaluate $y = \int_0^{\pi/2} \sqrt{1 - 0.25 \sin^2 x}\, dx$.

Try, first, $n = 2$, $h = \pi/4$. Thus, $f_0 = f(0) = 1$, $f_1 = f(\pi/4) = 0.9354$, $f_2 = f(\pi/2) = 0.8660$.

$$y_1 \approx h(\tfrac{1}{2}f_0 + f_1 + \tfrac{1}{2}f_2) = \frac{\pi}{4}\,(\tfrac{1}{2} \times 1 + 0.9354 + \tfrac{1}{2} \times 0.8660) = 1.4674$$

Next try $n = 4$, $h = \pi/8$. Hence, $f_0 = 1$, $f_1 = 0.9804$, $f_2 = 0.9354$, $f_3 = 0.8868$, and $f_4 = 0.8660$.

$$y_2 \approx h(\tfrac{1}{2}f_0 + f_1 + f_2 + f_3 + \tfrac{1}{2}f_4)$$

$$= \frac{\pi}{8}\,(\tfrac{1}{2} \times 1 + 0.9804 + 0.9354 + 0.8868 + \tfrac{1}{2} \times 0.8660) = 1.4670$$

259

More accurately,

$$y \approx \frac{4^2}{4^2 - 2^2} 1.4670 - \frac{2^2}{4^2 - 2^2} 1.4674 = 1.4669$$

11-2. Simpson's Rule for Integration. (See also Art. 11-1.)   Another method of evaluating

$$y = \int_a^b f(x) \, dx$$

when a general solution cannot be found is to divide the interval into n parts, each of length

$$h = \frac{b - a}{n}$$

Then compute, as a first approximation,

$$y_1 \approx \frac{h}{3} \left(f_0 + 4f_1 + 2f_2 + 4f_3 + \cdots + 2f_{n-2} + 4f_{n-1} + f_n \right)$$

where

$$f_0 = f(a), \ f_1 = f(a + h), \ f_2 = f(a + 2h), \ \cdots, \ f_n = f(a + nh) = f(b)$$

When n is an odd number, apply this formula to the first $n - 1$ parts and apply the trapezoidal rule (Art. 11-1) to the last part.

The accuracy of the result sometimes may be improved by repeating the calculation with another value of n, say m, and obtaining another approximation of y, say y_2.   Then, more accurately,

$$y \approx \frac{m^4}{m^4 - n^4} y_2 - \frac{n^4}{m^4 - n^4} y_1$$

• Evaluate $y = \int_0^{\pi/2} \sqrt{1 - 0.25 \sin^2 x} \, dx$.

Try first $n = 2$, $h = \pi/4$.   Thus, $f_0 = f(0) = 1$, $f_1 = f(\pi/4) = 0.9354$, and $f_2 = f(\pi/2) = 0.8660$.

$$y_1 \approx \frac{h}{3} \left(f_0 + 4f_1 + f_2 \right) = \frac{\pi}{12} \left(1 + 4 \times 0.9354 + 0.8660 \right) = 1.4681$$

Next try $n = 4$, $h = \pi/8$.   Hence, $f_0 = 1$, $f_1 = 0.9804$, $f_2 = 0.9354$, $f_3 = 0.8868$, and $f_4 = 0.8660$.

$$y_2 \approx \frac{h}{3} \left(f_0 + 4f_1 + 2f_2 + 4f_3 + f_4 \right)$$

$$= \frac{\pi}{24} \left(1 + 4 \times 0.9804 + 2 \times 0.9354 + 4 \times 0.8868 + 0.8660 \right) = 1.4668$$

More accurately,

$$y \approx \frac{4^4}{4^4 - 2^4} 1.4668 - \frac{2^4}{4^4 - 2^4} 1.4681 = 1.4667$$

11-3. Forward Differences. If $y = f(x)$, consider the values of y when x is given successive increments equal to h. Let $y = y_n$ when $x = x_n$, and $y = y_{n+1}$ when $x = x_n + h$. Then the first forward difference of y at x_n is defined by

$$\Delta y_n = y_{n+1} - y_n$$

The second forward difference is defined by

$$\Delta(\Delta y_n) = \Delta^2 y_n = y_{n+2} - 2y_{n+1} + y_n$$

The kth forward difference is given by

$$\Delta^k y_n = y_{n+k} + \sum_{r=1}^{k} (-1)^r C(k,r) y_{n+k-r}$$

where
$$C(k,r) = \frac{k(k - 1)(k - 2) \cdots (k - r + 1)}{r!}$$

Δ may be used as an operator in the same manner as $D = \dfrac{d}{dx}$ (Art. 9-7). Thus, y at any point $x = x_n + kh$ equals

$$y_{n+k} = (1 + \Delta)^k y_n$$

D is related to Δ by

$$D = \frac{1}{h}\left(\Delta - \frac{\Delta^2}{2} + \frac{\Delta^3}{3} - \frac{\Delta^4}{4} + \cdots\right)$$

Higher derivatives can be obtained by taking successive powers of D.

$$D^2 = \frac{1}{h^2}(\Delta^2 - \Delta^3 + 11\!/\!12\Delta^4 - 5\!/\!6\Delta^5 + \cdots)$$

$$D^3 = \frac{1}{h^3}(\Delta^3 - 3\!/\!2\Delta^4 + 7\!/\!4\Delta^5 - \cdots)$$

$$D^4 = \frac{1}{h^4}(\Delta^4 - 2\Delta^5 + 17\!/\!6\Delta^6 - \cdots)$$

Hence, Dy_n at $x = x_n$ is given approximately by $(1/h)\,\Delta y_n$, $D^2 y_n$ is given by $(1/h^2)\Delta^2 y_n$, . . . , $D^m y_n$ is given by $(1/h^m)\Delta^m y_n$.

11-4. Backward Differences. If $y = f(x)$, consider the values of y when x is given successive decrements equal to h. Let $y = y_n$ when $x = x_n$, and $y = y_{n-1}$ when $x = x_n - h$. Then the first backward difference of y at x_n is defined by

$$\nabla y_n = y_n - y_{n-1}$$

The second backward difference is defined by

$$\nabla(\nabla y_n) = \nabla^2 y_n = y_n - 2y_{n-1} + y_{n-2}$$

The kth backward difference is given by

$$\nabla^k y_n = y_n + \sum_{r=1}^{k} (-1)^r C(k,r) y_{n-r}$$

where $$C(k,r) = \frac{k(k-1)(k-2) \cdots (k-r+1)}{r!}$$

∇ (del) may be used as an operator in the same manner as $D = \dfrac{d}{dx}$ (Art. 9-7). Thus, at any point $x = x_n - kh$, y equals

$$y_{n-k} = (1 - \nabla)^k y_n$$

D is related to ∇ by

$$D = \frac{1}{h}\left(\nabla + \frac{\nabla^2}{2} + \frac{\nabla^3}{3} + \frac{\nabla^4}{4} + \cdots\right)$$

Higher derivatives can be obtained by taking successive powers of D.

$$D^2 = \frac{1}{h^2}\left(\nabla^2 + \nabla^3 + 11\!\!/_{12}\nabla^4 + 5\!\!/_6\nabla^5 + \cdots\right)$$

$$D^3 = \frac{1}{h^3}\left(\nabla^3 + 3\!\!/_2\nabla^4 + 7\!\!/_4\nabla^5 + \cdots\right)$$

$$D^4 = \frac{1}{h^4}\left(\nabla^4 + 2\nabla^5 + 17\!\!/_6\nabla^6 + \cdots\right)$$

Hence, Dy_n at $x = x_n$ is given approximately by $(1/h)\,\nabla y_n$, $D^2 y_n$ is given by $(1/h^2)\,\nabla^2 y_n$, $\ldots$, $D^m y_n$ is given by $(1/h^m)\,\nabla^m y_n$.

11-5. Central Differences. If $y = f(x)$, consider the values of y when x is given successive increments and decrements equal to $h/2$. Let $y = y_n$ when $x = x_n$, $y = y_{n\pm1}$ when $x = x_n \pm h$, and $y = y_{n\pm\frac{1}{2}}$ when $x = x_n \pm h/2$. Then, the first central difference of y at x_n is defined by

$$\delta y_n = y_{n+\frac{1}{2}} - y_{n-\frac{1}{2}}$$

The second central difference is defined by

$$\delta(\delta y_n) = \delta^2 y_n = y_{n+1} - 2y_n + y_{n-1}$$

The kth central difference is given by

$$\delta^k y_n = y_{n+k/2} + \sum_{r=1}^{k} (-1)^r C(k,r) y_{n-r+k/2}$$

where $$C(k,r) = \frac{k(k-1)(k-2) \cdots (k-r+1)}{r!}$$

To eliminate the need for computing y at $x \pm rh/2$, where r is an odd number, averaged odd differences are used. They are defined as the averages of odd central differences at $x \pm h/2$. Then, the first averaged difference at x_n is

$$\mu\delta y_n = \tfrac{1}{2}(y_{n+1} - y_{n-1})$$

where μ is an operator called the averager.

$$\mu y_n = \tfrac{1}{2}(y_{n+\frac{1}{2}} + y_{n-\frac{1}{2}})$$

Symbolically, $\mu^2 = 1 + \delta^2/4$. δ and μ may be used as operators in the same manner as $D = \dfrac{d}{dx}$ (Art. 9-7).

D is related to δ and μ by

$$D = \frac{\mu}{h}\left(\delta - \frac{\delta^3}{6} + \frac{\delta^5}{30} - \cdots\right)$$
$$D^2 = \frac{1}{h^2}\left(\delta^2 - \frac{\delta^4}{12} + \frac{\delta^6}{90} - \cdots\right)$$
$$D^3 = \frac{\mu}{h^3}\left(\delta^3 - \frac{\delta^5}{4} + \frac{7\delta^7}{120} - \cdots\right)$$
$$D^4 = \frac{1}{h^4}\left(\delta^4 - \frac{\delta^6}{6} + \frac{7\delta^8}{240} - \cdots\right)$$

Hence, Dy_n at $x = x_n$ is given approximately by

$$\frac{\mu\delta}{h}\, y_n = \frac{1}{2h}\,(y_{n+1} - y_{n-1})$$

$D^2 y_n$ is given by

$$\frac{1}{h^2}\,\delta^2 y_n = \frac{1}{h^2}\,(y_{n+1} - 2y_n + y_{n-1})$$

$D^3 y_n$ is given by

$$\frac{\mu}{h^3}\,\delta^3 y_n = \frac{1}{2h^3}\,(y_{n+2} - 2y_{n+1} + 2y_{n-1} - y_{n-2})$$

$D^4 y_n$ is given by

$$\frac{1}{h^4}\,(y_{n+2} - 4y_{n+1} + 6y_n - 4y_{n-1} + y_{n-2}), \text{ etc.}$$

When m is an even number,

$$D^m y_n \approx \frac{\delta^m}{h^m}\, y_n = \frac{1}{h^m}\left[y_{n+m/2} + \sum_{r=1}^{m} (-1)^r C(m,r) y_{n-r+m/2}\right]$$

When m is an odd number,

$$D^m y_n \approx \frac{\mu\delta^m}{h^m}\, y_n = \frac{\mu\delta}{h^m}\,(\delta^{m-1} y_n) = \frac{1}{2h^m}\,\delta^{m-1}(y_{n+1} - y_{n-1})$$

11-6. Milne Method for First-order Equations. The first-order differential equation

$$y' = f(x,y)$$

with a given initial condition, $y = y_0$ when $x = x_0$, can be solved numerically with Milne's recurrence formula.

$$y_{m+1} = y_{m-3} + \frac{4h}{3}(2y'_{m-2} - y'_{m-1} + 2y'_m)$$

where y_{m+1} = value of y at $x_m + h$
y_{m-3} = value of y at $x_m - 3h$
$\quad y'_m$ = value of y' at x_m
$\quad y'_{m-1}$ = value of y' at $x_m - h$
$\quad y'_{m-2}$ = value of y' at $x_m - 2h$

With y_{m+1} known, y'_{m+1} can be computed from the given differential equation.

The values of y and y' needed in the recurrence formula can be calculated by expanding f in a Taylor series about a value of x and y satisfying the initial condition. As a check, after y'_{m+1} has been found, substitute in the formula

$$y_{m+1} = y_{m-1} + \frac{h}{3}(y'_{m-1} + 4y'_m + y'_{m+1})$$

If the result checks with the previously obtained value, continue by determining y_{m+2} from the recurrence formula. If there is a disagreement, use the second value to compute a new y'_{m+1} and substitute in the check formula. Continue until the difference between two successive values of y_{m+1} is negligible.

In solving a differential equation numerically, it is convenient to transform the variables into nondimensional form.

• Solve $y' = x + y$ in the interval $0 \le x \le 1$ when $(x_0,y_0) = (0,1)$.

Divide the interval into ten equal parts ($h = 0.1$). To obtain the values of y and y' needed for the recurrence formula, expand y in a Taylor series about $(0,1)$.

$$y(x) = y_0 + y'_0(x - x_0) + \frac{y''_0}{2!}(x - x_0)^2 + \frac{y'''_0}{3!}(x - x_0)^3 + \cdots$$

$$= 1 + mhy'_0 + \frac{m^2h^2}{2!}y''_0 + \frac{m^3h^3}{3!}y''' + \cdots$$

where $x - x_0 = mh$ and m is an integer.

The value of y'_0 is obtained by substituting $(0,1)$ for (x_0,y_0) in the differential equation.

$$y'_0 = x_0 + y_0 = 0 + 1 = 1$$

y''_0 can be obtained by differentiating the above equation.

$$y''_0 = 1 + y'_0 = 1 + 1 = 2$$

Successive differentiation yields $y^{(n)} = y^{(n-1)} = \cdots = y''' = y'' = 2$. Substitution of these values of the derivatives in the Taylor series with $h = 0.1$ results in

$$y(x) = 1 + 0.1m + 0.01m^2 + (0.001/3)m^3 + (0.0001/3 \cdot 4)m^4 + \cdots$$

If m is set equal to 1, 2, 3, then, $y_1 = 1.1103$, $y_2 = 1.2428$, and $y_3 = 1.3997$. Substitution in the differential equation now gives

$$y_1' = 0.1 + 1.1103 = 1.2103$$
$$y_2' = 0.2 + 1.2428 = 1.4428$$
$$y_3' = 0.3 + 1.3997 = 1.6997$$

Substitution of these values in the recurrence formula with $m = 3$ yields

$$y_4 = y_0 + (0.4/3)(2y_1' - y_2' + 2y_3')$$
$$= 1 + 0.1333(2 \times 1.2103 - 1.4428 + 2 \times 1.6997)$$
$$= 1.5836$$
$$y_4' = 0.4 + 1.5836 = 1.9836$$

Check: $y_4 = y_2 + (0.1/3)(y_2' + 4y_3' + y_4') = 1.2428 + 0.03333(1.4428 + 4 \times 1.6997 + 1.9836) = 1.5836$.

Since this checks closely the previous value, continue by computing y_5 from the recurrence formula with $m = 4$.

$$y_5 = y_1 + (0.4/3)(2y_2' - y_3' + 2y_4') = 1.7974$$

11-7. Adams's Method for First-order Equations. The first-order differential equation

$$y' = f(x,y)$$

with a given initial condition $y = y_0$ when $x = x_0$, can be solved numerically with Adams's recurrence formula:

$$y_{m+1} = y_m + h(1 + \tfrac{1}{2}\nabla + \tfrac{5}{12}\nabla^2 + \tfrac{3}{8}\nabla^3 + \tfrac{251}{720}\nabla^4 + \tfrac{95}{288}\nabla^5 + \cdots)f_m$$

where y_m = value of y at x_m
y_{m+1} = value of y at $x_m + h$
f_m = value of y' at x_m
$\nabla, \nabla^2, \nabla^3, \ldots$ = backward difference operators del
(See Art. 11-4.)

To obtain the backward differences of f_m for use in the recurrence formula, successive values of f_m for increments of h in x must be known. These can be estimated by expanding f_m in a Taylor series about a value of x and y satisfying the initial condition.

In solving a differential equation numerically it is convenient to transform the variables into nondimensional form.

• Solve the equation $y' = x + y$ of Art. 11-6, given $(x_0,y_0) = (0,1)$.

Divide the interval into ten equal parts $(h = 0.1)$. Assume that sufficient accuracy will be attained if terms involving the fourth backward difference and higher are dropped from the recurrence formula. Use a Taylor-series expansion about $(0,1)$ to obtain three values of y' needed to compute the third backward difference at the start. The calculation is the same as in Art. 11-6.

The differences now can be obtained by constructing a table.

TABLE 11-1

x	y	y'	$\nabla y'$	$\nabla^2 y'$	$\nabla^3 y'$
0	1.0000	1.0000			
0.1	1.1103	1.2103	0.2103		
0.2	1.2428	1.4428	0.2325	0.0222	
0.3	1.3997	1.6997	0.2569	0.0244	0.0022
0.4	1.5836	1.9836	0.2839	0.0270	0.0026
0.5	1.7974				

The recurrence formula is used to expand the table, starting with $m = 3$.

$y_4 = 1.3997 + 0.1[1.6997 + \frac{1}{2}(0.2569) + \frac{5}{12}(0.0244) + \frac{3}{8}(0.0022)] = 1.5836$
$y_4' = 0.4 + 1.5836 = 1.9836$

After these are entered in the table, another line of backward differences is computed. Then y_5 is calculated from the recurrence formula and the computation is continued.

11-8. Euler-Fox Method for First-order Linear Equations. The differential equation

$$y' = f(x)y + g(x)$$

with a given initial condition $y = y_0$ when $x = x_0$, can be solved by the Euler-Fox recurrence equation:

$$y_{m+1} = \frac{1}{1 - (h/2)f_{m+1}}\left[\left(1 + \frac{h}{2}f_m\right)y_m + \frac{h}{2}(g_m + g_{m+1}) + e_{m+1}\right]$$

where $e_{m+1} = -(\frac{1}{12}\delta^3 - \frac{1}{120}\delta^5 + \frac{1}{840}\delta^7 - \cdots)y_{m+\frac{1}{2}}$
$\delta^3, \delta^5, \delta^7, \ldots$ = third, fifth, seventh, . . . central differences
       f_m = value of $f(x)$ at x_m
       y_m = value of y at x_m
    y_{m+1} = value of y at $x_m + h$
    $y_{m+\frac{1}{2}}$ = value of y at $x_m + h/2$

As a first approximation, e is ignored in calculating y. Then the central differences are computed to determine e, after which y is corrected.

• Solve the equation $y' = x + y$ of Art. 11-6 when $(x_0,y_0) = (0,1)$.

Divide the interval into ten equal parts $(h = 0.1)$. Construct a table giving values of x, f_m, first approximation of y_m, central differences, corrections, and corrected values of y_m.

TABLE 11-2

x	f_m	y_m	$\delta y_{m+\frac{1}{2}}$	$\delta^2 y_m$	$\delta^3 y_{m+\frac{1}{2}}$	$\dfrac{e}{0.95}$	y_m
0	1	1.0000	......	......	......	....	1.0000
0.05	1	......	0.1105	......	(0.0023)	−0.0002	
0.10	1	1.1105	......	0.0222	......	....	1.1103
0.15	1	......	0.1327	......	0.0023	−0.0002	
0.20	1	1.2432	......	0.0245	......	....	1.2430
0.25	1	......	0.1572	......	(0.0023)	−0.0002	
0.30	1	1.4004	......	......	......	....	1.4002

In this problem, f_m has the constant value 1 and $g(x) = x$. The first approximation ignores e and starts with $x_0 = 0$, $y_0 = 1$ in the recurrence formula.

$$y_1 \approx \frac{1}{1 - 0.05(1)} \{[1 + 0.05(1)]y_0 + 0.05(x_0 + x_0 + 0.1)\}$$
$$= (1/0.95)(1.05 + 0.005) = 1.1105$$
$$y_2 \approx (1/0.95)[1.05y_1 + 0.05(0.1 + 0.2)] = \frac{(1.05 \times 1.1105 + 0.015)}{0.95} = 1.2432$$

Similarly, $y_3 = 1.4004$. These values are entered in the table for $x = 0.1$, 0.2, and 0.3, respectively.

In computing e, assume that sufficient accuracy will be attained with only the first term, $-\frac{1}{12}\delta^3$.

The first central differences are

$$\delta y_{\frac{1}{2}} = 1.1105 - 1.0000 = 0.1105$$
$$\delta y_{1\frac{1}{2}} = 1.2432 - 1.1105 = 0.1327$$
$$\delta y_{2\frac{1}{2}} = 1.4004 - 1.2432 = 0.1572$$

The second and third differences are computed next. The third difference is found only at $x = 0.15$ and is equal to 0.0023. For a second approximation of y_m, assume the third difference equal to 0.0023 for $x = 0.05$, 0.15, and 0.25. Then

$$e \approx -\frac{1}{12} \times 0.0023 = -0.00019$$

and the correction to y equals $-0.00019/0.95 = -0.0002$. The second approximation of y is given in the table.

11-9. Solution of Equations of Second and Higher Order. An equation of order higher than first can be replaced by a system of simultaneous first-order equations. These can then be solved by one of the methods given in Arts. 11-6 to 11-8.

• Solve $\dfrac{d^2y}{dt^2} = -25y - y^3$ in the interval $0 \leq t \leq 0.20$ if $y = 1$ and $y' = 0$ when $t = 0$.

Write the equation in the form

$$\frac{dy}{dt} = z$$

$$\frac{dz}{dt} = -25y - y^3$$

and use Milne's method (Art. 11-6). Divide the interval into ten equal parts ($h = 0.02$). The third and higher derivatives of y needed for expansion of y in a Taylor series can be obtained by successive differentiation of the given equation, noting that $y_0 = 1$ and $y_0' = 0$.

$$y'' = -25y - y^3 \qquad\qquad y_0'' = -25 - 1 = -26$$
$$y''' = -25y' - 3y^2y' \qquad\quad y_0''' = 0$$
$$y^{(4)} = -25y'' - 3y^2y'' - 6yy'^2 \quad y_0^{(4)} = 728$$

$$y(t) = y_0 + y_0'(t - t_0) + \frac{y_0''}{2!}(t - t_0)^2 + \frac{y_0'''}{3!}(t - t_0)^3 + \frac{y_0^{(4)}}{4!}(t - t_0)^4 + \cdots$$

$$= 1 - 26\tfrac{1}{2}t^2 + \frac{728}{4!}t^4 - \cdots$$

Substitute $t = 0.02$, 0.04, and 0.06 in this equation to obtain $y_1 = 0.9948$, $y_2 = 0.9793$, and $y_3 = 0.9536$.

Since $z = y'$, differentiate the series to obtain z.

$$z(t) = -26t + \frac{728}{3!}t^3 - \cdots$$

from which $z_1 = -0.5190$, $z_2 = -1.0323$, and $z_3 = -1.5340$.

Values of $y'' = z'$ can be found by substitution in the given equation: $y_1'' = -25y_1 - y_1^3 = -25 \times 0.9948 - (0.9948)^3 = -25.8545$, $y_2'' = -25.4212$, and $y_3'' = -24.7072$.

The Milne recurrence formula (Art. 11-6) with $m = 3$ is used to continue.

$$y_4 = y_0 + \frac{4h}{3}(2z_1 - z_2 + 2z_3)$$

$$= 1 + \frac{4 \times 0.02}{3}[2(-0.5190) + 1.0323 + 2(-1.5340)]$$

$$= 0.9180$$

$$z_4 = z_0 + \frac{4h}{3}(2z_1' - z_2' + 2z_3')$$

$$= 0 + \frac{0.08}{3}[2(-25.8545) + 25.4212 + 2(-24.7072)]$$

$$= 2.0185$$

$$z_4' = y_4'' = -25y_4 - y_4^3 = -23.7236$$

After these values have been checked, successive values of y and z can be computed with the recurrence formula in the same way.

11-10. Solution of Boundary-value Problems. (See also Art. 9-2.) Given a differential equation with conditions at more than one point, express the derivatives in terms of central differences (Art. 11-5) to integrate numerically. Solve by trial and error, by interpolation, or by setting up simultaneous equations.

• Solve $y'' - (18x + 4)y = -(18x + 4)e^x$ in the interval $0 \le x \le 1$, given $y = 0$ when $x = 0$, $y = 1$ when $x = 1$.

Divide the interval into four equal parts ($h = 0.25$). Thus, $x_1 = 0.25$, $x_2 = 0.50$, $x_3 = 0.75$, and $x_4 = 1$. The given equation, expressed as a difference equation, with D^2 as given in Art. 11-5 and $h^2 = 0.0625$, becomes

$$y_{m+1} - 2y_m + y_{m-1} - 0.0625(18x_m + 4)y_m = -0.0625(18x_m + 4)e^{x_m}$$

or $\qquad y_{m+1} - (1.125x_m + 2.25)y_m + y_{m-1} = -(1.125x_m + 0.25)e^{x_m}$

with $(x_0,y_0) = (0,0)$ and $(x_4,y_4) = (1,1)$. When m is set equal to 1, 2, and 3, three equations are determined.

$$y_2 - (1.125 \times 0.25 + 2.25)y_1 + 0 = -(1.125 \times 0.25 + 0.25)e^{0.25}$$
$$y_3 - (1.125 \times 0.50 + 2.25)y_2 + y_1 = -(1.125 \times 0.50 + 0.25)e^{0.50}$$
$$1 - (1.125 \times 0.75 + 2.25)y_3 + y_2 = -(1.125 \times 0.75 + 0.25)e^{0.75}$$

Their simultaneous solution yields $y_1 = 0.7752$, $y_2 = 1.2801$, and $y_3 = 1.4854$. More accurate results can be obtained with smaller values of h.

11-11. Solution of Characteristic-value Problems. (See also Art. 9-3.) Given a differential equation which involves a parameter and which must satisfy conditions at more than one point, express the derivatives in terms of central differences (Art. 11-5). If the resulting equations will have a nontrivial solution only for specific values of the parameter (characteristic values or eigenvalues), set the determinant of the coefficients in the equations equal to zero and solve for the parameter.

• If $y = y' = 0$ when $x = 0$, and $y = y'' = 0$ when $x = 1$, solve

$$\frac{d^4y}{dx^4} + K\frac{d^2y}{dx^2} = 0$$

where K is a constant.

Divide the interval $0 \le x \le 1$ into $n = 4$ parts ($h = 0.25$). Substitute $(1/h^4)\delta^4y_m$ for $y^{(4)}$ and $(1/h^2)\delta^2y_m$ for y''.

$$\frac{1}{h^4}\delta^4y_m + \frac{K}{h^2}\delta^2y_m = 0$$

Multiply by h^4, let $k = h^2K$, and expand.

$$y_{m-2} - 4y_{m-1} + 6y_m - 4y_{m+1} + y_{m+2} + k(y_{m-1} - 2y_m + y_{m+1}) = 0$$

Combine like terms.

$$y_{m-2} + (k - 4)y_{m-1} - (2k - 6)y_m + (k - 4)y_{m+1} + y_{m+2} = 0$$

The given conditions require that $y_0 = 0$, $y_1 - y_{-1} = 0$, $y_4 = 0$, $y_5 - 0 + y_3 = 0$. Then set $m = 1$, 2, and 3 to obtain the three equations

$$y_1 + 0 - (2k - 6)y_1 + (k - 4)y_2 + y_3 = 0$$
$$0 + (k - 4)y_1 - (2k - 6)y_2 + (k - 4)y_3 + 0 = 0$$
$$y_1 + (k - 4)y_2 - (2k - 6)y_3 + 0 - y_3 = 0$$

If these equations are not to have the trivial solution $y_m = 0$, the determinant of the coefficients must be zero.

$$\begin{vmatrix} -(2k - 7) & (k - 4) & 1 \\ (k - 4) & -(2k - 6) & (k - 4) \\ 1 & (k - 4) & -(2k - 5) \end{vmatrix} = 0$$

or $-4k^3 + 30k^2 - 68k + 44 = 0$

The smallest root of this equation is $k = 1.111$.   Hence, $K = k/h^2 = 17.776$.

11-12. Solution of Partial Differential Equations.   To integrate a partial differential equation numerically, express the derivatives in terms of central differences (Art. 11-5) and solve as in Art. 11-10. The derivatives with respect to x and y may be approximated by

$$D_x z_m = \frac{1}{2h} (z_r - z_l) \qquad\qquad D_y z_m = \frac{1}{2k} (z_u - z_d)$$

$$D_x^2 z_m = \frac{1}{h^2} (z_r - 2z_m + z_l) \qquad\qquad D_y^2 z_m = \frac{1}{k^2} (z_u - 2z_m + z_d)$$

$$D_x^3 z_m = \frac{1}{2h^3} (z_{rr} - 2z_r + 2z_l - z_{ll}) \qquad D_y^3 z_m = \frac{1}{2k^3} (z_{uu} - 2z_u$$
$$+ 2z_d - z_{dd})$$

$$D_x^4 z_m = \frac{1}{h^4} (z_{rr} - 4z_r + 6z_m - 4z_l + z_{ll}) \quad D_y^4 z_m = \frac{1}{k^4} (z_{uu} - 4z_u$$
$$+ 6z_m - 4z_d + z_{dd})$$

where h = length of x subinterval
         k = length of y subinterval
         z_r = value of z at $(x_m + h, y_m)$
         z_{rr} = value of z at $(x_m + 2h, y_m)$
         z_l = value of z at $(x_m - h, y_m)$
         z_{ll} = value of z at $(x - 2h, y_m)$
         z_u = value of z at $(x_m, y_m + k)$
         z_{uu} = value of z at $(x_m, y_m + 2k)$
         z_d = value of z at $(x_m, y_m - k)$
         z_{dd} = value of z at $(x_m, y_m - 2k)$

$D_{xy} z_m$ can be obtained by operating on D_y with D_x. When $h = k$,

$$D_{xy} z_m = \frac{1}{4h^2} [(z_r - z_l)_u - (z_r - z_l)_d] = \frac{1}{4h^2} (z_{ur} - z_{ul} - z_{dr} + z_{dl})$$

where z_{ur} = value of z at $(x_m + h, y_m + h)$
z_{ul} = value of z at $(x_m - h, y_m + h)$
z_{dr} = value of z at $(x_m + h, y_m - h)$
z_{dl} = value of z at $(x_m - h, y_m - h)$

The Laplace equation $\nabla^2 z = D_x^2 z + D_y^2 z = 0$ becomes, with $h = k$,

$$z_m = \tfrac{1}{4}(z_r + z_l + z_u + z_d)$$

Given boundary conditions, the equation can be solved by iteration. Assume values of z at the start, then correct them one at a time by averaging values of z at a distance of h in the x and y directions.

BIBLIOGRAPHY

Agnew, R. P.: "Differential Equations," 2d ed., McGraw-Hill Book Company, Inc., New York, 1960.

Hildebrand, F. B.: "Methods of Applied Mathematics," Prentice-Hall, Inc., Englewood Cliffs, N.J., 1952.

Milne, W. E.: "Numerical Solution of Differential Equations," John Wiley & Sons, Inc., New York, 1953.

Salvadori, M. G., and M. L. Baron: "Numerical Methods in Engineering," 2d ed., Prentice-Hall, Inc., Englewood Cliffs, N.J., 1961.

Sokolnikoff, I. S., and R. M. Redheffer: "Mathematics of Physics and Modern Engineering," McGraw-Hill Book Company, Inc., New York, 1958.

Wylie, C. R., Jr: "Advanced Engineering Mathematics," 2d ed., McGraw-Hill Book Company, Inc., New York, 1960.

Chapter 12

SERIES

12-1. Infinite Series. A succession of terms $a_1, a_2, a_3, \ldots, a_n, \ldots,$ formed according to a fixed law, is called a sequence. A series is the sum of the terms of a sequence. The term a_n, expressing the law of formation of the series, is called the general term.

An infinite series is the sum of the terms of a sequence with an unlimited number of terms.

A convergent series is an infinite series that approaches a limit as the number of terms approaches infinity. A necessary condition for convergence is that the general term of the series a_n approach zero as n approaches infinity (see also Arts. 12-7 to 12-11).

A series is said to be absolutely convergent if the series of absolute values of its terms is convergent. Other convergent series are said to be conditionally convergent.

$1 - (1/2!) + (1/3!) - (1/4!) + \cdots$ is absolutely convergent since $1 + (1/2!) + (1/3!) + (1/4!) + \cdots$ is convergent.

$1 - \frac{1}{2} + \frac{1}{3} - \frac{1}{4} + \cdots$ is conditionally convergent since $1 + \frac{1}{2} + \frac{1}{3} + \frac{1}{4} + \cdots$ is divergent.

A divergent series is one that increases without bound as the number of terms increases.

An oscillating series is one that neither converges nor increases without bound as the number of terms increases.

12-2. Arithmetic Progression. A sequence in which each term after the first is equal to the sum of the preceding term and a constant. For example, 2, 4, 6, 8, ... and 12, 8, 4, 0, -4, ... are arithmetic progressions.

Each term is the arithmetic mean between the preceding and following terms and is equal to half their sum.

If the first term of the arithmetic progression is a and the nth term is l, then

$$l = a + (n - 1)d$$

where d is the constant difference between successive terms.

• In the series 12, 16, 20, . . . , what is the fifth term?
Here, $a = 12$, $d = 4$, and $n = 5$. Hence, the fifth term is

$$l = 12 + (5 - 1)4 = 28$$

The sum of the first n terms of an arithmetic progression is given by

$$s = \frac{n}{2}(a + l) = \frac{n}{2}[2a + (n - 1)d]$$

• What is the sum of the first five terms of the sequence 12, 16, 20, . . . ?
As in the previous example, $a = 12$, $d = 4$, $n = 5$, and $l = 28$, so

$$s = \tfrac{5}{2}(12 + 28) = 100$$

12-3. Geometric Progression. A sequence in which each term after the first is equal to the product of the preceding term and a constant. For example, 2, 4, 8, 16, 32, . . . and 81, 27, 9, 3, 1, $\frac{1}{3}$, . . . are geometric progressions. They can be written in terms of the first term and the constant multiplier in the form a, ar, ar^2, ar^3, . . . , ar^n, ...

Each term is the geometric mean between the preceding and following terms.

If the first term is a and the constant multiplier is r, the nth term is

$$l = ar^{n-1}$$

• What is the sixth term of the sequence 3, 6, 12, . . . ?
Here, $a = 3$, $r = 2$, and $n = 6$. Hence, the sixth term is

$$l = 3(2)^{6-1} = 3 \times 32 = 96$$

The sum of the first n terms of a geometric progression is

$$s = \frac{ar^n - a}{r - 1} \qquad r > 1$$

$$s = \frac{a - ar^n}{1 - r} \qquad r < 1$$

• What is the sum of the first six terms of 3, 6, 12, . . . ?
Here, $a = 3$, $r = 2$, and $n = 6$. Hence,

$$s = \frac{3(2)^6 - 3}{2 - 1} = 189$$

• What is the sum of the first five terms of 81, 27, 9, . . . ?
In this case, $a = 81$, $r = \frac{1}{3}$, and $n = 5$, so

$$s = \frac{81 - 81(\frac{1}{3})^5}{1 - \frac{1}{3}} = 121$$

When $|r| < 1$, the sum of the geometric progression approaches as a limit

$$s = \frac{a}{1 - r}$$

• What is the sum of the progression 1, $\frac{1}{2}$, $\frac{1}{4}$, $\frac{1}{8}$, . . . ?
Since $a = 1$ and $r = \frac{1}{2} < 1$,

$$s = \frac{1}{1 - \frac{1}{2}} = 2$$

The last equation also can be used to convert a repeating decimal to a fraction.

$$0.6370370370 \cdots = 0.6 + \frac{370}{10,000} + \frac{370}{10,000,000} + \cdots$$

$$= \frac{6}{10} + \frac{370/10,000}{1 - 1/1,000} = \frac{6}{10} + \frac{370}{9,990} = \frac{6}{10} + \frac{1}{27} = \frac{86}{135}$$

If, in a geometric progression, $r = 1$ or $|r| > 1$, the series diverges. If $r = -1$, the series oscillates.

12-4. Integer Series. If the variable x represents only successive integers, then an integer series has the form

$$\sum_1^n f(x) = f(1) + f(2) + f(3) + \cdots + f(n)$$

If the lower limit is omitted, 1 is understood; if the upper limit is omitted, n is understood. For example,

$$\Sigma(1 + a)^y = (1 + a) + (1 + a)^2 + (1 + a)^3 + \cdots + (1 + a)^n$$

Integer series satisfy the following relationships:

$$\Sigma[f(x) + g(x)] = \Sigma f(x) + \Sigma g(x)$$
$$\Sigma mf(x) = m\Sigma f(x)$$

Thus, $\Sigma(3x^4 + 4x^3) = \Sigma 3x^4 + \Sigma 4x^3 = 3\Sigma x^4 + 4\Sigma x^3$.

If the term after the summation sign is a constant, all terms of the sum are equal. $\Sigma 5 = 5 + 5 + 5 + \cdots = 5n$.

A useful formula for evaluating certain integer series is

$$\Sigma[f(x) - f(x - 1)] = f(n) - f(0)$$

• Find the sum of the first n terms of the sequence 1, 2, 3, 4, ...
Let $f(x) = x^2$. Then

$$\Sigma[x^2 - (x - 1)^2] = n^2 - 0$$

Evaluation of the left-hand side of this equation yields

$$\Sigma(x^2 - x^2 + 2x - 1) = \Sigma(2x - 1) = 2\Sigma x - n = n^2$$

Finally, solution for Σx gives

$$\sum x = 1 + 2 + 3 + \cdots + n = \frac{n^2 + n}{2} = \frac{n(n + 1)}{2}$$

If $f(x) = x^3$, the formula yields

$$\sum x^2 = \frac{n(n+1)(2n+1)}{6}$$

Similarly,

$$\sum x^3 = \frac{n^2(n+1)^2}{4}$$

If $f(x)$ is a rational integer function, $\Sigma f(x)$ can be represented by

$$\Sigma f(x) = \Sigma a_n + a_{n-1}\Sigma x + a_{n-2}\Sigma x^2 + \cdots$$
$$= na_n + \frac{n(n+1)}{2} a_{n-1} + \frac{n(n+1)(2n+1)}{6} a_{n-2} + \cdots$$

• Evaluate $1 \cdot 2 \cdot 3 + 2 \cdot 3 \cdot 4 + 3 \cdot 4 \cdot 5 + \cdots + n(n+1)(n+2)$.
Let $f(x) = x(x+1)(x+2)$. Then

$$\Sigma x(x+1)(x+2) = \Sigma(x^3 + 3x^2 + 2x) = \Sigma x^3 + 3\Sigma x^2 + 2\Sigma x$$
$$= \frac{n^2(n+1)^2}{4} + \frac{3n(n+1)(2n+1)}{6} + \frac{2n(n+1)}{2}$$
$$= \frac{1}{4}n(n+1)(n+2)(n+3)$$

12-5. Fractional Series. The relationship given in Art. 12-4

$$\sum_{1}^{n} [f(x) - f(x-1)] = f(n) - f(0)$$

also holds for fractions. But expressions with x as a factor in the denominator should be avoided, since $f(x-1)$ has no meaning when $x = 1$.

• Evaluate $1/1 \cdot 2 + 1/2 \cdot 3 + 1/3 \cdot 4 + \cdots + 1/[n(n+1)]$.
Let $f(x) = 1/(x+1)$. Then

$$\sum \left(\frac{1}{x+1} - \frac{1}{x}\right) = \frac{1}{n+1} - 1 = -\frac{n}{n+1}$$

Evaluation of the first term on the left of this equation yields, after multiplication by -1,

$$\sum \frac{1}{x(x+1)} = \frac{1}{1 \cdot 2} + \frac{1}{2 \cdot 3} + \cdots + \frac{1}{n(n+1)} = \frac{n}{n+1}$$

The series converges to 1 when n approaches infinity.

12-6. Harmonic Series. The infinite series called harmonic,

$$1 + \frac{1}{2} + \frac{1}{3} + \cdots + \frac{1}{n} + \cdots$$

is divergent. Divergence is slow, however; the sum of one million terms is less than 15.

12-7. Integral Test for Convergence. Let a be a positive integer and let $f(x)$ be a positive nonincreasing function in the interval $a \le x < \infty$.

Thus, $f(x) \to 0$ as $x \to +\infty$. Then the series $\displaystyle\sum_{n=1}^{\infty} f(n)$ converges if $\displaystyle\int_{a}^{\infty} f(x)\, dx$ converges.

- Test for convergence $1 + 1/2^k + 1/3^k + \cdots + 1/n^k + \cdots$. Take $a = 1$ and $f(x) = 1/x^k$. When $k \neq 1$,

$$\int_{1}^{\infty} f(x)\, dx = \int_{1}^{\infty} \frac{dx}{x^k} = \frac{1}{(1-k)x^{k-1}}\Big]_{1}^{\infty}$$

If $k > 1$, the last term converges to $1/(k-1)$; hence, the given series converges. If $k < 1$, the last term increases without bound, so the given series diverges. When $k = 1$ (the series is harmonic),

$$\int_{1}^{\infty} f(x)\, dx = \int_{1}^{\infty} \frac{dx}{x} = \log_e x\Big]_{1}^{\infty} = \infty$$

and the series diverges.

12-8. Comparison Test for Convergence. If $u_1 + u_2 + u_3 + \cdots$ is a series of positive terms to be tested and $v_1 + v_2 + v_3 + \cdots$ is a series known to be convergent, then if each term of the first series is less than the corresponding term of the second series, the first series is convergent and its limiting value cannot be greater than that of the second.

Here are some useful series for testing convergence:

$$a + ar + ar^2 + ar^3 + \cdots + ar^n + \cdots \qquad \text{for } -1 < r < 1$$

$$\frac{1}{1\cdot 2} + \frac{1}{2\cdot 3} + \frac{1}{3\cdot 4} + \cdots + \frac{1}{n(n+1)} + \cdots$$

$$1 + \frac{1}{2^p} + \frac{1}{3^p} + \cdots + \frac{1}{n^p} + \cdots \qquad \text{for } p > 1$$

- Is the series $1/2! + 1/3! + 1/4! + \cdots + 1/n! + \cdots$ convergent?

Compare the series with $1/(1\cdot 2) + 1/(2\cdot 3) + 1/(3\cdot 4) + \cdots + 1/[n(n+1)] + \cdots$, which is known to be convergent. The first two terms of both series are equal. The third term of the given series, however, is less than the third term of the test series. Now compare the nth terms for $n > 2$.

$$\frac{1}{n!} = \frac{1}{n(n-1)(n-2)\cdots 1} < \frac{1}{n(n+1)}$$

Since the nth term of the test series is larger than the nth term of the given series when $n > 2$, the given series is convergent.

12-9. Ratio Test for Convergence. Compute the ratio of the $(n+1)$th term of the series to the nth term. Allow n to approach infinity. If the absolute value of the limit is S and

$S < 1$ the series is absolutely convergent
$S > 1$ the series is divergent
$S = 1$ the series may be either convergent or divergent

• Test $2^2/2^2 + 3^2/2^3 + 4^2/2^4 + \cdots + (n+1)^2/2^{n+1} + \cdots$ for convergence.
The $(n+1)$th term of this series is $(n+2)^2/2^{n+2}$. The ratio of this term to the nth term is

$$\frac{(n+2)^2/2^{n+2}}{(n+1)^2/2^{n+1}} = \frac{1}{2}\left(\frac{n+2}{n+1}\right)^2 = \frac{1}{2}\left(\frac{1+2/n}{1+1/n}\right)^2$$

As n approaches infinity, this ratio approaches $\frac{1}{2}$, so the series is convergent.

• Is the harmonic series $1 + \frac{1}{2} + \frac{1}{3} + \frac{1}{4} + \cdots + 1/n + \cdots$ convergent?
The $(n+1)$th term is $1/(n+1)$. The ratio of this term to the nth term is

$$\frac{1/(n+1)}{1/n} = \frac{n}{n+1} = \frac{1}{1+1/n}$$

The ratio is less than 1, but this condition, while necessary, is not sufficient for convergence. When n approaches infinity, the ratio approaches 1, and this test fails. Actually, this series is divergent (Art. 12-6).

12-10. Convergence of Alternating Series. Consider the alternating series made up of the positive terms $u_1, u_2, u_3, \ldots$

$$S = u_1 - u_2 + u_3 - u_4 + \cdots (-1)^{n-1}u_n + \cdots$$

The series is convergent if the absolute value of each term is less than that of its predecessor and the limit of the general term as n approaches infinity is zero.

• Test for convergence $1 - \frac{1}{2} + \frac{1}{3} - \frac{1}{4} + \cdots (-1)^{n-1}(1/n) + \cdots$.
The absolute value of each term is less than that of its predecessor, and

$$\lim \frac{1}{n}\bigg]_{n \to \infty} = 0$$

Hence, the series is convergent. Yet, $1 + \frac{1}{2} + \frac{1}{3} + \cdots + 1/n + \cdots$ is not convergent (Art. 12-6).

12-11. Convergence of Power Series. To determine the range of x in which the series

$$a_0 + a_1x + a_2x^2 + a_3x^3 + \cdots + a_nx^n + \cdots$$

converges, determine the ratio a_{n+1}/a_n. Allow n to approach infinity. If this ratio has a limit r, and

$$|x| < \left|\frac{1}{r}\right| \qquad \text{the series converges}$$

$$|x| > \left|\frac{1}{r}\right| \qquad \text{the series diverges}$$

$$|x| = \left|\frac{1}{r}\right| \qquad \text{the test fails}$$

If $r = 0$, the series converges for all values of x.

• For what values of x is the following series convergent:

$$S = \frac{1}{2}x - \frac{1}{2^2 2!}x^2 + \frac{1 \cdot 3}{2^3 3!}x^3 - \frac{1 \cdot 3 \cdot 5}{2^4 4!}x^4 + \cdots$$

Analysis of the sequence indicates that

$$a_n = (-1)^{n-1}\frac{(1)(3)(5) \cdots [2(n-1)-1]}{2^n n!}$$

$$a_{n+1} = (-1)^n \frac{(1)(3)(5) \cdots (2n-1)}{2^{n+1}(n+1)!}$$

Hence, the ratio of a_{n+1} to a_n is

$$\frac{a_{n+1}}{a_n} = -\frac{1}{2}\frac{1}{n+1}\frac{2n-1}{1} = -\frac{1}{2}\left(\frac{2-1/n}{1+1/n}\right)$$

The limit of this ratio as n approaches infinity is -1, so the series converges for $-1 < x < 1$.

12-12. Binomial Series. (See also Arts. 3-37 and 3-38.) If n is a fraction or a negative number in the binomial expansion

$$(a + x)^n = a^n + na^{n-1}x + \frac{n(n-1)}{2!}a^{n-2}x^2$$

$$+ \frac{n(n-1)(n-2)}{3!}a^{n-3}x^3 + \cdots$$

an infinite (power) series results. It is convergent when x is numerically less than a. (If $x > a$, expand $x + a$ to get a convergent series.)

It is usually convenient to write $(a + x)^n$ as $a^n(1 + x/a)^n$ and to expand $(1 + x/a)^n$.

• Write $\sqrt{1+x}$ as a power series. Compute its value when $x = 0.44$.

$$(1 + x)^{1/2} = 1 + \frac{1}{2}x - \frac{1}{2^2 2!}x^2 + \frac{1 \cdot 3}{2^3 3!}x^3 - \frac{1 \cdot 3 \cdot 5}{2^4 4!}x^4 + \cdots$$

$$(1.44)^{1/2} = 1 + 0.22 - 0.0242 + 0.005324 - 0.001464 \approx 1.19966$$

12-13. Trigonometric Series. For x in radians ($1° = 0.0174533$ radian)

$$\sin x = x - \frac{x^3}{3!} + \frac{x^5}{5!} - \frac{x^7}{7!} + \cdots$$

$$\cos x = 1 - \frac{x^2}{2!} + \frac{x^4}{4!} - \frac{x^6}{6!} + \cdots$$

$$\tan x = x + \frac{x^3}{3} + \frac{2x^5}{15} + \frac{17x^7}{315} + \cdots \qquad -\frac{\pi}{2} < x < \frac{\pi}{2}$$

$$\cot x = \frac{1}{x} - \frac{x}{3} - \frac{x^3}{45} - \frac{2x^5}{945} - \cdots \qquad -\pi < x < \pi$$

$$\sec x = 1 + \frac{x^2}{2} + \frac{5x^4}{24} + \frac{61x^6}{720} + \cdots \qquad -\frac{\pi}{2} < x < \frac{\pi}{2}$$

$$\csc x = \frac{1}{x} + \frac{x}{6} + \frac{7x^3}{360} + \cdots \qquad\qquad -\pi < x < \pi$$

$$\arcsin x = x + \frac{x^3}{6} + \frac{3x^5}{40} + \frac{15x^7}{336} + \cdots \qquad -1 < x < 1$$

$$\arccos x = \frac{\pi}{2} - x - \frac{x^3}{6} - \frac{3x^5}{40} - \frac{15x^7}{336} - \cdots \qquad -1 < x < 1$$

$$\arctan x = x - \frac{x^3}{3} + \frac{x^5}{5} - \frac{x^7}{7} + \cdots \qquad\qquad -1 < x < 1$$

12-14. Hyperbolic Series

$$\sinh x = x + \frac{x^3}{3!} + \frac{x^5}{5!} + \frac{x^7}{7!} + \cdots$$

$$\cosh x = 1 + \frac{x^2}{2!} + \frac{x^4}{4!} + \frac{x^6}{6!} + \cdots$$

$$\tanh x = x - \frac{x^3}{3} + \frac{2x^5}{15} - \frac{17x^7}{315} + \cdots \qquad -\frac{\pi}{2} < x < \frac{\pi}{2}$$

$$\coth x = \frac{1}{x} + \frac{x}{3} - \frac{x^3}{45} + \frac{2x^5}{945} - \frac{x^7}{4{,}725} + \cdots \qquad -\pi < x < \pi$$

$$\operatorname{sech} x = 1 - \frac{x^2}{2!} + \frac{5x^4}{4!} - \frac{61x^6}{6!} + \frac{1{,}385x^8}{8!} - \cdots \qquad -\frac{\pi}{2} < x < \frac{\pi}{2}$$

12-15. Logarithmic Series

$$\log_e x = 2\left[\frac{x-1}{x+1} + \frac{1}{3}\left(\frac{x-1}{x+1}\right)^3 \right.$$
$$\left. + \frac{1}{5}\left(\frac{x-1}{x+1}\right)^5 + \cdots \right] \qquad x > 0$$

$$\log_e(1+x) = x - \frac{x^2}{2} + \frac{x^3}{3} - \frac{x^4}{4} + \cdots \qquad -1 < x \leq 1$$

$$\log_e(1-x) = -x - \frac{x^2}{2} - \frac{x^3}{3} - \frac{x^4}{4} - \cdots \qquad -1 \leq x < 1$$

12-16. Exponential Series

$$e = 2 + \frac{1}{2!} + \frac{1}{3!} + \frac{1}{4!} + \cdots \approx 2.71828 \qquad \text{(Art. 1-7)}$$

$$e^x = 1 + x + \frac{x^2}{2!} + \frac{x^3}{3!} + \frac{x^4}{4!} + \cdots$$

$$a^x = 1 + x \log_e a + \frac{x^2}{2!} \log_e^2 a + \frac{x^3}{3!} \log_e^3 a + \cdots$$

12-17. Maclaurin's Series.

If $f(x)$ possesses derivatives of all orders in an interval including $x = 0$, it can be represented in that interval by the power series

$$f(x) = f(0) + f'(0)x + \frac{f''(0)}{2!} x^2 + \frac{f'''(0)}{3!} x^3 + \cdots + \frac{f^{(n)}(0)}{n!} x^n + \cdots$$

for those values of x for which

$$\lim \frac{f^{(n)}(\epsilon)}{n!} x^n \bigg]_{n \to \infty} = 0 \qquad 0 < \epsilon < x$$

In this series, $f(0)$, $f'(0)$, $f''(0)$, $f'''(0)$, $\ldots$, $f^{(n)}(0)$ represent, respectively, the value of the function, the first derivative, second derivative, third derivative, $\ldots$, nth derivative at $x = 0$.

• Expand e^x by a Maclaurin series.

$$f(x) = e^x \qquad f(0) = 1$$
$$f^{(n)}(x) = e^x \qquad f^{(n)}(0) = 1$$

Hence $\qquad e^x = 1 + x + \dfrac{x^2}{2!} + \dfrac{x^3}{3!} + \cdots + \dfrac{x^n}{n!} + \cdots$

12-18. Taylor's Series. If $f(x)$ possesses derivatives of all orders in an interval including $x = a$, it can be represented in that interval by the power series

$$f(x) = f(a) + f'(a)(x - a) + \frac{f''(a)}{2!}(x - a)^2$$
$$+ \cdots + \frac{f^{(n)}(a)}{n!}(x - a)^n + \cdots$$

for those values of x for which

$$\lim \frac{f^{(n)}(\epsilon)}{n!}(x - a)^n \bigg]_{n \to \infty} = 0 \qquad a < \epsilon < x$$

In this series, $f(a)$, $f'(a)$, $f''(a)$, $\ldots$, $f^{(n)}(a)$ represent, respectively, the value of the function, its first derivative, second derivative, $\ldots$ nth derivative at $x = a$. [Maclaurin's series (Art. 12-16) is the special case of Taylor's series for which $a = 0$.]

Taylor's series may also be written in the form

$$f(a + h) = f(a) + f'(a)h + \frac{f''(a)}{2!}h^2 + \cdots + \frac{f^{(n)}(a)}{n!}h^n + \cdots$$

obtained by setting $x - a = h$.

• Expand $\log_e x$ about $x = 1$.
Let $h = x - 1$ and $\log_e x = \log_e(1 + h)$. Here, $a = 1$.

$$f(x) = \log_e x \qquad f(1) = \log_e 1 = 0$$
$$f'(x) = \frac{1}{x} \qquad f'(1) = 1$$
$$f''(x) = -x^{-2} \qquad f''(1) = -1$$
$$f'''(x) = 2x^{-3} \qquad f'''(1) = 2!$$
$$f^{(4)}(x) = -3!x^{-4} \qquad f^{(4)}(1) = -3!$$

$$\log_e(1 + h) = 0 + h - \frac{1}{2!}h^2 + \frac{2!}{3!}h^3 - \frac{3!}{4!}h^4 + \cdots$$
$$= h - \tfrac{1}{2}h^2 + \tfrac{1}{3}h^3 - \tfrac{1}{4}h^4 + \cdots$$

$$\log_e x = (x - 1) - \tfrac{1}{2}(x - 1)^2 + \tfrac{1}{3}(x - 1)^3 - \tfrac{1}{4}(x - 1)^4 + \cdots \qquad 0 < x < 2$$

12-19. Fourier Series. Any single-valued function $f(x)$ that is continuous, except possibly for a finite number of finite discontinuities in an interval of length 2π, and that has a finite number of maxima and minima in this interval may be represented by a convergent Fourier series.

$$f(x) = \tfrac{1}{2}a_0 + a_1 \cos x + a_2 \cos 2x + \cdots + a_n \cos nx + \cdots$$
$$+ b_1 \sin x + b_2 \sin 2x + \cdots + b_n \sin nx + \cdots$$

$$= \tfrac{1}{2}a_0 + \sum_{n=1}^{\infty} (a_n \cos nx + b_n \sin nx)$$

where
$$a_n = \frac{1}{\pi} \int_c^{c+2\pi} f(x) \cos nx \, dx$$

$$b_n = \frac{1}{\pi} \int_c^{c+2\pi} f(x) \sin nx \, dx$$

• Express as a Fourier series $f(x)$ defined by:

$$f(x) = 1 \qquad 0 < x < \pi$$
$$f(x) = 2 \qquad \pi < x < 2\pi$$

[$f(x)$ is not defined at the midpoint of the interval. The Fourier series will yield $f(\pi) = \tfrac{3}{2}$. Also, $f(0) = f(2\pi) = \tfrac{3}{2}$ because of the periodicity of the series.]
The coefficients of the trigonometric functions in the series are:

$$a_0 = \frac{1}{\pi} \int_0^{2\pi} f(x) \, dx = \frac{1}{\pi} \int_0^{\pi} 1 \, dx + \frac{1}{\pi} \int_\pi^{2\pi} 2 \, dx = 1 + 2 = 3$$

$$a_n = \frac{1}{\pi} \int_0^{\pi} 1 \cos nx \, dx + \frac{1}{\pi} \int_\pi^{2\pi} 2 \cos nx \, dx = 0$$

$$b_n = \frac{1}{\pi} \int_0^{\pi} 1 \sin nx \, dx + \frac{1}{\pi} \int_\pi^{2\pi} 2 \sin nx \, dx = \frac{1}{n\pi} (\cos n\pi - 1)$$

Hence $\qquad f(x) = \tfrac{3}{2} - \dfrac{2}{\pi} (\sin x + \tfrac{1}{3} \sin 3x + \tfrac{1}{5} \sin 5x + \cdots).$

When an even function $[f(-x) \equiv f(x)]$ is expanded in a Fourier series in the interval $-\pi < x < \pi$, the series will not contain sine terms, and

$$a_n = \frac{2}{\pi} \int_0^{\pi} f(x) \cos nx \, dx$$

When an odd function $[f(-x) \equiv -f(x)]$ is expanded in this interval, the series will not contain cosine terms, and

$$b_n = \frac{2}{\pi} \int_0^{\pi} f(x) \sin nx \, dx$$

The Fourier series for an interval $-L < x < L$ is given by

$$f(x) = \tfrac{1}{2}a_0 + a_1 \cos \frac{\pi x}{L} + a_2 \cos \frac{2\pi x}{L} + \cdots + a_n \cos \frac{n\pi x}{L} + \cdots$$
$$+ b_1 \sin \frac{\pi x}{L} + b_2 \sin \frac{2\pi x}{L} + \cdots + b_n \sin \frac{n\pi x}{L} + \cdots$$

where
$$a_n = \frac{1}{L} \int_{-L}^{L} f(x) \cos \frac{n\pi x}{L} \, dx$$

$$b_n = \frac{1}{L} \int_{-L}^{L} f(x) \sin \frac{n\pi x}{L} \, dx$$

For the half interval $0 < x < L$, the half-range sine series is

$$f(x) = b_1 \sin \frac{\pi x}{L} + b_2 \sin \frac{2\pi x}{L} + \cdots + b_n \sin \frac{n\pi x}{L} + \cdots$$

where
$$b_n = \frac{2}{L} \int_{0}^{L} f(x) \sin \frac{n\pi x}{L} \, dx$$

The half-range cosine series is

$$f(x) = \tfrac{1}{2}a_0 + a_1 \cos \frac{\pi x}{L} + a_2 \cos \frac{2\pi x}{L} + \cdots + a_n \cos \frac{n\pi x}{L} + \cdots$$

where
$$a_n = \frac{2}{L} \int_{0}^{L} f(x) \cos \frac{n\pi x}{L} \, dx$$

BIBLIOGRAPHY

Agnew, R. P.: "Differential Equations," 2d ed., McGraw-Hill Book Company, Inc., New York, 1960.

Bacon, H. M.: "Differential and Integral Calculus," 2d ed., McGraw-Hill Book Company, Inc., New York, 1955.

Burington, R. S.: "Handbook of Mathematical Tables and Formulas," 3d ed., McGraw-Hill Book Company, Inc., New York, 1948.

Dwight, H. B.: "Tables of Integrals and Other Mathematical Data," The Macmillan Company, New York, 1947.

Peirce, B. O., and R. M. Foster: "A Short Table of Integrals," Ginn & Company, Boston, 1956.

Reddick, H. W., and F. H. Miller: "Advanced Mathematics for Engineers," 3d ed., John Wiley & Sons, Inc., New York, 1955.

Sokolnikoff, I. S., and R. M. Redheffer: "Mathematics of Physics and Modern Engineering," McGraw-Hill Book Company, Inc., New York, 1958.

Chapter 13

MATRICES AND DETERMINANTS

13-1. Basic Properties of Matrices. A matrix is a group of real or complex numbers in a rectangular array.   In the remainder of this chapter a pair of brackets enclosing an array will signify that the array can be operated on in accordance with the rules of matrix algebra that follow.

$$A = \begin{bmatrix} a_{11} & a_{12} & \cdots & a_{1n} \\ a_{21} & a_{22} & \cdots & a_{2n} \\ \cdots & \cdots & \cdots & \cdots \\ a_{m1} & a_{m2} & \cdots & a_{mn} \end{bmatrix}$$

The numbers a_{ij} are called the elements of the matrix. The subscript i denotes the row and subscript j denotes the column in which the element is located.

If a matrix contains m rows and n columns, it is called an $m \times n$ matrix, or is said to be of order m,n.

A square matrix with n rows and n columns is of order n. The elements a_{11}, a_{22}, . . . , a_{nn} comprise the main diagonal. The elements a_{1n}, a_{2n-1}, . . . a_{n1} form the secondary diagonal.

Two matrices, $A = [a_{ij}]$ and $B = [b_{ij}]$, are equal if they are of the same order and corresponding elements are equal: $a_{ij} = b_{ij}$ for all values of i and j.

To add two matrices of the same order, add corresponding elements.

$$\begin{bmatrix} 3x & 3 \\ 2 & k \end{bmatrix} + \begin{bmatrix} -x & -2 \\ 4 & 2k \end{bmatrix} = \begin{bmatrix} 2x & 1 \\ 6 & 3k \end{bmatrix}$$

The negative of a matrix is obtained by changing the sign of every element.   Hence, to subtract two matrices of the same order, subtract corresponding elements.

$$\begin{bmatrix} 3x & 3 \\ 2 & k \end{bmatrix} - \begin{bmatrix} -x & -2 \\ 4 & 2k \end{bmatrix} = \begin{bmatrix} 4x & 5 \\ -2 & -k \end{bmatrix}$$

If A, B, and C are matrices of the same order, $A + B = B + A$ and $A + (B + C) = (A + B) + C$.

To multiply a matrix by a scalar, multiply every element of the matrix by the scalar.

$$2\begin{bmatrix} 1 & 3 \\ x & k \end{bmatrix} = \begin{bmatrix} 2 & 6 \\ 2x & 2k \end{bmatrix}$$

The product of an $m \times p$ matrix A and a $p \times n$ matrix B is an $m \times n$ matrix AB. The element in the ith row and jth column of AB is found by multiplying corresponding elements of the ith row of A and the jth column of B, and then adding the products. Thus, if c_{ij} are the elements of AB,

$$c_{ij} = a_{i1}b_{1j} + a_{i2}b_{2j} + \cdots + a_{ip}b_{pj} = \sum_{k=1}^{p} a_{ik}b_{kj}$$

A matrix A is conformable for multiplication to a matrix B if the number of columns of A equals the number of rows of B. The product AB indicates that B is premultiplied by A and A is postmultiplied by B. In general, $AB \neq BA$.

• Postmultiply $A = \begin{bmatrix} 1 & 2 & 3 \\ -2 & 4 & -1 \end{bmatrix}$ by $B = \begin{bmatrix} 5 & -3 & 6 \\ 0 & -4 & -5 \\ -1 & 0 & -6 \end{bmatrix}$.

A is a 2×3 matrix, and B is a 3×3 matrix, so AB will be a 2×3 matrix.

$$AB = \begin{bmatrix} 1(5) + 2(0) + 3(-1) & 1(-3) + 2(-4) + 3(0) & 1(6) + 2(-5) + 3(-6) \\ -2(5) + 4(0) - 1(-1) & -2(-3) + 4(-4) - 1(0) & -2(6) + 4(-5) - 1(-6) \end{bmatrix}$$

$$= \begin{bmatrix} 2 & -11 & -22 \\ -9 & -10 & -26 \end{bmatrix}$$

If A, B, and C are matrices conformable for multiplication,

$$AB(C) = A(BC) \qquad A(B + C) = AB + AC \qquad (A + B)C = AC + BC$$

However, if $AB = 0$, it is not necessary that either A or B is zero (composed of elements all equal to zero). If $AB = AC$ or $BA = CA$, B is not necessarily equal to C.

13-2. Linear Equations in Matrix Notation. In a system of m linear equations in n unknowns

$$a_{11}x_1 + a_{12}x_2 + \cdots + a_{1n}x_n = b_1$$
$$a_{21}x_1 + a_{22}x_2 + \cdots + a_{2n}x_n = b_2$$
$$\cdots\cdots\cdots\cdots\cdots\cdots\cdots$$
$$a_{m1}x_1 + a_{m2}x_2 + \cdots + a_{mn}x_n = b_n$$

the left-hand side can be written as the product of two matrices AX, where

$$A = \begin{bmatrix} a_{11} & a_{12} & \cdots & a_{1n} \\ a_{21} & a_{22} & \cdots & a_{2n} \\ \cdots\cdots\cdots\cdots\cdots \\ a_{m1} & a_{m2} & \cdots & a_{mn} \end{bmatrix} \qquad X = \begin{bmatrix} x_1 \\ x_2 \\ \cdots \\ x_n \end{bmatrix}$$

Note that the elements of A are the coefficients of the unknowns and the elements of X are the unknowns.

If B is a column matrix whose elements are the constants on the right-hand side of the system of equations, then the system can be written in matrix notation as

$$AX = B$$

(See also Art. 13-15.)

13-3. Transpose of a Matrix. The matrix obtained by interchanging rows and columns of a given matrix.

• Find the transpose of $A = \begin{bmatrix} 1 & 2 & 3 \\ -2 & 4 & -1 \end{bmatrix}$.

The first row becomes the first column and the second row becomes the second column of the transpose.

$$A^T = \begin{bmatrix} 1 & -2 \\ 2 & 4 \\ 3 & -1 \end{bmatrix}$$

If A^T and B^T are the transposes of A and B, then

$$(A^T)^T = A \qquad (A + B)^T = A^T + B^T \qquad (kA)^T = kA^T$$

(where k is a scalar), and $(AB)^T = B^T A^T$.

If A is a square matrix and $A = A^T$, A is symmetric and $a_{ij} = a_{ji}$. If $A = -A^T$, A is skew-symmetric, $a_{ij} = -a_{ji}$, and the main diagonal consists of zeros.

13-4. Hermitian Matrices. If A is a matrix the elements of which are complex numbers, its conjugate $\bar{A}$ is a matrix whose corresponding elements are the conjugates of the elements of A. The transpose of $\bar{A}$, written $A^* = (\bar{A})^T$, is called the tranjugate or adjoint of A.

If $A = A^*$, A is a Hermitian matrix. Its diagonal is composed of real numbers.

• Show that $A = \begin{bmatrix} 3 & 1 - i \\ 1 + i & 2 \end{bmatrix}$ is Hermitian.

$$\bar{A} = \begin{bmatrix} 3 & 1 + i \\ 1 - i & 2 \end{bmatrix}$$

$$A^* = (\bar{A})^T = \begin{bmatrix} 3 & 1 - i \\ 1 + i & 2 \end{bmatrix} = A$$

13-5. Scalar Matrices. Matrices that behave in multiplication as scalars. Thus, they are square matrices with all elements zero except those in the main diagonal. Each element of the main diagonal is the equivalent scalar.

$$k\begin{bmatrix} a_{11} & a_{12} & \cdots & a_{1n} \\ a_{21} & a_{22} & \cdots & a_{2n} \\ \cdots & \cdots & \cdots & \cdots \\ a_{m1} & a_{m2} & \cdots & a_{mn} \end{bmatrix} = \begin{bmatrix} k & 0 & \cdots & 0 \\ 0 & k & \cdots & 0 \\ \cdots & \cdots & \cdots & \cdots \\ 0 & 0 & \cdots & k \end{bmatrix} \begin{bmatrix} a_{11} & a_{21} & \cdots & a_{1n} \\ a_{21} & a_{22} & \cdots & a_{2n} \\ \cdots & \cdots & \cdots & \cdots \\ a_{m1} & a_{m2} & \cdots & a_{mn} \end{bmatrix}$$

If the scalar k is unity, the scalar matrix is the identity or unit matrix I. Thus, a scalar matrix can be written kI, where

$$I = \begin{bmatrix} 1 & 0 & \cdots & 0 \\ 0 & 1 & \cdots & 0 \\ \cdots & \cdots & \cdots & \cdots \\ 0 & 0 & \cdots & 1 \end{bmatrix}$$

13-6. Inversion of Subscripts. In any term consisting of literal factors with numerical subscripts, the occurrence of a larger subscript to the left of a smaller one is called an inversion. Thus, $a_1b_3c_2$ contains one inversion, b_3c_2; $a_3b_1c_2$ contains two inversions, a_3b_1 and a_3c_2; $a_3b_2c_1$ contains three inversions, a_3b_2, a_3c_1, and b_2c_1.

13-7. Determinant. A square array of numbers. The value of a determinant is the sum of products each of which is formed by multiplying together one element from each row and each column; the sign of each product is positive if it contains an even number of inversions and negative if it contains an odd number of inversions (Art. 13-6). Vertical bars at the sides of the array are symbols indicating that it is a determinant.

$$\bullet \text{ Evaluate } D = \begin{vmatrix} a_{11} & a_{12} & a_{13} \\ a_{21} & a_{22} & a_{23} \\ a_{31} & a_{32} & a_{33} \end{vmatrix}$$

Follow the definition; select one element from each row and each column. Keep the second subscripts of the elements in the products in the order 1, 2, 3, and determine the inversions by the order of the first subscripts. The solution contains six terms.

$$D = a_{11}a_{22}a_{33} + a_{21}a_{32}a_{13} + a_{31}a_{12}a_{23} - a_{31}a_{22}a_{13} - a_{11}a_{32}a_{23} - a_{21}a_{12}a_{33}$$

The order of a determinant is equal to the number of rows or columns. Thus the determinant in the preceding example is of the third order.

The principal diagonal begins at the upper left-hand corner and ends at the lower right-hand corner; it consists of the elements a_{11}, a_{22}, $\ldots$, a_{nn}. The secondary diagonal begins at the upper right and extends to the lower left; it consists of the elements a_{1n}, $a_{2,n-1}$, $\ldots$, a_{n1}.

13-8. Minor of a Determinant. A determinant obtained from a given determinant when the row and column containing a specific element are eliminated.

In determinant D of Art. 13-7, the minor of element a_{22} is obtained by deleting the second row and the second column.

$$M_{22} = \begin{vmatrix} a_{11} & a_{13} \\ a_{31} & a_{33} \end{vmatrix}$$

13-9. Cofactors of a Determinant. The cofactor of any element a_{ij} equals its minor multiplied by $(-1)^{i+j}$ (see also Art. 13-8).

In determinant D of Art. 13-7, the cofactor of a_{31} is

$$A_{31} = (-1)^{3+1}M_{31} = + \begin{vmatrix} a_{12} & a_{13} \\ a_{22} & a_{23} \end{vmatrix}$$

13-10. General Properties of Determinants. The value of a determinant is not changed when the rows are transposed to columns or the columns to rows.

$$\begin{vmatrix} a_{11} & a_{12} \\ a_{21} & a_{22} \end{vmatrix} = \begin{vmatrix} a_{11} & a_{21} \\ a_{12} & a_{22} \end{vmatrix}$$

Any theorem that holds for the rows of a determinant is also true for its columns.

If any two columns (or rows) of a determinant are interchanged, its sign is changed.

$$\begin{vmatrix} a_{11} & a_{12} \\ a_{21} & a_{22} \end{vmatrix} = - \begin{vmatrix} a_{12} & a_{11} \\ a_{22} & a_{21} \end{vmatrix} = + \begin{vmatrix} a_{22} & a_{21} \\ a_{12} & a_{11} \end{vmatrix}$$

If all the elements of a row or column are zero, or if two columns (or rows) are identical, the determinant equals zero.

If the elements of any column (or row) are multiplied by the cofactors of the corresponding elements of any other column (or row), the sum of the products is zero.

If all the elements in any column (or row) are multiplied by any scalar, the determinant is multiplied by that scalar.

$$k\begin{vmatrix} a_{11} & a_{12} \\ a_{21} & a_{22} \end{vmatrix} = \begin{vmatrix} ka_{11} & a_{12} \\ ka_{21} & a_{22} \end{vmatrix} = \begin{vmatrix} a_{11} & ka_{12} \\ a_{21} & ka_{22} \end{vmatrix} = \begin{vmatrix} ka_{11} & ka_{12} \\ a_{21} & a_{22} \end{vmatrix}$$

A determinant is not altered in value if each element of a column (or row) is multiplied by the same number and added to (or subtracted from) another column (or row).

$$\begin{vmatrix} 6 & 3 \\ 4 & 5 \end{vmatrix} = \begin{vmatrix} 6 & 3 + 6 \times 2 \\ 4 & 5 + 4 \times 2 \end{vmatrix} = \begin{vmatrix} 6 & 15 \\ 4 & 13 \end{vmatrix} = 18$$

If each element of the jth column (or ith row) of a determinant A is expressed as the sum of p terms, determinant A may be expressed as the sum of the determinants of p matrices. In these matrices, the elements of the jth column (or ith row) are, respectively, the first, second, . . . , pth terms of the elements of the jth column (or ith row) of A. All other columns (or rows) are the same throughout.

$$\begin{vmatrix} 7 & -1 & 5 \\ 12 & 6 & 4 \\ 3 & -3 & 0 \end{vmatrix} = \begin{vmatrix} 3+3+1 & -1 & 5 \\ 6+5+1 & 6 & 4 \\ 2+1+0 & -3 & 0 \end{vmatrix}$$

$$= \begin{vmatrix} 3 & -1 & 5 \\ 6 & 6 & 4 \\ 2 & -3 & 0 \end{vmatrix} + \begin{vmatrix} 3 & -1 & 5 \\ 5 & 6 & 4 \\ 1 & -3 & 0 \end{vmatrix} + \begin{vmatrix} 1 & -1 & 5 \\ 1 & 6 & 4 \\ 0 & -3 & 0 \end{vmatrix}$$

13-11. Reduction of a Determinant to Lower Order. If all elements but one in a row (or column) are zero, the determinant is equal to the product of the nonzero element and its cofactor. Since the cofactor is a determinant of the next lower order, the order of a determinant is reduced by one when all but one element of any row (or column) is made zero.

To reduce an element to zero, multiply another column (or row) by a factor that makes a corresponding element in that line equal in absolute value to the given element. Then add the multiplied column (or row) or its negative to the column (or row) containing the given element.

• Evaluate by successive reduction

$$D = \begin{vmatrix} 5 & -13 & 2 \\ 1 & 5 & 1 \\ 2 & -12 & 7 \end{vmatrix}$$

Convert the second row to 1, 0, 0. To do this, multiply the first column by 5 and subtract the result from the second column, then subtract the first column from the third. The result is

$$D = \begin{vmatrix} 5 & -38 & -3 \\ 1 & 0 & 0 \\ 2 & -22 & 5 \end{vmatrix} = - \begin{vmatrix} -38 & -3 \\ -22 & 5 \end{vmatrix}$$

Now multiply the second column by -13 and add to the first column to get unity for the first element.

$$D = - \begin{vmatrix} 1 & -3 \\ -87 & 5 \end{vmatrix}$$

Finally, either multiply the top row by 87 and add to the bottom row to obtain a zero there, or multiply the first column by 3 and add to the second column to get a zero there. Both lead to $D = 256$.

Another, and usually more efficient, method of reducing a determinant to the next lower order is pivotal condensation. Each element of the lower-order determinant is computed from a second-order determinant containing elements of the given determinant.

$$
\begin{vmatrix} a_{11} & a_{12} & \cdots & a_{1n} \\ a_{21} & a_{22} & \cdots & a_{2n} \\ \cdots & \cdots & \cdots & \cdots \\ a_{n1} & a_{n2} & \cdots & a_{nn} \end{vmatrix} = \frac{1}{a_{11}{}^{n-2}} \begin{vmatrix} \begin{vmatrix} a_{11} & a_{12} \\ a_{21} & a_{22} \end{vmatrix} & \begin{vmatrix} a_{11} & a_{13} \\ a_{21} & a_{23} \end{vmatrix} & \cdots & \begin{vmatrix} a_{11} & a_{1n} \\ a_{21} & a_{2n} \end{vmatrix} \\ \begin{vmatrix} a_{11} & a_{12} \\ a_{31} & a_{32} \end{vmatrix} & \begin{vmatrix} a_{11} & a_{13} \\ a_{31} & a_{33} \end{vmatrix} & \cdots & \begin{vmatrix} a_{11} & a_{1n} \\ a_{31} & a_{3n} \end{vmatrix} \\ \cdots & \cdots & \cdots & \cdots \\ \begin{vmatrix} a_{11} & a_{12} \\ a_{n1} & a_{n2} \end{vmatrix} & \begin{vmatrix} a_{11} & a_{13} \\ a_{n1} & a_{n3} \end{vmatrix} & \cdots & \begin{vmatrix} a_{11} & a_{1n} \\ a_{n1} & a_{nn} \end{vmatrix} \end{vmatrix}
$$

• Solve determinant D in the previous example by pivotal condensation.

$$
D = \frac{1}{5^{3-2}} \begin{vmatrix} \begin{vmatrix} 5 & -13 \\ 1 & 5 \end{vmatrix} & \begin{vmatrix} 5 & 2 \\ 1 & 1 \end{vmatrix} \\ \begin{vmatrix} 5 & -13 \\ 2 & -12 \end{vmatrix} & \begin{vmatrix} 5 & 2 \\ 2 & 7 \end{vmatrix} \end{vmatrix} = \frac{1}{5} \begin{vmatrix} 38 & 3 \\ -34 & 31 \end{vmatrix} = \frac{1,280}{5} = 256
$$

13-12. Multiplication of Determinants. If a_{ij} is the element in the ith row and jth column of determinant A, and b_{ij} is the corresponding element of determinant B, both A and B having n rows and columns, then c_{ij}, the element in the ith row and jth column of the product AB, is

$$
c_{ij} = a_{i1}b_{1j} + a_{i2}b_{2j} + \cdots + a_{in}b_{nj} = \sum_{k}^{n} a_{ik}b_{kj}
$$

That is, c_{ij} is the sum of the products of the elements in the ith row of A and the elements in the jth column of B. The product AB is an nth-order determinant.

• Multiply $A = \begin{vmatrix} 3 & 3 & 5 \\ 3 & 1 & 1 \\ 1 & 4 & 1 \end{vmatrix}$ by $B = \begin{vmatrix} 4 & 3 & 1 \\ 3 & 1 & 2 \\ 3 & 1 & 5 \end{vmatrix}$.

$$
AB = \begin{vmatrix} 3(4) + 3(3) + 5(3) & 3(3) + 3(1) + 5(1) & 3(1) + 3(2) + 5(5) \\ 3(4) + 1(3) + 1(3) & 3(3) + 1(1) + 1(1) & 3(1) + 1(2) + 1(5) \\ 1(4) + 4(3) + 1(3) & 1(3) + 4(1) + 1(1) & 1(1) + 4(2) + 1(5) \end{vmatrix}
$$

$$
= \begin{vmatrix} 36 & 17 & 34 \\ 18 & 11 & 10 \\ 19 & 8 & 14 \end{vmatrix}
$$

13-13. Evaluation of a Determinant. A second-order determinant can be evaluated rapidly by subtracting the product of the elements in the secondary diagonal from the product of the elements in the principal diagonal.

$$
\begin{vmatrix} a_{11} & a_{12} \\ a_{21} & a_{22} \end{vmatrix} = a_{11}a_{22} - a_{21}a_{12}
$$

A third-order determinant can be evaluated as follows: Add (1) the products of the elements of the principal diagonal $a_{11}a_{22}a_{33}$, (2) the products of the elements of the next parallel diagonal $a_{21}a_{32}a_{13}$, including the upper right corner element a_{13}, and (3) the parallel diagonal $a_{31}a_{23}a_{12}$, including the lower left corner element a_{31}. Subtract from this sum (1) the product of the elements of the secondary diagonal $a_{13}a_{22}a_{31}$, (2) the products of the elements of the next parallel diagonal $a_{23}a_{32}a_{11}$, including the element in the upper left corner, and (3) the product of the elements in the next parallel diagonal $a_{33}a_{21}a_{12}$, including the element in the lower right corner.

$$\begin{vmatrix} 5 & -13 & 2 \\ 1 & 5 & 1 \\ 2 & -12 & 7 \end{vmatrix} = \begin{aligned} & 5(5)(7) + 1(-12)(2) + 2(1)(-13) \\ & - 2(5)(2) - 1(-12)(5) - 7(1)(-13) \end{aligned}$$
$$= 175 - 24 - 26 - 20 + 60 + 91 = 256$$

Usually, a determinant of the fourth or higher order can be most easily and rapidly evaluated by successively reducing its order (Art. 13-11). The process can be carried out to complete evaluation or can be stopped when a second- or third-order determinant, which can be evaluated as just indicated, is reached. A determinant also is equal to the sum of the products of the elements of any row (or column) and their respective cofactors (see Art. 13-9). This method is useful for proofs and derivations. Of course, a determinant may be evaluated in accordance with its definition (Art. 13-7), but this is seldom the best method.

13-14. Derivative of a Determinant. If the elements of a determinant are differentiable functions of a variable x,

$$\frac{d}{dx}\begin{vmatrix} a_{11} & a_{12} & \cdots & a_{1n} \\ a_{21} & a_{22} & \cdots & a_{2n} \\ \cdots & \cdots & \cdots & \cdots \\ a_{n1} & a_{n2} & \cdots & a_{nn} \end{vmatrix} = \begin{vmatrix} a'_{11} & a'_{12} & \cdots & a'_{1n} \\ a_{21} & a_{22} & \cdots & a_{2n} \\ \cdots & \cdots & \cdots & \cdots \\ a_{n1} & a_{n2} & \cdots & a_{nn} \end{vmatrix} + \begin{vmatrix} a_{11} & a_{12} & \cdots & a_{1n} \\ a'_{21} & a'_{22} & \cdots & a'_{2n} \\ \cdots & \cdots & \cdots & \cdots \\ a_{n1} & a_{n2} & \cdots & a_{nn} \end{vmatrix}$$

$$+ \cdots + \begin{vmatrix} a_{11} & a_{12} & \cdots & a_{1n} \\ a_{21} & a_{22} & \cdots & a_{2n} \\ \cdots & \cdots & \cdots & \cdots \\ a'_{n1} & a'_{n2} & \cdots & a'_{nn} \end{vmatrix}$$

where the primes denote differentiation with respect to x.

• Evaluate $y' = \dfrac{d}{dx}\begin{vmatrix} x^2 & 3x + 2 \\ -2x & x^3 \end{vmatrix}$.

$$y' = \begin{vmatrix} 2x & 3 \\ -2x & x^2 \end{vmatrix} + \begin{vmatrix} x^2 & 3x + 2 \\ -2 & 3x^2 \end{vmatrix}$$

13-15. Inverse of a Square Matrix. If A^{-1} is the inverse of A, then $AA^{-1} = I$, where I is the identity matrix. A square matrix A has an inverse only if determinant $A \neq 0$. (An inverse is not defined for a nonsquare matrix.)

To find the inverse of A, replace each element a_{ij} by its cofactor divided by determinant A and transpose the resulting matrix.

• Find the inverse of $A = \begin{bmatrix} 2 & -1 & 4 \\ 1 & 0 & -3 \\ 3 & -2 & -4 \end{bmatrix}$.

The matrix of cofactors (Art. 13-9) is

$$\begin{bmatrix} -6 & -5 & -2 \\ -12 & -20 & 1 \\ 3 & 10 & 1 \end{bmatrix}$$

Determinant $A = -15$ (Art. 13-13). Hence,

$$A^{-1} = \begin{bmatrix} \frac{6}{15} & \frac{5}{15} & \frac{2}{15} \\ \frac{12}{15} & \frac{20}{15} & -\frac{1}{15} \\ -\frac{3}{15} & -\frac{10}{15} & -\frac{1}{15} \end{bmatrix}^T = \begin{bmatrix} \frac{2}{5} & \frac{4}{5} & -\frac{1}{5} \\ \frac{1}{3} & \frac{4}{3} & -\frac{2}{3} \\ \frac{2}{15} & -\frac{1}{15} & -\frac{1}{15} \end{bmatrix}$$

The process of finding the inverse of a matrix may be used to solve a system of linear simultaneous equations. Let $\mathbf{AX} = \mathbf{Y}$ represent in matrix notation the equations

$$\begin{array}{llll}
a_{11}x_1 + a_{12}x_2 + \cdots + a_{1n}x_n = y_{11} & \text{or} & y_{12} \cdots & \text{or} & y_{1m} \\
a_{21}x_1 + a_{22}x_2 + \cdots + a_{2n}x_n = y_{21} & \text{or} & y_{22} \cdots & \text{or} & y_{2m} \\
\cdots\cdots\cdots\cdots\cdots\cdots\cdots\cdots\cdots\cdots\cdots\cdots\cdots\cdots \\
a_{n1}x_1 + a_{n2}x_2 + \cdots + a_{nn}x_n = y_{n1} & \text{or} & y_{n2} \ldots & \text{or} & y_{nm}
\end{array}$$

$\mathbf{A}$ is the matrix formed by the coefficients and $\mathbf{X}$ a column matrix composed of the unknowns. Note that $\mathbf{Y}$ may be taken as an m-column matrix; that is, the left-hand side of each equation may be equal to several sets of constants, each set represented by the vectors, or columns, comprising $\mathbf{Y}$. To solve the equations, multiply both sides of the matrix equation by the inverse of $\mathbf{A}$:

$$\mathbf{A^{-1}AX = A^{-1}Y}$$
$$\mathbf{X = A^{-1}Y}$$

For computer solution, the following method, based on solution of linear equations by elimination (Art. 3-83), permits compact representation and can be executed routinely:

Augment coefficient matrix $\mathbf{A}$ by adding matrix $\mathbf{Y}$ on the right. On the right of **Y** add a unit vector, or column matrix with one in the first row and zero for all other elements.

$$\begin{bmatrix} a_{11} & a_{12} & \cdots & a_{1n} & \vdots & y_{11} & y_{12} & \cdots & y_{1m} & \vdots & 1 \\ a_{21} & a_{22} & \cdots & a_{2n} & \vdots & y_{21} & y_{22} & \cdots & y_{2m} & \vdots & 0 \\ \cdots\cdots\cdots\cdots\cdots & & & & & \cdots\cdots\cdots\cdots\cdots & & & & & \cdot \\ a_{n1} & a_{n2} & \cdots & a_{nn} & \vdots & y_{n1} & y_{n2} & \cdots & y_{nm} & \vdots & 0 \end{bmatrix}$$

Add a row at the bottom—called the pivot row—with each element equal to

$$a_{pj} = \frac{a_{1k}}{a_{11}}$$

where $p = n + 1$ and $k = j + 1 = 2, 3. \ldots$ Let j represent an element's column, i its row.

Replace all the elements of the augmented matrix above the pivot row with new values b_{ij} computed from

$$b_{ij} = a_{ik} - a_{i1}a_{pj}$$

The new values for the first row are all zero. Drop this row and renumber the remaining rows from 1 to n, the elements of the last row now being represented by $b_{nj} = a_{pj}$.

Augment the new matrix with a unit vector and repeat the steps. If this is done n times, the result is a matrix

$$\begin{bmatrix} x_{11} & x_{12} & \cdots & x_{1m} & \vdots & o_{11} & o_{12} & \cdots & o_{1n} \\ x_{21} & x_{22} & \cdots & x_{2m} & \vdots & o_{21} & o_{22} & \cdots & o_{2n} \\ \cdots\cdots\cdots\cdots\cdots & & & & & \cdots\cdots\cdots\cdots\cdots \\ x_{n1} & x_{n2} & \cdots & x_{nm} & \vdots & o_{n1} & o_{n2} & \cdots & o_{nn} \end{bmatrix}$$

where the o_{ij} are the elements of the inverse of the original coefficient matrix and the x_{ij} are the solutions for each of the y vectors comprising the **Y** matrix.

• Solve the following system of equations and find the inverse of the coefficient matrix.

$$\begin{aligned} x - 2y + 3z &= 2 \\ 2x + 0y - 3z &= 3 \\ x + \ y + \ z &= 6 \end{aligned}$$

Augment the coefficient matrix with the vector (2,3,6) and the unit vector, to get

$$\begin{bmatrix} 1 & -2 & 3 & \vdots & 2 & \vdots & 1 \\ 2 & 0 & -3 & \vdots & 3 & \vdots & 0 \\ 1 & 1 & 1 & \vdots & 6 & \vdots & 0 \end{bmatrix}$$

To obtain the elements of the pivot row a_{4j}, divide by $a_{11} = 1$ the elements of the first row, beginning with $a_{12} = -2$. The pivot row, then, is

$$[-2 \quad 3 \quad 2 \quad 1]$$

Replace the rest of the augmented matrix with new elements computed from $b_{ij} = a_{ik} - a_{i1}a_{4j}$. For example,

$$b_{21} = a_{22} - a_{21}a_{41} = 0 - 2(-2) = 4$$
$$b_{22} = a_{23} - a_{21}a_{42} = -3 - 2(3) = -9$$
$$b_{31} = a_{32} - a_{31}a_{41} = 1 - 1(-2) = 3$$

So the new matrix, after dropping the former first row, becomes (with a new unit vector added)

$$\begin{bmatrix} 4 & -9 & -1 & -2 & \vdots & 1 \\ 3 & -2 & 4 & -1 & \vdots & 0 \\ -2 & 3 & 2 & 1 & \vdots & 0 \end{bmatrix}$$

Note that the third row is the former pivot row, with a zero at the right.

The new pivot row is obtained by dividing by $b_{11} = 4$ the elements of the first row, beginning with the second column.

$$\begin{bmatrix} -\dfrac{9}{4} & -\dfrac{1}{4} & -\dfrac{2}{4} & \dfrac{1}{4} \end{bmatrix}$$

Replace the rest of the matrix by new elements obtained from $c_{ij} = b_{ik} - b_{i1}b_{4j}$. For example, $c_{21} = -2 - 3(-\frac{9}{4}) = 1\frac{9}{4}$. The new matrix, then, is

$$\begin{bmatrix} 1\frac{9}{4} & 1\frac{9}{4} & \frac{2}{4} & -\frac{3}{4} & \vdots & 1 \\ -\frac{6}{4} & \frac{6}{4} & 0 & \frac{2}{4} & \vdots & 0 \\ -\frac{9}{4} & -\frac{1}{4} & -\frac{2}{4} & \frac{1}{4} & \vdots & 0 \end{bmatrix}$$

This time the new pivot row is obtained by dividing the first row by $1\frac{9}{4}$:

$$[1 \quad \frac{2}{19} \quad -\frac{3}{19} \quad \frac{4}{19}]$$

Finally, replace the rest of the matrix by new elements obtained from $d_{ij} = c_{ik} - c_{i1}c_{4j}$, to get the final solution:

$$\begin{bmatrix} 3 & \vdots & \frac{3}{19} & \frac{5}{19} & \frac{6}{19} \\ 2 & \vdots & -\frac{5}{19} & -\frac{2}{19} & \frac{9}{19} \\ 1 & \vdots & \frac{2}{19} & -\frac{3}{19} & \frac{4}{19} \end{bmatrix}$$

The first column gives the solution $x = 3; y = 2; z = 1$. The rest of the matrix is the inverse of the coefficient matrix.

To check the results, multiply the inverse matrix by the vector, or column, matrix $(2,3,6)$. To get x, sum the products of the elements of the first row of the inverse matrix by corresponding elements of the vector:

$$x = \frac{3}{19} \times 2 + \frac{5}{19} \times 3 + \frac{6}{19} \times 6 = \frac{6 + 15 + 36}{19} = \frac{57}{19} = 3$$

Similarly

$$y = -\frac{5}{19} \times 2 - \frac{2}{19} \times 3 + \frac{9}{19} \times 6 = \frac{-10 - 6 + 54}{19} = \frac{38}{19} = 2$$

$$z = \frac{2}{19} \times 2 - \frac{3}{19} \times 3 + \frac{4}{19} \times 6 = \frac{4 - 9 + 24}{19} = \frac{19}{19} = 1$$

13-16. Rank of a Matrix. A submatrix of a given matrix is the array remaining after certain rows and columns have been deleted. The determinant of a square submatrix of order r is called the determinant of order r of the matrix.

A matrix has rank r if it has at least one determinant of order r that is not zero and has no determinant of order more than r that is not zero. A matrix is of zero rank only if all its elements are zero.

If the determinants of all submatrices of order k equal zero, then so do the determinants of all submatrices of order $k + 1$ or higher.

• What is the rank of $A = \begin{bmatrix} 2 & -1 & 1 \\ 0 & -2 & 3 \\ 2 & -3 & 4 \end{bmatrix}$?

Determinant $A = 0$, so rank $r < 3$. Next, test submatrices of the second order formed by deleting a column and a row. Since the submatrix formed by dropping the third row and second column has determinant

$$\begin{vmatrix} 2 & 1 \\ 0 & 3 \end{vmatrix} \neq 0$$

A has rank 2.

13-17. Elementary Transformations of Matrices. Following are operations that do not alter the rank or order of a matrix:

1. Interchange of any two rows (or columns)

2. Multiplication of all the elements of a row (or column) by the same nonzero constant

3. Addition to any row (or column) of an arbitrary multiple of any other row (or column)

The inverse of an elementary transformation is an operation that restores the original matrix. It is an elementary transformation of the same type.

Any elementary transformation on a given matrix can be effected by first performing the same elementary transformation on an identity matrix of appropriate order, then premultiplying the given matrix by the result if the operation is on rows, postmultiplying if it is on columns.

• By a matrix multiplication, interchange the second and third rows of

$$\begin{bmatrix} 1 & 2 & -1 & -2 \\ 3 & 0 & 2 & 4 \\ -4 & -3 & 5 & 0 \end{bmatrix}$$

Interchange the second and third rows of a third-order identity matrix and premultiply the given matrix by it.

$$\begin{bmatrix} 1 & 0 & 0 \\ 0 & 0 & 1 \\ 0 & 1 & 0 \end{bmatrix} \begin{bmatrix} 1 & 2 & -1 & -2 \\ 3 & 0 & 2 & 4 \\ -4 & -3 & 5 & 0 \end{bmatrix} = \begin{bmatrix} 1 & 2 & -1 & -2 \\ -4 & -3 & 5 & 0 \\ 3 & 0 & 2 & 4 \end{bmatrix}$$

• Use matrix multiplication to multiply the third column of the matrix in the preceding example by k.

Multiply the third column of the fourth-order identity matrix by k and postmultiply the given matrix by the result.

$$\begin{bmatrix} 1 & 2 & -1 & -2 \\ 3 & 0 & 2 & 4 \\ -4 & -3 & 5 & 0 \end{bmatrix} \begin{bmatrix} 1 & 0 & 0 & 0 \\ 0 & 1 & 0 & 0 \\ 0 & 0 & k & 0 \\ 0 & 0 & 0 & 1 \end{bmatrix} = \begin{bmatrix} 1 & 2 & -k & -2 \\ 3 & 0 & 2k & 4 \\ -4 & -3 & 5k & 0 \end{bmatrix}$$

• Use matrix multiplication to add double the second column to the first column of the matrix in the preceding examples.

Add double the second column of the fourth-order identity matrix to the first column, then postmultiply the given matrix by the result.

$$\begin{bmatrix} 1 & 2 & -1 & -2 \\ 3 & 0 & 2 & 4 \\ -4 & -3 & 5 & 0 \end{bmatrix} \begin{bmatrix} 1 & 0 & 0 & 0 \\ 2 & 1 & 0 & 0 \\ 0 & 0 & 1 & 0 \\ 0 & 0 & 0 & 1 \end{bmatrix} = \begin{bmatrix} 5 & 2 & -1 & -2 \\ 3 & 0 & 2 & 4 \\ -10 & -3 & 5 & 0 \end{bmatrix}$$

13-18. Nomograms. A nomogram, or alignment chart, is a graphical solution of a formula in which values of the variables are plotted along lines in such fashion that a straight line cuts these lines in a set of points that satisfy the formula.

For example, the equation $x + y = z$ can be solved with three parallel scales (Fig. 13-1). Values of x are plotted along the left-hand line, y along the right-hand line, and z along the center line. A straight line between any point on the x scale and any point on the y scale cuts the z scale at a point satisfying $x + y = z$. Thus, the line joining 1 on the x scale and 3 on the y scale meets the z scale at $z = 4$.

A general method for plotting a nomogram for a formula with three variables is to write a third-order determinant whose value is identical with the given formula. Each row of the determinant should contain only one

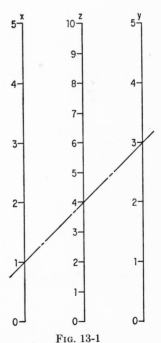

FIG. 13-1

of the variables. Then operate on the determinant without changing its value (Art. 13-10) to obtain unity for every element of the third column. Thus, the determinant will be in the form

$$\begin{vmatrix} U_1 & U_2 & 1 \\ V_1 & V_2 & 1 \\ W_1 & W_2 & 1 \end{vmatrix} = 0$$

where U_1, U_2 = functions of a variable u

V_1, V_2 = functions of a second variable v

W_1, W_2 = functions of a third variable w

Now, using cartesian coordinates, plot first $x = U_1$, $y = U_2$; next plot $x = V_1$, $y = V_2$; finally plot $x = W_1$, $y = W_2$. (Sometimes, it is convenient to use skew coordinates instead of rectangular.) In general, the scales will be curved lines.

• Construct a nomogram for $u = vw$.

The design determinant could be obtained by trial and error. However, a more direct approach in this case is to set two new variables $z_1 = u$ and $z_2 = v$; or better still, let $z_1 = k_u u$ and $z_2 = k_v v$, where k_u and k_v are scale moduli that permit adjustments in the plotting of $f(u)$ and $f(v)$, respectively. Then, three equations can be written in the unknowns z_1 and z_2.

$$z_1 + 0 - k_u u = 0$$
$$0 + z_2 - k_v v = 0$$
$$\frac{z_1}{k_u} - \frac{z_2}{k_v} w + 0 = 0$$

(The last equation was obtained by substituting $z_1 = k_u u$ and $z_2 = k_v v$ in the equation $u - vw = 0$.) Since these equations are consistent, the determinant of the coefficients and constants must equal zero (Art. 3-88).

$$\begin{vmatrix} 1 & 0 & -k_u u \\ 0 & 1 & -k_v v \\ \dfrac{1}{k_u} & -\dfrac{w}{k_v} & 0 \end{vmatrix} = 0$$

To convert this to the proper form, first multiply columns 2 and 3 and row 2 by -1, then add the second column to the first.

$$\begin{vmatrix} 1 & 0 & k_u u \\ 1 & 1 & -k_v v \\ \dfrac{k_v + k_u w}{k_u k_v} & \dfrac{w}{k_v} & 0 \end{vmatrix} = 0$$

Next, divide the third row by $(k_v + k_u w)/k_u k_v$ and interchange columns.

$$\begin{vmatrix} 0 & k_u u & 1 \\ 1 & -k_v v & 1 \\ \dfrac{k_u w}{k_v + k_u w} & 0 & 1 \end{vmatrix} = 0$$

The resulting scales, $x = 0$, $y = k_u u$; $x = 1$, $y = -k_v v$; $x = k_u w/(k_v + k_u w)$, $y = 0$, are straight lines. Plot the y axis at 45° to the x axis to get a more compact chart.

If the u and v scales are to be 10 in. long, and if the range of u is 0 to 10 and the range of v is 0 to 5, then

$$k_u = \frac{10}{10 - 0} = 1 \qquad k_v = \frac{10}{5 - 0} = 2$$

The final design determinant is

$$\begin{vmatrix} 0 & u & 1 \\ 1 & -2v & 1 \\ \dfrac{w}{2 + w} & 0 & 1 \end{vmatrix} = \begin{vmatrix} 0 & u & 1 \\ 14.1 & -2v & 1 \\ \dfrac{14.1w}{2 + w} & 0 & 1 \end{vmatrix} = 0$$

after the first column is multiplied by $10\sqrt{2} = 14.1$ to get a convenient diagram. Figure 13-2 was drawn in three steps: (1) Plotting $y = u$ along the vertical line

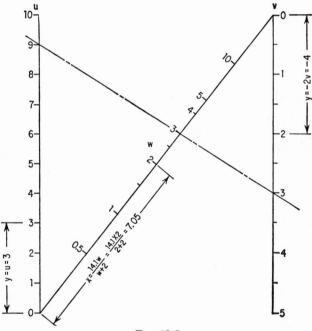

FIG. 13-2

$x = 0$. (2) Plotting $y = -2v$ along the vertical line $x = 14.1$ and writing the value of v at each point determined by $f(v)$. (3) Plotting $x = 14.1w/(2 + w)$ along the 45° line $y = 0$ and writing the value of w at each point determined by $f(w)$.

When four variables are involved in a formula, pair two of the variables on one row of the design determinant. Assign one value at a time to one of this pair of variables and plot the resulting scales. The result will be a network of scales to be used with the scales for the other two variables.

BIBLIOGRAPHY

Bellman, Richard: "Introduction to Matrix Analysis," McGraw-Hill Book Company, Inc., New York, 1960.

Hohn, Franz E.: "Elementary Matrix Algebra," The Macmillan Company, New York, 1958.

Levens, A. S.: "Nomography," 2d ed., John Wiley & Sons, Inc., New York, 1959.

Chapter 14

VECTORS AND TENSORS

14-1. Vector and Tensor Notation. Vectors and tensors generally are represented by boldface letters. Scalars are indicated by normal face.

Subscripts and superscripts are used with letters to denote different quantities. Hence, to indicate a power, a quantity is either repeated or enclosed in parentheses before a superscript indicating a power. For example, x_1, x_2, x_3, . . . and x^1, x^2, x^3 are different ways of writing sets of variables, whereas $(x)^3$ represents x cubed. x^1, x^2, x^3 would usually be written x, y, z in algebra. Some quantities may have both subscripts and superscripts; the order in which they are written generally is important. Thus, the components of a tensor may be written T_{23}, T^{23} or $T^2_{.3}$, etc.; T_{23} may differ from T_{32}.

For abbreviation, letters may be used as subscripts or superscripts, with a range of numerical values indicated.

In most texts, a repeated letter, subscript or superscript, calls for summation over all values of that index. For example,

$$a^i b_i = \sum_1^3 a^i b_i = a^1 b_1 + a^2 b_2 + a^3 b_3$$

14-2. Kronecker Delta. Symbols for the numbers 0, $+1$, and -1. The first-order delta is defined by

$$\delta^i_j = \begin{cases} 0 & i \neq j \\ 1 & i = j \end{cases}$$

The second-order delta δ^{ij}_{rs} is zero if $i = j$ or $r = s$ or ij is not the same set as rs. It is $+1$ if r, s is an even permutation of i, j and -1 if an odd permutation. Thus, $\delta^{11}_{21} = \delta^{21}_{22} = 0$; $\delta^{12}_{12} = +1$; $\delta^{12}_{21} = -1$.

14-3. Admissible Coordinate Transformations. Those satisfying the following rules in a given region R:

If $x^i = x^1$, x^2, . . . , x^n are coordinates in a given system and $y^i = y^1$, y^2, . . . , y^n are the coordinates after transformation, then the y^i are given by the n equations

$$y^i = f^i(x^1, x^2, \ldots, x^n)$$

These equations should be solvable for x^i.

$$x^i = g^i(y^1, y^2, \ldots, y^n)$$

Both functions f^i and g^i should be single valued throughout R.
In R, the jacobian determinant is not equal to zero.

$$\begin{vmatrix} \dfrac{\partial y^1}{\partial x^1} & \cdots & \dfrac{\partial y^1}{\partial x^n} \\ \cdots \cdots \cdots \\ \dfrac{\partial y^n}{\partial x^1} & \cdots & \dfrac{\partial y^n}{\partial x^n} \end{vmatrix} \neq 0$$

An affine transformation is the special case in which the functions f^i are linear and homogeneous in x^i.

14-4. Invariant. An entity that remains unchanged while certain of its constituent elements or associated quantities change under a transformation of coordinates.

Thus, the distance between two points is an invariant when a cartesian coordinate system is rotated.

Under admissible coordinate transformations (Art. 14-3) vector and tensor equations remain invariant in form.

14-5. Scalar. A quantity completely specified when its magnitude is given. Thus it can be represented by a single real number.

Mass, density, temperature, energy, length of a line or curve, and the angle between two lines are scalars.

14-6. Vector. A quantity that is completely defined in n-dimensional space only when n components are given, and that remains invariant

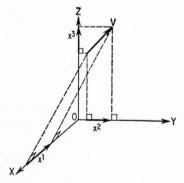

under an admissible coordinate transformation (Art. 14-3).

In a plane or in three-dimensional space, a vector can be represented by an arrow of specific length. Thus it is completely defined when its length, direction, and sense are given.

Such quantities as velocity, acceleration, force, and electric and magnetic field intensities are vectors.

If cartesian coordinates are chosen, the projections of the arrow on the x, y, and

Fig. 14-1

z axes, x^1, x^2, and x^3, respectively, are a set of components of the vector (Fig. 14-1) (see also Art. 14-8). These components may change in value under an admissible transformation of coordinates, but the vector remains invariant.

A zero vector has no length and is devoid of direction. If a vector is zero in one coordinate system, it is zero in all coordinate systems.

A unit vector is a vector of unit length.

• If a vector has components 3, 4, and 12 in a cartesian system, what is its magnitude? What are its direction cosines?

Its magnitude is $\sqrt{3^2 + 4^2 + 12^2} = 13$. Its direction cosines are $\frac{3}{13}$, $\frac{4}{13}$, and $\frac{12}{13}$.

• If a vector $\mathbf{V}$ lies in the xy plane, has a magnitude of 24, and makes an angle of 30° with the x axis, what are its x and y components?

If $\mathbf{V}_x$ and $\mathbf{V}_y$ are the components parallel, respectively, to the x and y axes, then

$$\mathbf{V}_x = 24 \cos 30 = 12 \sqrt{3}$$
$$\mathbf{V}_y = 24 \sin 30 = 12$$

14-7. Fundamental Properties of Vectors. Two vectors are equal if corresponding components on the same bases (or axes) are equal or if the

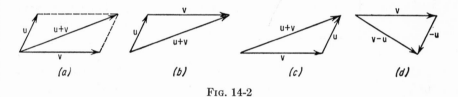

(a)   *(b)*   *(c)*   *(d)*

FIG. 14-2

arrows representing them have the same length, direction, and sense. If two vectors are equal in one coordinate system, they are equal in all coordinate systems.

The negative of a three-dimensional vector is represented by an arrow of the same length oppositely directed.

To add or subtract two vectors, add or subtract corresponding components. Geometrically, two vectors can be added by the parallelogram law:

To add $\mathbf{u}$ to $\mathbf{v}$, draw $\mathbf{u}$ and $\mathbf{v}$ from a common origin and complete a parallelogram by drawing a line through the terminal of $\mathbf{u}$ parallel to $\mathbf{v}$ and one through the terminal of $\mathbf{v}$ parallel to $\mathbf{u}$. Then the diagonal of the parallelogram through the origin equals $\mathbf{u} + \mathbf{v}$ (Fig. 14-2a).

This construction can be shortened by first drawing $\mathbf{u}$, then projecting $\mathbf{v}$ from its terminal; $\mathbf{u} + \mathbf{v}$ equals the arrow from the origin of $\mathbf{u}$ to the terminal of $\mathbf{v}$ (Fig. 14-2b). The construction also can start with $\mathbf{v}$ (Fig. 14-2c).

To subtract a vector, reverse its direction and add (Fig. 14-2d).

The product $k\mathbf{u}$, or $\mathbf{u}k$, of a vector $\mathbf{u}$ and a scalar k is a vector each of whose components is k times those of $\mathbf{u}$. It can be represented by an arrow k times as long but having the same direction as **u**. (If k is negative, the direction is reversed.)

Vectors can be treated formally in accordance with the rules of algebra for addition and subtraction, and for multiplication and division by numbers. (For vector multiplication, see Arts. 14-14 and 14-15.)

The n vectors $\mathbf{u}_1$, $\mathbf{u}_2$, . . . , $\mathbf{u}_n$ are linearly dependent if n real numbers k_1, k_2, . . . , k_n exist, not all zero, such that

$$k_1\mathbf{u}_1 + k_2\mathbf{u}_2 + \cdots k_n\mathbf{u}_n = 0$$

If m vectors are linearly dependent, any greater number of vectors including these also are dependent. Two vectors are linearly dependent if they are collinear; three vectors are linearly dependent if they are coplanar. In three-dimensional space, any four vectors are linearly dependent.

A set of vectors not linearly dependent is linearly independent.

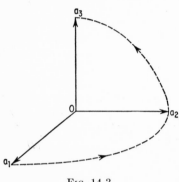

FIG. 14-3

14-8. Base Vectors. Let $\mathbf{a}_1$, $\mathbf{a}_2$, and $\mathbf{a}_3$ be three linearly independent (noncoplanar) vectors with a common origin. These vectors form a right-handed, or dextral, system if, when $\mathbf{a}_1$ is rotated through the shortest distance to coincide with $\mathbf{a}_2$ and then with $\mathbf{a}_3$, the terminal of $\mathbf{a}_1$ rotates clockwise when viewed from the origin (Fig. 14-3). (Thumb, index finger, and middle finger of your right hand can point in the directions of $\mathbf{a}_1$, $\mathbf{a}_2$, and $\mathbf{a}_3$, respectively.) If the rotation is counterclockwise, the set is left-handed or sinistral.

Any vector **u** can be expressed as the sum of three vectors parallel to $\mathbf{a}_1$, $\mathbf{a}_2$, and $\mathbf{a}_3$.

$$\mathbf{u} = u^1\mathbf{a}_1 + u^2\mathbf{a}_2 + u^3\mathbf{a}_3$$

(The superscripts are not exponents.) The scalars u^i are components of **u** with respect to the vectors $\mathbf{a}_i$ ($i = 1, 2, 3$).

Consider a generalized coordinate system in n-dimensional space with the radius vector **r** to any point a function of the coordinates u^1, u^2, u^3, . . . , u^n. When a set of vectors $\mathbf{a}_1$, $\mathbf{a}_2$, . . . , $\mathbf{a}_n$ are so chosen that

$$\mathbf{a}_1 = \frac{\partial \mathbf{r}}{\partial u^1} \qquad \mathbf{a}_2 = \frac{\partial \mathbf{r}}{\partial u^2} \qquad \cdots \qquad \mathbf{a}_n = \frac{\partial \mathbf{r}}{\partial u^n}$$

the vectors $\mathbf{a}_i$ are called unitary base vectors. They are tangent to the u^i curves at each point, directed toward increasing values of the coordinates, and, in general, vary from point to point.

$$d\mathbf{r} = \mathbf{a}_1 du^1 + \mathbf{a}_2 du^2 + \cdots + \mathbf{a}_n du^n$$

Note also that

$$\frac{\partial \mathbf{a}_i}{\partial u^j} = \frac{\partial^2 \mathbf{r}}{\partial u^i \, \partial u^j} = \frac{\partial \mathbf{a}_j}{\partial u^i}$$

where $i \neq j$.

Mutually perpendicular dextral unit vectors usually are represented by **i**, **j**, and **k**. If rectangular coordinate axes are chosen to coincide with these vectors, the x axis lying along **i**, the y axis along **j**, and the z axis along **k**, then the coordinates of any point P are the components on the axes of the radius vector **r** to P.

$$\mathbf{r} = x\mathbf{i} + y\mathbf{j} + z\mathbf{k}$$

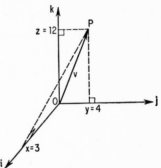

FIG. 14-4

• What is the magnitude of the vector $\mathbf{v} = 3\mathbf{i} + 4\mathbf{j} + 12\mathbf{k}$ (Fig. 14-4)?

The components of **v** on the rectangular axes are 3, 4, and 12. Hence,

$$|\mathbf{v}| = \sqrt{3^2 + 4^2 + 12^2} = 13$$

14-9. Reciprocal Base Vectors. Two sets of base vectors $\mathbf{a}_i$ and $\mathbf{a}^j$, where i, $j = 1, 2, 3$, are reciprocal if they satisfy the nine equations given by their scalar products (Art. 14-14):

$$\mathbf{a}_i \cdot \mathbf{a}^j = \delta_i^j = \begin{cases} 0 & i \neq j \\ 1 & i = j \end{cases}$$

Either set can be computed from its reciprocal.

$$\mathbf{a}^1 = \frac{\mathbf{a}_2 \times \mathbf{a}_3}{[\mathbf{a}_1\mathbf{a}_2\mathbf{a}_3]} \qquad \mathbf{a}^2 = \frac{\mathbf{a}_3 \times \mathbf{a}_1}{[\mathbf{a}_1\mathbf{a}_2\mathbf{a}_3]} \qquad \mathbf{a}^3 = \frac{\mathbf{a}_1 \times \mathbf{a}_2}{[\mathbf{a}_1\mathbf{a}_2\mathbf{a}_3]}$$

$$\mathbf{a}_1 = \frac{\mathbf{a}^2 \times \mathbf{a}^3}{[\mathbf{a}^1\mathbf{a}^2\mathbf{a}^3]} \qquad \mathbf{a}_2 = \frac{\mathbf{a}^3 \times \mathbf{a}^1}{[\mathbf{a}^1\mathbf{a}^2\mathbf{a}^3]} \qquad \mathbf{a}_3 = \frac{\mathbf{a}^1 \times \mathbf{a}^2}{[\mathbf{a}^1\mathbf{a}^2\mathbf{a}^3]}$$

where the multiplication sign indicates vector product (Art. 14-15) and the brackets indicate scalar triple product (Art. 14-17).

When a set of base vectors and its reciprocal are identical, the sets are self-reciprocal. This can happen only when the vectors of each set are mutually perpendicular.

A vector has the same value whether its components are given for a base set of vectors or the reciprocal set, but corresponding components may not necessarily be equal.

$$\mathbf{u} = u^1\mathbf{a}_1 + u^2\mathbf{a}_2 + u^3\mathbf{a}_3 = u_1\mathbf{a}^1 + u_2\mathbf{a}^2 + u_3\mathbf{a}^3$$

The components $u^i = \mathbf{u} \cdot \mathbf{a}^i$ are called contravariant. The components $u_i = \mathbf{u} \cdot \mathbf{a}_i$ are called covariant.

The right-handed set of mutually orthogonal unit vectors, $\mathbf{i}$, $\mathbf{j}$, $\mathbf{k}$, is self-reciprocal. Components of vectors in that system may be treated as contravariant or covariant.

14-10. Reciprocal Differentials. The differential of the radius vector to any point $d\mathbf{r}$ can be expressed in terms of either the unitary base vectors $\mathbf{a}_i$ (Art. 14-8), or their reciprocals $\mathbf{a}^i$ (Art. 14-9).

$$d\mathbf{r} = \mathbf{a}_1 \, du^1 + \mathbf{a}_2 \, du^2 + \cdots + \mathbf{a}_n \, du^n = \mathbf{a}^1 \, du_1 + \mathbf{a}^2 \, du_2 + \cdots + \mathbf{a}^n \, du_n$$

where the du^i are the differentials of the coordinates. The differentials du_i are called reciprocal differentials. They can be determined from

$$du_i = \sum_{j=1}^{n} \mathbf{a}_i \cdot \mathbf{a}_j \, du^j$$

Similarly, the differentials of the coordinates can be obtained from the reciprocal differentials.

$$du^i = \sum_{j=1}^{n} \mathbf{a}^i \cdot \mathbf{a}^j \, du_j$$

14-11. Vector Field. If a vector varies with position and time, the function representing it is called a vector field. If the components in rectangular coordinates of a vector field $\mathbf{V}$ are u, v, and w (scalar functions of the coordinates), the field is described by

$$\mathbf{V} = u(x,y,z,t)\mathbf{i} + v(x,y,z,t)\mathbf{j} + w(x,y,z,t)\mathbf{k}$$

where t is the time coordinate.

A steady-state vector field does not vary with time. It is described by

$$\mathbf{V} = u(x,y,z)\mathbf{i} + v(x,y,z)\mathbf{j} + w(x,y,z)\mathbf{k}$$

If $\mathbf{V}$ is a constant, the field is called uniform.

14-12. Cogredient and Contragredient Transformations. Assume an admissible coordinate transformation (Art. 14-3) given by

$$v^i = v^i(u^1,u^2, \ldots ,u^n) \qquad u^i = u^i(v^1,v^2, \ldots ,v^n) \qquad i = 1, 2, \ldots , n$$

Differentiation of these equations yields

$$dv^i = \sum_{j=1}^{n} \frac{\partial v^i}{\partial u^j} du^j \qquad du^i = \sum_{j=1}^{n} \frac{\partial u^i}{\partial v^j} dv^j$$

The result is the law of transformation of differentials of coordinates. Note that the partial derivatives are taken with respect to the given system of coordinates in each case.

For the same transformation of coordinates, the unitary base vectors at any point $\mathbf{a}_i$ change into the set $\mathbf{b}_i$ in accordance with the equations

$$\mathbf{b}_i = \sum_{j=1}^{n} \frac{\partial u^j}{\partial v^i} \mathbf{a}_j \qquad \mathbf{a}_i = \sum_{j=1}^{n} \frac{\partial v^j}{\partial u^i} \mathbf{b}_j$$

Since the partial derivatives are taken with respect to the new system of coordinates in each case, this law of transformation differs from that for the differentials of coordinates.

Quantities that transform in accordance with the same law are said to transform cogrediently.

Differentials of coordinates and quantities that transform as they do are said to transform contragrediently to the unitary base vectors.

The matrix of the coefficients for the transformation of du^i to dv^i is the reciprocal of the transpose of the matrix of the coefficients for the transformation of $\mathbf{a}_i$ to $\mathbf{b}_i$. For an affine transformation, the coefficients (partial derivatives) are constants.

14-13. Covariant, Contravariant, and Mixed Quantities. Quantities that transform cogrediently (Art. 14-12) with the unitary base vectors are covariant. They are denoted by subscripts. Thus, the reciprocal differentials (Art. 14-10) are covariant.

$$dv_i = \sum_{j=1}^{n} \frac{\partial u^j}{\partial v^i} du_j \qquad du_i = \sum_{j=1}^{n} \frac{\partial v^j}{\partial u^i} du_j$$

Quantities that transform contragrediently (Art. 14-12) to the unitary base vectors are contravariant. They are denoted by superscripts. Thus, the reciprocal unitary vectors are contravariant.

$$\mathbf{b}^i = \sum_{j=1}^{n} \frac{\partial v^i}{\partial u^j} \mathbf{a}^j \qquad \mathbf{a}^i = \sum_{j=1}^{n} \frac{\partial u^i}{\partial v^j} \mathbf{b}^j$$

where $\mathbf{b}^i$ = reciprocal base vectors in v system of coordinates
$\mathbf{a}^i$ = reciprocal base vectors in u system
$i = 1, 2, \ldots, n$

Tensors may be contravariant or covariant with respect to more than one index. When they have both contravariant and covariant components, they are called mixed tensors. For example, the metric coefficients g_{ij} (Art. 14-28) transform in accordance with the law

$$h_{lm} = \sum_{i,j=1}^{n} \frac{\partial u^i}{\partial v^l} \frac{\partial u^j}{\partial v^m} g_{ij}$$

where u^i are the given coordinates, v^l the new coordinates, and h_{lm} are the metric coefficients in the v system. Hence, g_{ij} is covariant. But

$$h^{lm} = \sum_{i,j=1}^{n} \frac{\partial v^l}{\partial u^i} \frac{\partial v^m}{\partial u^j} g^{ij}$$

Therefore, g^{ij} is contravariant. $g^{i}_{.j}$ is mixed.

$$h^{l.}_{.m} = \sum_{i,j=1}^{n} \frac{\partial v^l}{\partial u^i} \frac{\partial u^j}{\partial v^m} g^{i.}_{.j}$$

Note that the metric coefficients act as operators to raise or lower indices. For example,

$$A_i = \sum_{j=1}^{n} g_{ij} A^j \quad \text{and} \quad A^{ij} = \sum_{l=1}^{n} \sum_{m=1}^{n} g^{im} g^{jl} A_{lm}$$

14-14. Scalar (Dot or Inner) Product. The scalar product of two vectors **u** and **v** equals the product of their lengths u and v and the cosine of the included angle. It is written **u · v**.

$$\mathbf{u} \cdot \mathbf{v} = uv \cos (\mathbf{u},\mathbf{v}) = \mathbf{v} \cdot \mathbf{u}$$

When **u** is perpendicular to **v**, **u · v** = 0. When **u** is parallel to **v**, **u · v** = uv. Hence, **u · u** = $(u)^2$.

If **i**, **j**, and **k** are mutually perpendicular unit vectors,

$$\mathbf{i} \cdot \mathbf{i} = \mathbf{j} \cdot \mathbf{j} = \mathbf{k} \cdot \mathbf{k} = 1$$
$$\mathbf{i} \cdot \mathbf{j} = \mathbf{j} \cdot \mathbf{k} = \mathbf{k} \cdot \mathbf{i} = 0$$

The scalar product of two vectors is an invariant equal to the sum of the products of their corresponding components in a rectangular coordinate system.

$$\mathbf{u} \cdot \mathbf{v} = (u_1\mathbf{i} + u_2\mathbf{j} + u_3\mathbf{k}) \cdot (v_1\mathbf{i} + v_2\mathbf{j} + v_3\mathbf{k}) = u_1v_1 + u_2v_2 + u_3v_3$$

When the base vectors are not mutually perpendicular, the scalar product is simplified when **u** and **v** are referred to reciprocal bases.

$$\mathbf{u} \cdot \mathbf{v} = (u^1\mathbf{a}_1 + u^2\mathbf{a}_2 + u^3\mathbf{a}_3) \cdot (v_1\mathbf{a}^1 + v_2\mathbf{a}^2 + v_3\mathbf{a}^3)$$
$$= u^1v_1 + u^2v_2 + u^3v_3 = u_1v^1 + u_2v^2 + u_3v^3$$

• Evaluate **a · b** where **a** = 3**i** + 4**j** and **b** = 5**i** − 12**j**.

$$\mathbf{a} \cdot \mathbf{b} = 3(5) + 4(-12) = -33$$

Check: $a = \sqrt{3^2 + 4^2} = 5$, $b = \sqrt{5^2 + (-12)^2} = 13$, $\cos (\mathbf{a},\mathbf{b}) = \frac{3}{5}(\frac{5}{13}) + \frac{4}{5}(-\frac{12}{13}) = -\frac{33}{65}$.

$$\mathbf{a} \cdot \mathbf{b} = ab \cos (\mathbf{a},\mathbf{b}) = 5 \times 13(-\tfrac{33}{65}) = -33$$

14-15. Vector (Cross) Product. The vector product of two vectors **u** and **v** is a vector normal to the plane of **u** and **v** and forming a right-handed set with them (Fig. 14-3). The length of this vector equals the product of the lengths u and v of the given vectors and the absolute value of the sine of the included angle.

$$\mathbf{u} \times \mathbf{v} = uv \, |\sin (\mathbf{u},\mathbf{v})| \, \mathbf{a} = -\mathbf{v} \times \mathbf{u}$$

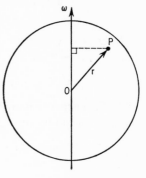

FIG. 14-5

where **a** is a unit vector normal to **u** and **v** and forming a right-handed system with them. If **u** is parallel to **v**, **u** × **v** = 0.

If **i**, **j**, and **k** are a right-handed set of mutually perpendicular unit vectors,

$$\mathbf{i} \times \mathbf{j} = \mathbf{k} \qquad \mathbf{j} \times \mathbf{k} = \mathbf{i} \qquad \mathbf{k} \times \mathbf{i} = \mathbf{j}$$
$$\mathbf{i} \times \mathbf{i} = \mathbf{j} \times \mathbf{j} = \mathbf{k} \times \mathbf{k} = 0$$

Hence, in a right-handed rectangular system, the vector product can be written for any two vectors **u** and **v** as

$$\begin{aligned}
\mathbf{u} \times \mathbf{v} &= (u_1\mathbf{i} + u_2\mathbf{j} + u_3\mathbf{k}) \times (v_1\mathbf{i} + v_2\mathbf{j} + v_3\mathbf{k}) \\
&= (u_2v_3 - u_3v_2)\mathbf{i} + (u_3v_1 - u_1v_3)\mathbf{j} + (u_1v_2 - u_2v_1)\mathbf{k} \\
&= \begin{vmatrix} \mathbf{i} & \mathbf{j} & \mathbf{k} \\ u_1 & u_2 & u_3 \\ v_1 & v_2 & v_3 \end{vmatrix}
\end{aligned}$$

The vector product represents the area of a parallelogram formed with the given vectors as sides.

When the base vectors are not mutually orthogonal, the vector product is given by

$$\mathbf{u} \times \mathbf{v} = [\mathbf{a}^1\mathbf{a}^2\mathbf{a}^3] \begin{vmatrix} \mathbf{a}_1 & \mathbf{a}_2 & \mathbf{a}_3 \\ u_1 & u_2 & u_3 \\ v_1 & v_2 & v_3 \end{vmatrix} = [\mathbf{a}_1\mathbf{a}_2\mathbf{a}_3] \begin{vmatrix} \mathbf{a}^1 & \mathbf{a}^2 & \mathbf{a}^3 \\ u^1 & u^2 & u^3 \\ v^1 & v^2 & v^3 \end{vmatrix}$$

where the brackets denote scalar triple product and $\mathbf{a}^1$, $\mathbf{a}^2$, $\mathbf{a}^3$ and $\mathbf{a}_1$, $\mathbf{a}_2$, $\mathbf{a}_3$ are reciprocal bases (Art. 14-9).

• Determine the velocity **v** of a point P in a rigid body rotating with angular velocity $\dfrac{d\theta}{dt}$ about a fixed axis (Fig. 14-5).

Let $\boldsymbol{\omega} = \dfrac{d\theta}{dt}\,\mathbf{a}$, where **a** is a unit vector directed along the axis of rotation in the direction a right-handed screw would move when turned in the same direction as the body. The velocity **v** is normal to the plane of **ω** and **r** (the radius vector to P) and its magnitude is $r\dfrac{d\theta}{dt}\,|\sin (\mathbf{a},\mathbf{r})|$. Hence,

$$\mathbf{v} = \boldsymbol{\omega} \times \mathbf{r}$$

• Evaluate $\mathbf{u} \times \mathbf{v}$ if $\mathbf{u} = 3\mathbf{i} + 4\mathbf{j}$ and $\mathbf{v} = 5\mathbf{i} - 12\mathbf{j}$.

$$\mathbf{u} \times \mathbf{v} = \begin{vmatrix} \mathbf{i} & \mathbf{j} & \mathbf{k} \\ 3 & 4 & 0 \\ 5 & -12 & 0 \end{vmatrix} = -36\mathbf{k} - 20\mathbf{k} = -56\mathbf{k}$$

Check: $u = \sqrt{3^2 + 4^2} = 5,\ v = \sqrt{5^2 + (-12)^2} = 13,\ \sin\ (\mathbf{u},\mathbf{v}) = \frac{4}{5}(\frac{5}{13})$
$-\frac{3}{5}(-\frac{12}{13}) = \frac{56}{65}.$

$$\mathbf{u} \times \mathbf{v} = uv \sin\ (\mathbf{u},\mathbf{v})\mathbf{k} = -5 \times 13(\frac{56}{65})\mathbf{k} = -56\mathbf{k}$$

14-16. Vector Triple Product.

The vector product of a vector $\mathbf{w}$ and the vector product of two other vectors $\mathbf{u} \times \mathbf{v}$.

$$(\mathbf{u} \times \mathbf{v}) \times \mathbf{w} = \mathbf{u} \cdot \mathbf{wv} - \mathbf{v} \cdot \mathbf{wu}$$
$$\mathbf{w} \times (\mathbf{u} \times \mathbf{v}) = \mathbf{w} \cdot \mathbf{vu} - \mathbf{w} \cdot \mathbf{uv}$$

14-17. Scalar Triple (Box) Product.

The scalar product of a vector $\mathbf{w}$ and the cross product of two other vectors $\mathbf{u} \times \mathbf{v}$.

In three-dimensional space, a scalar product is numerically equal to the volume of a parallelepiped formed with $\mathbf{u}$, $\mathbf{v}$, and $\mathbf{w}$ as sides. The sign is positive if $\mathbf{u}$, $\mathbf{v}$, and $\mathbf{w}$ form a right-handed system (Art. 14-8).

If the scalar triple product is zero, the vectors are coplanar, unless one of them is zero.

If the same vector appears twice, the scalar triple product is zero.

The value of the product is not altered by interchanging dot and cross. Since there is no ambiguity in omitting the dot and cross, the scalar triple product may be indicated by three vectors written in the proper order and enclosed in brackets.

$$\mathbf{u} \times \mathbf{v} \cdot \mathbf{w} = [\mathbf{uvw}] = \mathbf{u} \cdot \mathbf{v} \times \mathbf{w}$$

If $\mathbf{i}$, $\mathbf{j}$, and $\mathbf{k}$ are a right-handed set of mutually perpendicular unit vectors,

$$[\mathbf{ijk}] = [\mathbf{jki}] = [\mathbf{kij}] = 1 \qquad [\mathbf{ikj}] = [\mathbf{kji}] = [\mathbf{jik}] = -1$$

All other permutations of $\mathbf{i}$, $\mathbf{j}$, and $\mathbf{k}$ make the product zero. Hence,

$$[\mathbf{uvw}] = (u_1\mathbf{i} + u_2\mathbf{j} + u_3\mathbf{k}) \times (v_1\mathbf{i} + v_2\mathbf{j} + v_3\mathbf{k}) \cdot (w_1\mathbf{i} + w_2\mathbf{j} + w_3\mathbf{k})$$
$$= \begin{vmatrix} u_1 & u_2 & u_3 \\ v_1 & v_2 & v_3 \\ w_1 & w_2 & w_3 \end{vmatrix}$$

• Evaluate $\mathbf{a} \cdot \mathbf{b} \times \mathbf{c}$ if $\mathbf{a} = \sqrt{3}\,\mathbf{i} + \mathbf{j}$, $\mathbf{b} = -2\mathbf{i} + 2\sqrt{3}\,\mathbf{j}$, and $\mathbf{c} = 3\mathbf{k}$.

$$\mathbf{a} \cdot \mathbf{b} \times \mathbf{c} = \begin{vmatrix} \sqrt{3} & 1 & 0 \\ -2 & 2\sqrt{3} & 0 \\ 0 & 0 & 3 \end{vmatrix} = 18 + 6 = 24$$

When the base vectors a_1, a_2, and a_3 are not mutually perpendicular, the scalar triple product is given by

$$[\mathbf{uvw}] = [\mathbf{a}_1\mathbf{a}_2\mathbf{a}_3] \begin{vmatrix} u^1 & u^2 & u^3 \\ v^1 & v^2 & v^3 \\ w^1 & w^2 & w^3 \end{vmatrix} = [\mathbf{a}^1\mathbf{a}^2\mathbf{a}^3] \begin{vmatrix} u_1 & u_2 & u_3 \\ v_1 & v_2 & v_3 \\ w_1 & w_2 & w_3 \end{vmatrix}$$

where $\mathbf{a}^1$, $\mathbf{a}^2$, and $\mathbf{a}^3$ are reciprocal base vectors.

14-18. Products of Four Vectors

$$(\mathbf{a} \times \mathbf{b}) \cdot (\mathbf{c} \times \mathbf{d}) = \begin{vmatrix} \mathbf{a} \cdot \mathbf{c} & \mathbf{a} \cdot \mathbf{d} \\ \mathbf{b} \cdot \mathbf{c} & \mathbf{b} \cdot \mathbf{d} \end{vmatrix}$$

$$(\mathbf{a} \times \mathbf{b}) \times (\mathbf{c} \times \mathbf{d}) = [\mathbf{acd}]\mathbf{b} - [\mathbf{bcd}]\mathbf{a} = [\mathbf{abd}]\mathbf{c} - [\mathbf{abc}]\mathbf{d}$$

where brackets indicate scalar triple product (Art. 14-17).

14-19. Derivative of a Vector. If a vector $\mathbf{v}$ is a continuous function of a scalar variable s, the derivative of $\mathbf{v}(s)$ with respect to s is defined by

$$\frac{d\mathbf{v}}{ds} = \lim \frac{\Delta\mathbf{v}}{\Delta s}\bigg]_{\Delta s \to 0} = \lim \frac{\mathbf{v}(s + \Delta s) - \mathbf{v}(s)}{\Delta s}\bigg]_{\Delta s \to 0}$$

If the radius vector from the origin to a curve given as a function of s is $\mathbf{r}(s)$, the derivative $\dfrac{d\mathbf{r}}{ds}$ is a vector tangent to the curve and having the direction of increasing s. If s denotes arc length, $\dfrac{d\mathbf{r}}{ds}$ is a unit tangent vector.

Vector calculus is analogous to scalar calculus.

$$\frac{d\mathbf{u}}{ds} = 0 \qquad \mathbf{u} = \text{constant}$$

$$\frac{d}{ds}(\mathbf{u} + \mathbf{v}) = \frac{d\mathbf{u}}{ds} + \frac{d\mathbf{v}}{ds}$$

$$\frac{d}{ds}(k\mathbf{u}) = k\frac{d\mathbf{u}}{ds} + \frac{dk}{ds}\mathbf{u}$$

$$\frac{d}{ds}(\mathbf{u} \cdot \mathbf{v}) = \mathbf{u} \cdot \frac{d\mathbf{v}}{ds} + \frac{d\mathbf{u}}{ds} \cdot \mathbf{v}$$

$$\frac{d}{ds}(\mathbf{u} \times \mathbf{v}) = \mathbf{u} \times \frac{d\mathbf{v}}{ds} + \frac{d\mathbf{u}}{ds} \times \mathbf{v}$$

As in scalar calculus, integration is the inverse of differentiation. Partial differentiation also is analogous.

• If $\mathbf{u} = (x)^2\mathbf{i} + xy\mathbf{j} + xz\mathbf{k}$, determine $\dfrac{\partial\mathbf{u}}{\partial x}$.

Since $\mathbf{i}$, $\mathbf{j}$, and $\mathbf{k}$ are constants,

$$\frac{\partial\mathbf{u}}{\partial x} = \frac{\partial(x)^2}{\partial x}\mathbf{i} + \frac{\partial xy}{\partial x}\mathbf{j} + \frac{\partial xz}{\partial x}\mathbf{k} = 2x\mathbf{i} + y\mathbf{j} + z\mathbf{k}$$

(If the base vectors also are point functions, they too must be differentiated.)

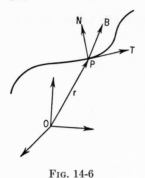

14-20. Frenet's Formulas. Let the radius vector **r** from the origin to a space curve be given as a function of arc length s. Then

$$\mathbf{T} = \frac{d\mathbf{r}}{ds} \qquad (14\text{-}1)$$

is a unit tangent vector (Fig. 14-6). The principal normal is a unit vector **N** perpendicular to **T**. It is given by

$$\frac{d\mathbf{T}}{ds} = \kappa\mathbf{N} \qquad (14\text{-}2)$$

Fɪɢ. 14-6

where κ is a scalar called the curvature. The plane of **T** and **N** is the osculating plane.

The binormal is a unit vector **B** defined by

$$\mathbf{B} = \mathbf{T} \times \mathbf{N} \qquad (14\text{-}3)$$

B, **T**, and **N** form a right-handed set of mutually perpendicular vectors.
The set of equations

$$\frac{d\mathbf{T}}{ds} = \kappa\mathbf{N} \qquad \frac{d\mathbf{N}}{ds} = -\kappa\mathbf{T} + \tau\mathbf{B} \qquad \frac{d\mathbf{B}}{ds} = -\tau\mathbf{N} \qquad (14\text{-}4)$$

where τ is a scalar called the torsion, are known as the Frenet-Serret formulas. For a straight line, $\kappa = 0$ and $\tau = 0$. For a plane curve, $\tau = 0$.

• Calculate the acceleration of a moving particle at time t, given the radius vector $\mathbf{r}(t)$.

If **v** is the velocity at time t and s is the length of the path, then from Eq. (14-1),

$$\mathbf{v} = \frac{d\mathbf{r}}{dt} = \frac{d\mathbf{r}}{ds}\frac{ds}{dt} = \mathbf{T}v$$

From Eq. (14-2),

$$\frac{d\mathbf{T}}{dt} = \frac{d\mathbf{T}}{ds}\frac{ds}{dt} = \kappa\mathbf{N}v$$

Then the acceleration, obtained by differentiating the velocity equation and substituting the last result, is

$$\frac{dv}{dt} = \mathbf{T}\frac{dv}{dt} + v\frac{d\mathbf{T}}{dt} = \mathbf{T}\frac{dv}{dt} + \kappa\mathbf{N}(v)^2$$

The equation indicates that the acceleration lies in the osculating plane. The magnitude of the tangential component of acceleration is $\dfrac{dv}{dt} = \dfrac{d^2s}{dt^2}$ and of the normal component is $\kappa(v)^2$, or the square of the speed divided by the radius of curvature.

14-21. Gradient. A vector representing at a given point both the direction and magnitude of the greatest space rate of increase of a continuous function. It is written grad f, where f is the function.

When f is a scalar function,

$$\operatorname{grad} f = \nabla f = \mathbf{i}\frac{\partial f}{\partial x} + \mathbf{j}\frac{\partial f}{\partial y} + \mathbf{k}\frac{\partial f}{\partial z} = \frac{df}{dn}\,\mathbf{n}$$

where $\mathbf{i}$, $\mathbf{j}$, and $\mathbf{k}$ are a set of mutually perpendicular unit vectors lying along the x, y, and z axes, respectively, $\mathbf{n}$ is a unit vector normal to the surface consisting of the points for which f has a constant value (level surface), and $\frac{df}{dn}$ is the derivative of f in the direction of $\mathbf{n}$.

$$\nabla = \mathbf{i}\frac{\partial}{\partial x} + \mathbf{j}\frac{\partial}{\partial y} + \mathbf{k}\frac{\partial}{\partial z}$$

is an operator called del, nabla, or atled.

In curvilinear coordinates u, v, and w,

$$\nabla = \nabla u\frac{\partial}{\partial u} + \nabla v\frac{\partial}{\partial v} + \nabla w\frac{\partial}{\partial w}$$

$$\nabla u = \frac{\mathbf{r}_v \times \mathbf{r}_w}{[\mathbf{r}_u\mathbf{r}_v\mathbf{r}_w]} \qquad \nabla v = \frac{\mathbf{r}_w \times \mathbf{r}_u}{[\mathbf{r}_u\mathbf{r}_v\mathbf{r}_w]} \qquad \nabla w = \frac{\mathbf{r}_u \times \mathbf{r}_v}{[\mathbf{r}_u\mathbf{r}_v\mathbf{r}_w]}$$

where $\mathbf{r}$ = radius vector

$$\mathbf{r}_u = \frac{\partial \mathbf{r}}{\partial u}, \; \mathbf{r}_v = \frac{\partial \mathbf{r}}{\partial v}, \; \text{and } \mathbf{r}_w = \frac{\partial \mathbf{r}}{\partial w}$$

Brackets indicate scalar triple product.

If r is the distance from the origin to the terminal of the radius vector, then ∇r is a unit radial vector.

When $\mathbf{f}$ is a vector function, grad $\mathbf{f}$ is the dyadic $\nabla\mathbf{f}$.

• Evaluate grad f at $(1,2,3)$ if $f = xyz$.

Since $\frac{\partial f}{\partial x} = yz$, $\frac{\partial f}{\partial y} = xz$, and $\frac{\partial f}{\partial z} = xy$,

$$\operatorname{grad} f = yz\mathbf{i} + xz\mathbf{j} + xy\mathbf{k} = (2)(3)\mathbf{i} + (1)(3)\mathbf{j} + (1)(2)\mathbf{k} = 6\mathbf{i} + 3\mathbf{j} + 2\mathbf{k}$$

Some useful relationships follow:

$$\operatorname{grad}(u + v) = \nabla u + \nabla v$$
$$\operatorname{grad}(uv) = v\,\nabla u + u\,\nabla v$$
$$\operatorname{grad}(\mathbf{u}\cdot\mathbf{v}) = (\mathbf{u}\cdot\nabla)\mathbf{v} + (\mathbf{v}\cdot\nabla)\mathbf{u} + \mathbf{u}\times(\nabla\times\mathbf{v}) + \mathbf{v}\times(\nabla\times\mathbf{u})$$

14-22. Directional Derivative. The component of the gradient of a function at a given point in a given direction.

If **e** is a unit vector acting in the given direction, the directional derivative of a continuous scalar or vector function **f** is

$$\frac{d\mathbf{f}}{ds} = \mathbf{e} \cdot \nabla \mathbf{f}$$

where $\nabla \mathbf{f}$ is the gradient (Art. 14-21).

14-23. Divergence of a Vector. This is given in rectangular coordinates for a vector function **f** by

$$\operatorname{div} \mathbf{f} = \nabla \cdot \mathbf{f} = \frac{\partial f_1}{\partial x} + \frac{\partial f_2}{\partial y} + \frac{\partial f_3}{\partial z}$$

where f_1, f_2, and f_3 are the components of **f** in the **i**, **j**, **k** system of unit vectors parallel to the coordinate axes.

• Evaluate div **r**, where **r** is the radius vector.
Since $\mathbf{r} = x\mathbf{i} + y\mathbf{j} + z\mathbf{k}$,

$$\operatorname{div} \mathbf{r} = \frac{\partial x}{\partial x} + \frac{\partial y}{\partial y} + \frac{\partial z}{\partial z} = 1 + 1 + 1 = 3$$

• Evaluate div grad f.

$$\operatorname{div} \operatorname{grad} f = \nabla \cdot \nabla f = \nabla^2 f = \frac{\partial f^2}{\partial x^2} + \frac{\partial^2 f}{\partial y^2} + \frac{\partial^2 f}{\partial z^2}$$

The operator $\nabla^2 = \dfrac{\partial^2}{\partial x^2} + \dfrac{\partial^2}{\partial y^2} + \dfrac{\partial^2}{\partial z^2}$. It is called the Laplacian.

If f is a scalar function and **u** and **v** are vector functions,

$$\operatorname{div} (\mathbf{u} + \mathbf{v}) = \operatorname{div} \mathbf{u} + \operatorname{div} \mathbf{v}$$
$$\operatorname{div} (f\mathbf{v}) = (\nabla f) \cdot \mathbf{v} + f \operatorname{div} \mathbf{v}$$
$$\operatorname{div} (\mathbf{u} \times \mathbf{v}) = \mathbf{v} \cdot \operatorname{curl} \mathbf{u} - \mathbf{u} \cdot \operatorname{curl} \mathbf{v}$$
$$\operatorname{div} \operatorname{curl} \mathbf{u} = 0$$

In curvilinear coordinates, **u**, **v**, **w**,

$$\nabla \cdot \mathbf{f} = \frac{1}{[\mathbf{r}_u \mathbf{r}_v \mathbf{r}_w]} \left[\frac{\partial}{\partial u} (\mathbf{r}_v \times \mathbf{r}_w \cdot \mathbf{f}) + \frac{\partial}{\partial v} (\mathbf{r}_w \times \mathbf{r}_u \cdot \mathbf{f}) + \frac{\partial}{\partial w} (\mathbf{r}_u \times \mathbf{r}_v \cdot \mathbf{f}) \right]$$

where **r** = radius vector

$$\mathbf{r}_u = \frac{\partial \mathbf{r}}{\partial u}, \ \mathbf{r}_v = \frac{\partial \mathbf{r}}{\partial v}, \ \text{and} \ \mathbf{r}_w = \frac{\partial \mathbf{r}}{\partial w}$$

Brackets indicate scalar triple product.

14-24. Curl (Rotation) of a Vector. This is given in rectangular coordinates for a vector function **f** by

$$\text{curl } \mathbf{f} = \text{rot } \mathbf{f} = \nabla \times \mathbf{f} = \left(\frac{\partial f_3}{\partial y} - \frac{\partial f_2}{\partial z} \right) \mathbf{i} + \left(\frac{\partial f_1}{\partial z} - \frac{\partial f_3}{\partial x} \right) \mathbf{j} + \left(\frac{\partial f_2}{\partial x} - \frac{\partial f_1}{\partial y} \right) \mathbf{k}$$

$$= \begin{vmatrix} \mathbf{i} & \mathbf{j} & \mathbf{k} \\ \dfrac{\partial}{\partial x} & \dfrac{\partial}{\partial y} & \dfrac{\partial}{\partial z} \\ f_1 & f_2 & f_3 \end{vmatrix}$$

where f_1, f_2, f_3 are components of $\mathbf{f}$ in the $\mathbf{i}$, $\mathbf{j}$, $\mathbf{k}$ system of unit vectors parallel to the coordinate axes.

If $\mathbf{r}$ is the radius vector $x\mathbf{i} + y\mathbf{j} + z\mathbf{k}$, curl $\mathbf{r} = 0$.

If f is a scalar function and $\mathbf{u}$ and $\mathbf{v}$ are vector functions,

$$\text{curl } (\mathbf{u} + \mathbf{v}) = \text{curl } \mathbf{u} + \text{curl } \mathbf{v}$$
$$\text{div curl } \mathbf{v} = \nabla \cdot (\nabla \times \mathbf{v}) = 0$$
$$\text{curl curl } \mathbf{v} = \nabla \times (\nabla \times \mathbf{v}) = \nabla(\nabla \cdot \mathbf{v}) - \nabla^2 \mathbf{v}$$
$$\text{curl grad } f = \nabla \times \nabla f = 0$$
$$\text{curl } (f\mathbf{v}) = (\nabla f) \times \mathbf{v} + f \text{ curl } \mathbf{v}$$
$$\text{curl } (\mathbf{u} \times \mathbf{v}) = (\mathbf{v} \cdot \nabla)\mathbf{u} - (\mathbf{u} \cdot \nabla)\mathbf{v} + \mathbf{u} \text{ div } \mathbf{v} - \mathbf{v} \text{ div } \mathbf{u}$$

• Evaluate the constant angular velocity $\boldsymbol{\omega}$ of a rigid body rotating about a fixed axis given the velocities $\mathbf{v}$ of its points.

Since $\mathbf{v} = \boldsymbol{\omega} \times \mathbf{r}$, curl $\mathbf{v} = \text{curl } (\boldsymbol{\omega} \times \mathbf{r})$. But

$$\text{curl } (\boldsymbol{\omega} \times \mathbf{r}) = \mathbf{r} \cdot \nabla \boldsymbol{\omega} - \boldsymbol{\omega} \cdot \nabla \mathbf{r} + \boldsymbol{\omega} \text{ div } \mathbf{r} - \mathbf{r} \text{ div } \boldsymbol{\omega} = 0 - \boldsymbol{\omega} + 3\boldsymbol{\omega} - 0 = 2\boldsymbol{\omega}$$

Therefore, the angular rotation $\boldsymbol{\omega} = \frac{1}{2} \text{ curl } \mathbf{v}$.

In curvilinear coordinates u, v, and w,

$$\nabla \times \mathbf{f} = \frac{1}{[\mathbf{r}_u \mathbf{r}_v \mathbf{r}_w]} \begin{vmatrix} \mathbf{r}_u & \mathbf{r}_v & \mathbf{r}_w \\ \dfrac{\partial}{\partial u} & \dfrac{\partial}{\partial v} & \dfrac{\partial}{\partial w} \\ \mathbf{r}_u \cdot \mathbf{f} & \mathbf{r}_v \cdot \mathbf{f} & \mathbf{r}_w \cdot \mathbf{f} \end{vmatrix}$$

where $\mathbf{r} =$ radius vector

$$\mathbf{r}_u = \frac{\partial \mathbf{r}}{\partial u}, \ \mathbf{r}_v = \frac{\partial \mathbf{r}}{\partial v}, \text{ and } \mathbf{r}_w = \frac{\partial \mathbf{r}}{\partial w}$$

Brackets indicate scalar triple product.

14-25. Total Differential. If $\mathbf{r}$ is the radius vector and $\mathbf{f}$ is a vector or tensor function of $\mathbf{r}$, the total differential of $\mathbf{f}$ is

$$d\mathbf{f} = d\mathbf{r} \cdot \nabla \mathbf{f}$$

If $\mathbf{f}$ is a function $\mathbf{f}(u,v,w)$ of rectangular or curvilinear coordinates,

$$d\mathbf{f} = \frac{\partial \mathbf{f}}{\partial u} \, du + \frac{\partial \mathbf{f}}{\partial v} \, dv + \frac{\partial \mathbf{f}}{\partial w} \, dw$$

14-26. Irrotational (Lamellar) Vectors. Those vectors for which the curl is zero.

The gradient of a scalar function with continuous second derivatives in a given region is irrotational in that region. If curl $\mathbf{f} = 0$, $\mathbf{f}$ is the gradient of a scalar φ, called the scalar potential of $\mathbf{f}$. If

$$\mathbf{f} = f_1\mathbf{i} + f_2\mathbf{j} + f_3\mathbf{k} = \text{grad } \varphi$$

where $\mathbf{i}$, $\mathbf{j}$, and $\mathbf{k}$ are unit vectors parallel to the x, y, and z axes,

$$\frac{\partial \varphi}{\partial x} = f_1 \qquad \frac{\partial \varphi}{\partial y} = f_2 \qquad \frac{\partial \varphi}{\partial z} = f_3$$

14-27. Solenoidal Vectors. Those vectors for which divergence is zero.

The curl of a vector function with continuous second derivatives in a given region is solenoidal in that region. A vector $\mathbf{g}$ is the vector potential of a vector $\mathbf{f}$ if $\mathbf{f} = \text{curl } \mathbf{g}$.

14-28. Metric Coefficients. If $\mathbf{r}$ is the radius vector from the origin of any coordinate system to a point P, and u^1 and u^2 are parametric coordinates of a surface, then

$$d\mathbf{r} = \frac{\partial \mathbf{r}}{\partial u^1}\, du^1 + \frac{\partial \mathbf{r}}{\partial u^2}\, du^2 = \mathbf{a}_1\, du^1 + \mathbf{a}_2\, du^2$$

where $\mathbf{a}_1 = \dfrac{\partial \mathbf{r}}{\partial u^1}$, a vector tangent to u^1 curve (u^2 equals a constant)

$\mathbf{a}_2 = \dfrac{\partial \mathbf{r}}{\partial u^2}$, a vector tangent to u^2 curve

$\mathbf{a}_1, \mathbf{a}_2 = $ base vectors that vary from point to point on the surface

Metric coefficients are defined by

$$g_{ij} = \mathbf{a}_i \cdot \mathbf{a}_j \qquad i, j = 1, 2, \ldots, n$$

Thus, $g_{ij} = g_{ji}$. In a cartesian system, $g_{ij} = 1$ if $i = j$ and $g_{ij} = 0$ if $i \neq j$. The base vectors are orthogonal only if

$$g_{12} = g_{21} = g_{31} = g_{13} = g_{23} = g_{32} = 0$$

(see also Arts. 14-29 and 14-30).

Similar coefficients are defined for the reciprocal system of vectors $\mathbf{a}^i$.

$$g^{ij} = \mathbf{a}^i \cdot \mathbf{a}^j$$

In curvilinear coordinates $u^1, u^2, \ldots, u^n$,

$$du_j = \sum_{i=1}^{n} g_{ij}\, du^i \qquad du^i = \sum_{j=1}^{n} g^{ij}\, du_j$$

$$\mathbf{a}^j = \sum_{i=1}^{n} g^{ij}\mathbf{a}_i \qquad \mathbf{a}_i = \sum_{j=1}^{n} g_{ij}\mathbf{a}^j$$

Note that the metric coefficients are operators that raise or lower the indices of the terms by which they are multiplied in accordance with the above equations.

If g is the determinant of the metric coefficients g_{ij} and g' is the determinant of the coefficients g^{ji}, then

$$g^{ji} = G^{ij} \qquad g_{ij} = G_{ji}$$

where G^{ij} is the cofactor of the element g_{ij} of the determinant g divided by g and G_{ji} is the cofactor of the element g^{ji} of the determinant g' divided by g'.

If x^1, x^2, and x^3 are rectangular cartesian coordinates related to curvilinear coordinates u^1, u^2, and u^3 by the equations

$$x^1 = x^1(u^1,u^2,u^3) \qquad x^2 = x^2(u^1,u^2,u^3) \qquad x^3 = x^3(u^1,u^2,u^3)$$

$$dx^i = \sum_{j=1}^{3} \frac{\partial x^i}{\partial u^j} du^j$$

then the metric coefficients, which are known in the cartesian system, may be determined in the curvilinear system from

$$g_{ij} = \frac{\partial x^1}{\partial u^i}\frac{\partial x^1}{\partial u^j} + \frac{\partial x^2}{\partial u^i}\frac{\partial x^2}{\partial u^j} + \frac{\partial x^3}{\partial u^i}\frac{\partial x^3}{\partial u^j}$$

• Determine the metric coefficients in spherical coordinates, which are related to cartesian coordinates by the following equations:

$$x^1 = r \sin \theta \cos \varphi \qquad x^2 = r \sin \theta \sin \varphi \qquad x^3 = r \cos \theta$$

Let $u^1 = r$, $u^2 = \theta$, and $u^3 = \varphi$. Then

$$g_{11} = \left(\frac{\partial x^1}{\partial u^1}\right)^2 + \left(\frac{\partial x^2}{\partial u^1}\right)^2 + \left(\frac{\partial x^3}{\partial u^1}\right)^2 = \sin^2 \theta \cos^2 \varphi + \sin^2 \theta \sin^2 \varphi + \cos^2 \theta = 1$$

$$g_{22} = \left(\frac{\partial x^1}{\partial u^2}\right)^2 + \left(\frac{\partial x^2}{\partial u^2}\right)^2 + \left(\frac{\partial x^3}{\partial u^2}\right)^2 = (r)^2 \cos^2 \theta \cos^2 \varphi + (r)^2 \cos^2 \theta \sin^2 \varphi$$
$$+ (r)^2 \sin^2 \theta = (r)^2$$

$$g_{33} = \left(\frac{\partial x^1}{\partial u^3}\right)^2 + \left(\frac{\partial x^2}{\partial u^3}\right)^2 + \left(\frac{\partial x^3}{\partial u^3}\right)^2 = (r)^2 \sin^2 \theta \sin^2 \varphi + (r)^2 \sin^2 \theta \cos^2 \varphi$$
$$+ 0 = (r)^2 \sin^2 \theta$$

Since spherical coordinates are orthogonal, all other coefficients are zero. For example,

$$g_{12} = \frac{\partial x^1}{\partial u^1}\frac{\partial x^1}{\partial u^2} + \frac{\partial x^2}{\partial u^1}\frac{\partial x^2}{\partial u^2} + \frac{\partial x^3}{\partial u^1}\frac{\partial x^3}{\partial u^2}$$
$$= (\sin \theta \cos \varphi)(r \cos \theta \cos \varphi) + (\sin \theta \sin \varphi)(r \cos \theta \sin \varphi)$$
$$+ \cos \theta(-r \sin \theta) = 0$$

14-29. First Fundamental Form, Angles, Areas, and Volumes. If $\mathbf{r}$ is the radius vector to a surface with parametric coordinates u^1 and u^2, then the differential length of a curve on that surface is given by the first fundamental form:

$$(ds)^2 = d\mathbf{r} \cdot d\mathbf{r} = g_{11}\, du^1\, du^1 + 2g_{12}\, du^1\, du^2 + g_{22}\, du^2\, du^2$$

where the g_{ij} are metric coefficients (Art. 14-28). In orthogonal cartesian coordinates, $g_{12} = 0$ and $g_{11} = g_{22} = 1$.

In curvilinear coordinates, $u^1, u^2, \ldots, u^n$, or with reciprocal base vectors,

$$(ds)^2 = \Sigma g_{ij}\, du^i\, du^j = \Sigma g^{ij}\, du_i\, du_j = \Sigma\, du^i\, du_i$$

summed for values of i and j from 1 to n.

The angle θ between two differential surface vectors $d\mathbf{s}$ and $\delta\mathbf{s}$ is given by

$$\cos\theta = \sum g_{ij}\frac{du^i}{ds}\frac{\delta u^j}{\delta s} = \sum g^{ij}\frac{du_i}{ds}\frac{\delta u_j}{\delta s} \qquad i,j = 1,2,\ldots,n$$

where ds = magnitude of $d\mathbf{s}$

δs = magnitude of $\delta\mathbf{s}$

Let $d\mathbf{s}_1$, $d\mathbf{s}_2$, and $d\mathbf{s}_3$, at a point $P(u^1, u^2, u^3)$, represent elements of arc in the directions of the base vectors $\mathbf{a}_1$, $\mathbf{a}_2$, and $\mathbf{a}_3$, respectively. (For example, $d\mathbf{s}_1$ lies along the u^1 curve, $u^2 = 0$, $u^3 = 0$.) Then

$$ds_1 = \sqrt{g_{11}}\, du^1 \qquad ds_2 = \sqrt{g_{22}}\, du^2 \qquad ds_3 = \sqrt{g_{33}}\, du^3$$

The areas of the parallelograms constructed on ds_1, ds_2, and ds_3 are

$$dA_1 = \sqrt{g_{22}g_{33} - g_{23}g_{23}}\, du^2\, du^3$$
$$dA_2 = \sqrt{g_{33}g_{11} - g_{31}g_{31}}\, du^3\, du^1$$
$$dA_3 = \sqrt{g_{11}g_{22} - g_{12}g_{12}}\, du^1\, du^2$$

The volume of a parallelepiped constructed on the three vectors is

$$dV = \sqrt{g}\, du^1\, du^2\, du^3$$

where g is the determinant of the metric coefficients g_{ij}. In orthogonal coordinates, $dV = \sqrt{g_{11}g_{22}g_{33}}\, dx^1\, dx^2\, dx^3$.

Equations of the same form hold for the reciprocal system with g^{ij}.

• Find differential arc length and volume in cylindrical coordinates, which are related to cartesian coordinates by the following equations:

$$x^1 = r\cos\varphi \qquad x^2 = r\sin\varphi \qquad x^3 = z$$

Let $u^1 = r$, $u^2 = \varphi$, and $u^3 = z$. From the equations of Art. 14-28,

$$g_{11} = \left(\frac{\partial x^1}{\partial u^1}\right)^2 + \left(\frac{\partial x^2}{\partial u^1}\right)^2 + \left(\frac{\partial x^3}{\partial u^1}\right)^2 = \cos^2\varphi + \sin^2\varphi + 0 = 1$$

$$g_{22} = \left(\frac{\partial x^1}{\partial u^2}\right)^2 + \left(\frac{\partial x^2}{\partial u^2}\right)^2 + \left(\frac{\partial x^3}{\partial u^2}\right)^2 = (r)^2\sin^2\varphi + (r)^2\cos^2\varphi + 0 = (r)^2$$

$$g_{33} = \left(\frac{\partial x^1}{\partial u^3}\right)^2 + \left(\frac{\partial x^2}{\partial u^3}\right)^2 + \left(\frac{\partial x^3}{\partial u^3}\right)^2 = 0 + 0 + 1 = 1$$

The remaining metric coefficients equal zero.

$$g = \begin{vmatrix} g_{11} & g_{12} & g_{13} \\ g_{21} & g_{22} & g_{23} \\ g_{31} & g_{32} & g_{33} \end{vmatrix} = \begin{vmatrix} 1 & 0 & 0 \\ 0 & (r)^2 & 0 \\ 0 & 0 & 1 \end{vmatrix} = (r)^2$$

$$(ds)^2 = \Sigma g_{ij}\,du^i\,du^j = g_{11}(dr)(dr) + g_{22}(d\varphi)(d\varphi) + g_{33}(dz)(dz)$$
$$= (dr)^2 + r^2(d\varphi)^2 + (dz)^2$$
$$dV = \sqrt{g}\,du^1\,du^2\,du^3 = r\,dr\,d\varphi\,dz$$

14-30. Second Fundamental Form and Surface Curvatures. If **r** is the radius vector to a surface curve and **n** is a unit vector normal to the surface whose parametric coordinates are u^1 and u^2, then

$$d\mathbf{n} \cdot d\mathbf{r} = l\,du^1\,du^1 + 2m\,du^1\,du^2 + n\,du^2\,du^2$$

$$l = -\frac{\left[\dfrac{\partial \mathbf{r}}{\partial u^1}\ \dfrac{\partial \mathbf{r}}{\partial u^2}\ \dfrac{\partial^2 \mathbf{r}}{\partial u^1\,\partial u^1}\right]}{\sqrt{g_{11}g_{22} - (g_{12})^2}}$$

$$m = -\frac{\left[\dfrac{\partial \mathbf{r}}{\partial u^1}\ \dfrac{\partial \mathbf{r}}{\partial u^2}\ \dfrac{\partial^2 \mathbf{r}}{\partial u^1\,\partial u^2}\right]}{\sqrt{g_{11}g_{22} - (g_{12})^2}}$$

$$n = -\frac{\left[\dfrac{\partial \mathbf{r}}{\partial u^1}\ \dfrac{\partial \mathbf{r}}{\partial u^2}\ \dfrac{\partial^2 \mathbf{r}}{\partial u^2\,\partial u^2}\right]}{\sqrt{g_{11}g_{22} - (g_{12})^2}}$$

where brackets denote scalar triple product.

The normal curvature of the curve is

$$N = \frac{d\mathbf{n} \cdot d\mathbf{r}}{(ds)^2} = \frac{l\,du^1\,du^1 + 2m\,du^1\,du^2 + n\,du^2\,du^2}{g_{11}\,du^1\,du^1 + 2g_{12}\,du^1\,du^2 + g_{22}\,du^2\,du^2}$$

The total or Gaussian curvature of the surface is

$$K = \frac{ln - (m)^2}{g_{11}g_{22} - (g_{12})^2}$$

The mean curvature of the surface is

$$H = \frac{g_{11}n - 2g_{12}m + g_{22}l}{g_{11}g_{22} - (g_{12})^2}$$

14-31. Integral Transformations. If S is a surface bounding a volume V, u is a scalar point function, $\mathbf{v}$ is a vector function, dA is an infinitesimal element of S, $\mathbf{n}$ is an outward unit vector normal to dA, and dV is an infinitesimal element of volume, then

$$\iiint_V \operatorname{grad} u \, dV = \iint_S \mathbf{n} \, dA$$

$$\iiint_V \operatorname{div} \mathbf{v} \, dV = \iint_S \mathbf{n} \cdot \mathbf{v} \, dA \qquad \text{divergence theorem}$$

$$\iiint_V \operatorname{curl} \mathbf{v} \, dV = \iint_S \mathbf{n} \times \mathbf{v} \, dA$$

If C is a curve bounding a surface S, ds is an element of arc, and $\mathbf{T}$ is a unit vector tangent to C, measured positive when the curve is traversed with S on the left, then

$$\iint_S \mathbf{n} \times \operatorname{grad} u \, dA = \int_C u\mathbf{T} \, ds$$

$$\iint_S \mathbf{n} \cdot \operatorname{curl} \mathbf{v} \, dA = \int_C \mathbf{v} \cdot \mathbf{T} \, ds \qquad \text{Stokes' theorem}$$

If $\mathbf{r}$ is the radius vector from the origin O to a surface S enclosing a volume V, and $\mathbf{r}_1$ is the unit radial vector, then

$$\iint_S \frac{\mathbf{n} \cdot \mathbf{r}_1}{(r)^2} \, dA = \begin{cases} 0 & \text{for } O \text{ outside } S \\ 4\pi & \text{for } O \text{ inside } S \end{cases} \qquad \text{Gauss' theorem}$$

If u and w are two scalar functions,

$$\iiint_V (u\nabla^2 w + \nabla u \cdot \nabla w) \, dV = \iint_S u\mathbf{n} \cdot \nabla w \, dA$$

$$\text{Green's theorem in first form}$$

$$\iiint_V (u\nabla^2 w - w\nabla^2 u) \, dV = \iint_S \mathbf{n} \cdot (u\nabla w - w\nabla u) \, dA$$

$$\text{Green's theorem in second form}$$

14-32. Dyadic. A tensor of valence two. It can be represented by

$$\varphi = \mathbf{a}_1\mathbf{b}_1 + \mathbf{a}_2\mathbf{b}_2 + \cdots + \mathbf{a}_n\mathbf{b}_n$$

where $\mathbf{a}_i$ and $\mathbf{b}_i$ are vectors, called antecedents and consequents, respectively. Each term $\mathbf{a}_i\mathbf{b}_i$ is a dyad.

Two dyadics φ and ψ are equal if $\varphi \cdot \mathbf{r} = \psi \cdot \mathbf{r}$ or $\mathbf{s} \cdot \varphi = \mathbf{s} \cdot \psi$, for every vector $\mathbf{r}$ and $\mathbf{s}$.

A dyadic $\varphi = 0$ if $\varphi \cdot \mathbf{r} = 0$ or $\mathbf{s} \cdot \varphi = 0$.

The sum of two dyadics equals the sum of all the dyads of each.

In three-dimensional space, any dyadic can be reduced to the sum of three dyads. This can be done by expressing the antecedents or the consequents in terms of three linearly independent vectors (Arts. 14-7 and 14-8).

The unit dyadic or idemfactor is defined by

$$\mathbf{I} \cdot \mathbf{r} = \mathbf{r} \cdot \mathbf{I} = \mathbf{r}$$

for every vector $\mathbf{r}$. If $\mathbf{a}^1$, $\mathbf{a}^2$, $\mathbf{a}^3$ and $\mathbf{a}_1$, $\mathbf{a}_2$, $\mathbf{a}_3$ are reciprocal sets of base vectors,

$$\mathbf{I} = \mathbf{a}_1\mathbf{a}^1 + \mathbf{a}_2\mathbf{a}^2 + \mathbf{a}_3\mathbf{a}^3 = \mathbf{a}^1\mathbf{a}_1 + \mathbf{a}^2\mathbf{a}_2 + \mathbf{a}^3\mathbf{a}_3$$

If $\mathbf{i},\mathbf{j},\mathbf{k}$ is a right-handed set of mutually perpendicular unit vectors,

$$\mathbf{I} = \mathbf{ii} + \mathbf{jj} + \mathbf{kk}$$

Two dyadics φ and ψ are reciprocals if $\varphi \cdot \psi = \psi \cdot \varphi = \mathbf{I}$. If $\varphi = \mathbf{e}_1\mathbf{f}_1 + \mathbf{e}_2\mathbf{f}_2 + \mathbf{e}_3\mathbf{f}_3$, then $\varphi^{-1} = \mathbf{f}^1\mathbf{e}^1 + \mathbf{f}^2\mathbf{e}^2 + \mathbf{f}^3\mathbf{e}^3$, where $\mathbf{e}^1,\mathbf{e}^2,\mathbf{e}^3$ is the reciprocal of the set $\mathbf{e}_1,\mathbf{e}_2,\mathbf{e}_3$, and $\mathbf{f}^1,\mathbf{f}^2,\mathbf{f}^3$ is the reciprocal of the set $\mathbf{f}_1,\mathbf{f}_2,\mathbf{f}_3$.

• Solve $\varphi \cdot \mathbf{r} = \mathbf{v}$ for $\mathbf{r}$.

Multiply both sides of the equation by the reciprocal of φ. $\varphi^{-1} \cdot \varphi \cdot \mathbf{r} = \varphi^{-1} \cdot \mathbf{v}$. Hence, $\mathbf{r} = \varphi^{-1} \cdot \mathbf{v}$.

The nonian form of a dyadic is

$$\begin{aligned}
\varphi = \quad & a_{11}\mathbf{ii} + a_{12}\mathbf{ij} + a_{13}\mathbf{ik} \\
+ \; & a_{21}\mathbf{ji} + a_{22}\mathbf{jj} + a_{23}\mathbf{jk} \\
+ \; & a_{31}\mathbf{ki} + a_{32}\mathbf{kj} + a_{33}\mathbf{kk}
\end{aligned}$$

The nonian form of its reciprocal is

$$\begin{aligned}
\varphi^{-1} = \frac{1}{D} \; [& A_{11}\mathbf{ii} + A_{21}\mathbf{ij} + A_{31}\mathbf{ik} \\
+ \; & A_{12}\mathbf{ji} + A_{22}\mathbf{jj} + A_{32}\mathbf{jk} \\
+ \; & A_{13}\mathbf{ki} + A_{23}\mathbf{kj} + A_{33}\mathbf{kk}]
\end{aligned}$$

where $D = \begin{vmatrix} a_{11} & a_{12} & a_{13} \\ a_{21} & a_{22} & a_{23} \\ a_{31} & a_{32} & a_{33} \end{vmatrix}$

A_{rs} = cofactor of typical term a_{rs} (Art. 13-9)

Also, if $\mathbf{a}_1$, $\mathbf{a}_2$, $\mathbf{a}_3$ and $\mathbf{a}^1$, $\mathbf{a}^2$, $\mathbf{a}^3$ are reciprocal sets of base vectors, φ can be written in four different forms (each with nine components when $\mathbf{i},\mathbf{j} = 1, 2, 3$).

$$\varphi = \sum_i \sum_j b^{ij}\mathbf{a}_i\mathbf{a}_j = \sum_i \sum_j b_{ij}\mathbf{a}^i\mathbf{a}^j = \sum_i \sum_j b^i_{\cdot j}\mathbf{a}_i\mathbf{a}^j = \sum_i \sum_j b^{\cdot j}_{i\cdot}\mathbf{a}^i\mathbf{a}_j$$

The components b^{ij} are contravariant, b_{ij} are covariant, and $b^i_{.j}$ and $b^j_{i.}$ are mixed (Art. 14-13).

If $\mathbf{r}$ is the radius vector from the origin O to a point P, then

$$\mathbf{r} \cdot \boldsymbol{\varphi} \cdot \mathbf{r} = 1$$

is the equation of a general quadric surface with center at O, and

$$\mathbf{r} \cdot \boldsymbol{\varphi} \cdot \mathbf{r} = 0$$

is the equation of a cone with vertex at O.

Any dyadic $\boldsymbol{\varphi}$ acting as a prefactor (or postfactor) on $\mathbf{r}$ represents an affine transformation. Thus, $\mathbf{q} = \boldsymbol{\varphi} \cdot \mathbf{r}$ (or $\mathbf{r} \cdot \boldsymbol{\varphi}$) transforms a point P into a point Q, straight lines and planes being transformed into straight lines and planes.

14-33. Absolute Tensor. A tensor consists of two parts, scalars (called its components) and zero, one, two, three, or more base vectors associated with each scalar. The number of base vectors determines the rank or valence of the tensor.

An absolute tensor of valence m in an n-dimensional coordinate system is a set of n^m components associated with m base vectors that is invariant under an admissible coordinate transformation (Art. 14-3).

Thus, components of a tensor transform as the product of m vector components. Since each of these components may be covariant or contravariant, the components are of 2^m types. One of these is purely covariant ($T_{ij\ldots k}$), another is purely contravariant ($T^{ij\cdots k}$), and the remainder are mixed.

A scalar is a tensor of zero valence; a vector is a tensor of valence 1; a dyadic is a tensor of valence 2. A triadic in a three-dimensional system may be represented by

$$\mathbf{T} = \Sigma T_{ijk}\mathbf{a}^i\mathbf{a}^j\mathbf{a}^k = \Sigma T^{ijk}\mathbf{a}_i\mathbf{a}_j\mathbf{a}_k = \Sigma T^{ij}_{..k}\mathbf{a}_i\mathbf{a}_j\mathbf{a}^k = \cdots$$

summed with respect to i, j, and k over the range 1 to 3. This tensor may be written in $2^3 = 8$ different forms, each of which has $3^3 = 27$ components associated with the base vectors $\mathbf{a}_i$, $\mathbf{a}_j$, and $\mathbf{a}_k$ or their reciprocals $\mathbf{a}^i$, $\mathbf{a}^j$, and $\mathbf{a}^k$. The indices on the components occur in the same order as the indices on the base vectors. Contravariant indices on the components are associated with covariant indices on the vectors, and covariant indices on the components are associated with contravariant indices on the vectors.

Contravariant components A^{ijk} of a triadic transform into B^{lmn} under the coordinate transformation $v^s = v^s(u^1, u^2, \ldots, u^n)$ according to the law

$$B^{lmn} = \sum \frac{\partial v^l}{\partial u^i}\frac{\partial v^m}{\partial u^j}\frac{\partial v^n}{\partial u^k}A^{ijk} \qquad A^{ijk} = \sum \frac{\partial u^i}{\partial v^l}\frac{\partial u^j}{\partial v^m}\frac{\partial u^k}{\partial v^n}B^{lmn}$$

The left-hand set of equations is summed with respect to i,j,k; the right-hand set is summed with respect to l,m,n. The covariant components transform according to the law

$$B_{lmn} = \sum \frac{\partial u^i}{\partial v^l} \frac{\partial u^j}{\partial v^m} \frac{\partial u^k}{\partial v^n} A_{ijk} \qquad A_{ijk} = \sum \frac{\partial v^l}{\partial u^i} \frac{\partial v^m}{\partial u^j} \frac{\partial v^n}{\partial u^k} B_{lmn}$$

In these and mixed forms, there is a base vector corresponding to an index of the component with respect to which the component transforms contragrediently. For example, compare the partial derivatives in the transformation of the scalar part of a mixed tensor with those of the vector part.

$$B^{lm.}_{..n} = \sum \frac{\partial v^l}{\partial u^i} \frac{\partial v^m}{\partial u^j} \frac{\partial u^k}{\partial v^n} A^{ij.}_{..k}$$

$$\mathbf{b}_l \mathbf{b}_m \mathbf{b}^n = \sum \frac{\partial u^i}{\partial v^l} \frac{\partial u^j}{\partial v^m} \frac{\partial v^n}{\partial u^k} \mathbf{a}_i \mathbf{a}_j \mathbf{a}^k$$

If the components of a tensor vanish in one coordinate system, they vanish in all.

Tensor equations maintain their forms in all coordinate systems.

14-34. Relative Tensors. (See also Art. 14-33.) Invariants that transform in the same manner as corresponding contravariant, covariant, and mixed absolute tensors except that each component is multiplied by the same power of the jacobian of the coordinates.

The power of the jacobian is called the weight of the tensor. A tensor of weight zero is an absolute tensor; a tensor of weight one is a tensor density.

• What is the law of transformation of $A^i_{.j}$, a tensor of weight two, if its components are $B^l_{.m}$ under the coordinate transformation $v^s = v^s(u^1, u^2, \ldots, u^n)$?

If $\left| \dfrac{\partial u}{\partial v} \right|$ represents the jacobian of the transformation (Art. 8-14), the components transform in accordance with

$$B^l_{.m} = \left| \frac{\partial u}{\partial v} \right|^2 \sum A^i_{.j} \frac{\partial v^l}{\partial u^i} \frac{\partial u^j}{\partial v^m}$$

summed over the indices i,j.

14-35. Operations with Tensors. The sum of two tensors of the same valence and type is a new tensor with the same characteristics.

• Add $\mathbf{A} = \Sigma A^{ijk} \mathbf{a}_i \mathbf{a}_j \mathbf{a}_k$ and $\mathbf{B} = \Sigma B^{ij}_{..t} \mathbf{a}_i \mathbf{a}_j \mathbf{a}^t$.

Before the addition can be performed, $\mathbf{A}$ must be converted into the same type tensor as $\mathbf{B}$ (or vice versa). Use the metric coefficients as operators to do this (Art. 14-28).

$$A^{ij}_{..t} = \Sigma g_{tk} A^{ijk} \qquad \mathbf{a}_i \mathbf{a}_j \mathbf{a}^t = \Sigma g^{tk} \mathbf{a}_i \mathbf{a}_j \mathbf{a}_k$$

where the summation is with respect to k. Hence,

$$A = \Sigma A^{ij}_{\cdot\cdot i} \mathbf{a}_i \mathbf{a}_j \mathbf{a}^t$$
$$A + B = \Sigma A^{ij}_{\cdot\cdot i} \mathbf{a}_i \mathbf{a}_j \mathbf{a}^t + \Sigma B^{ij}_{\cdot\cdot i} \mathbf{a}_i \mathbf{a}_j \mathbf{a}^t = \Sigma (A^{ij}_{\cdot\cdot i} + B^{ij}_{\cdot\cdot i}) \mathbf{a}_i \mathbf{a}_j \mathbf{a}^t$$

The product of two tensors of valence m_1 and m_2, respectively, is a tensor of valence $m_1 + m_2$; the order of the base vectors is maintained.

• Premultiply $B = \Sigma B_{rs} \mathbf{a}^r \mathbf{a}^s$ by $A = \Sigma A^{ij}_{\cdot\cdot i} \mathbf{a}_i \mathbf{a}_j \mathbf{a}^t$.

$$AB = \Sigma A^{ij}_{\cdot\cdot i} B_{rs} \mathbf{a}_i \mathbf{a}_j \mathbf{a}^t \mathbf{a}^r \mathbf{a}^s$$

summed for all values of each index. (Note that $BA = \Sigma B_{rs} A^{ij}_{\cdot\cdot i} \mathbf{a}^r \mathbf{a}^s \mathbf{a}_i \mathbf{a}_j \mathbf{a}^t$, a different tensor.)

The direct or inner product is produced in the same way, except that the dot product is formed between the two adjoining base vectors of each tensor.

$$A \cdot B = \Sigma A^{ij}_{\cdot\cdot i} B_{rs} \mathbf{a}_i \mathbf{a}_j \mathbf{a}^t \cdot \mathbf{a}^r \mathbf{a}^s = \Sigma g^{tr} A^{ij}_{\cdot\cdot i} B_{rs} \mathbf{a}_i \mathbf{a}_j \mathbf{a}^s$$

The result is a tensor of valence $m_1 + m_2 - 2 = 3 + 2 - 2 = 3$.

A mixed tensor may be contracted from valence m to $m - 2$. This can be done by setting an upper index equal to a lower index and summing over the index range.

• Contract $T = \Sigma T^{ij}_{\cdot\cdot k} \mathbf{a}_i \mathbf{a}_j \mathbf{a}^k$.

Set $j = k$ (equivalent to replacing $\mathbf{a}_j \mathbf{a}^k$ by $\mathbf{a}_j \cdot \mathbf{a}^k = g^k_j = \delta^k_j$). Then,

$$V = \Sigma T^{ij}_{\cdot\cdot j} \mathbf{a}_i$$

is a tensor of valence two less than that of the given tensor.

To contract a tensor with respect to two indices on the same level, first raise or lower one by using the metric coefficients as operators.

14-36. Quotient Law for Tensors. If $A^i B^j_{k\cdot\cdot\cdot}$ are tensor components and A^i represents the components of an arbitrary vector, then $B^j_{k\cdot\cdot\cdot}$ represents the components of a tensor.

14-37. Christoffel Symbols. If $\mathbf{a}_1, \mathbf{a}_2, \ldots, \mathbf{a}_n$ are unitary base vectors and the metric coefficients $g_{ij} = \mathbf{a}_i \cdot \mathbf{a}_j$, then the Christoffel symbol of the first kind is defined by

$$\Gamma_{ij,k} = \frac{1}{2} \left(\frac{\partial g_{jk}}{\partial x^i} + \frac{\partial g_{ki}}{\partial x^j} - \frac{\partial g_{ij}}{\partial x^k} \right)$$

The Christoffel symbol of the second kind is defined by

$$\Gamma^k_{ij} = \sum_{r=1}^{n} g^{kr} \Gamma_{ij,r}$$

It is related to the derivatives of the base vectors and the reciprocal vectors.

$$\Gamma^k_{ij} = \mathbf{a}^k \cdot \frac{\partial \mathbf{a}_j}{\partial x^i}$$

The symbols are symmetric in the indices i,j. Also,

$$\Gamma_{ij,k} = \sum_{r=1}^{n} g_{kr}\Gamma_{ij}^{r} = \mathbf{a}_k \cdot \frac{\partial \mathbf{a}_j}{\partial x^i}$$

For the transformation $v^s = v^s(u^1,u^2, \ldots ,u^n)$, the Christoffel symbol Γ_{mn}^l in the u system is given in the v system by

$$G_{jk}^i = \sum \Gamma_{mn}^l \frac{\partial u^m}{\partial v^j} \frac{\partial u^n}{\partial v^k} \frac{\partial v^i}{\partial u^l} + \sum \frac{\partial^2 u^r}{\partial v^j \, \partial v^k} \frac{\partial v^i}{\partial u^r}$$

where the first term on the right is summed on the indices l, m, n and the second term is summed on the index r.

14-38. General Equation of a Geodesic. For a general coordinate system x^i in n-dimensional space ($i = 1,2, \ldots ,n$), the equation of a geodesic (shortest line between two points) is given by

$$\frac{d^2 x^i}{ds^2} + \sum \Gamma_{jk}^i \frac{dx^j}{ds} \frac{dx^k}{ds} = 0$$

summed on the indices $j,k = 1,2, \ldots ,n$, where s is arc length and Γ_{jk}^i is a Christoffel symbol of the second kind (Art. 14-37).

• What are the equations of the geodesics for cartesian coordinates in Euclidean space?

For this condition, the metric coefficients are constants; hence, their derivatives are zero, and the Christoffel symbols also are zero.   Therefore, the equations of the geodesics are

$$\frac{d^2 x^i}{ds^2} = 0$$

for which the solution is $x^i = a^i s + b^i$ (straight lines).

Any system of coordinates for which the Christoffel symbols $\Gamma_{jk}^i = 0$ is a geodesic coordinate system.

14-39. Covariant Differentiation. If $T_{ijk\ldots}^{abc\ldots}$ represents the components of a mixed tensor of weight N, then its covariant derivative in n-dimensional space is a tensor of the same weight but covariant rank one greater and is given by

$$T_{ijk\ldots,\alpha}^{abc\ldots} = \frac{\partial T_{ijk\ldots}^{abc\ldots}}{\partial x^\alpha} + \sum_{\beta=1}^{n} T_{ijk\ldots}^{\beta bc\ldots}\Gamma_{\beta\alpha}^{a} + \sum_{\beta=1}^{n} T_{ijk\ldots}^{a\beta c\ldots}\Gamma_{\beta\alpha}^{b} + \cdots$$

$$- \sum_{\beta=1}^{n} T_{\beta jk\ldots}^{abc\ldots}\Gamma_{i\alpha}^{\beta} - \sum_{\beta=1}^{n} T_{i\beta k\ldots}^{abc\ldots}\Gamma_{j\alpha}^{\beta} - \cdots - N T_{ijk\ldots}^{abc\ldots}\sum_{\beta=1}^{n} \Gamma_{\beta\alpha}^{\beta}$$

The last term vanishes for absolute tensors. The covariant derivative reduces to the ordinary derivative for Euclidean space or for geodesic coordinates.

In covariant differentiation:

The derivative of the sum or product of two tensors may be computed by the rules for ordinary differentiation.

The metric coefficients may be treated as constants.

The operations of contraction and raising or lowering an index are commutative.

14-40. Curvature Tensors. The curvature tensor is defined by

$$R^i_{.\alpha jk} = \frac{\partial \Gamma^i_{\alpha k}}{x^j} - \frac{\partial \Gamma^i_{\alpha j}}{\partial x^k} + \sum_{\beta=1}^{n} \Gamma^\beta_{\alpha k}\Gamma^i_{\beta j} - \sum_{\beta=1}^{n} \Gamma^\beta_{\alpha j}\Gamma^i_{\beta k}$$

A necessary and sufficient condition that a space be Euclidean is that the curvature tensor be zero.

The contracted curvature tensor

$$R_{ij} = R^\alpha_{.i\alpha j} = \frac{\partial \Gamma^\alpha_{ij}}{\partial x^\alpha} - \frac{\partial \Gamma^\alpha_{i\alpha}}{\partial x^j} + \sum_{\beta=1}^{n} \Gamma^\beta_{ij}\Gamma^\alpha_{\beta\alpha} - \sum_{\beta=1}^{n} \Gamma^\beta_{i\alpha}\Gamma^\alpha_{\beta j}$$

each term being summed over the range $\alpha = 1,2, \ldots ,n$, is called the Ricci tensor.

The scalar invariant

$$R = \Sigma g^{ij}R_{ij}$$

summed on the indices $i,j = 1,2, \ldots ,n$, is called the scalar curvature.

$$R_{hijk} = \sum_{\alpha=1}^{n} g_{h\alpha}R^\alpha_{.ijk}$$

is the Riemann-Christoffel or covariant curvature tensor.

BIBLIOGRAPHY

Craig, H. V.: "Vector and Tensor Analysis," McGraw-Hill Book Company, Inc., New York, 1943.

Lass, Harry: "Vector and Tensor Analysis," McGraw-Hill Book Company, Inc., New York, 1950.

Sokolnikoff, I. S.: "Tensor Analysis," John Wiley & Sons, Inc., New York, 1958.

Chapter 15

COMPLEX VARIABLES AND CONFORMAL MAPPING

15-1. Real, Imaginary, and Complex Numbers. A complex number is an ordered pair of real numbers (a,b). The first component a is called a real number and the second component b is an imaginary number. Complex numbers obey the following rules:

1. Two complex numbers (a_1,b_1) and (a_2,b_2) are equal only if $a_1 = a_2$ and $b_1 = b_2$. In particular, $(a,b) = 0$ only if $a = 0$ and $b = 0$.

2. The sum and product of two complex numbers are:

$$(a_1,b_1) + (a_2,b_2) = (a_1 + a_2, \ b_1 + b_2)$$
$$(a_1,b_1)(a_2,b_2) = (a_1a_2 - b_1b_2, \ a_1b_2 + a_2b_1)$$

As usual, subtraction is the inverse of addition and division is the inverse of multiplication.

These rules are satisfied if the imaginary number is given as the square root of a negative real number or the product of a real number and i, where

$$i = \sqrt{-1}$$

Note that $i^2 = (-i)^2 = -1$ (see also Art. 15-2).

15-2. Representation of Complex Numbers. (See also Art. 15-1.) Any complex number $z = (x,yi)$, where x and y are real numbers, can be written in the forms

$$x + iy \qquad r(\cos \theta + i \sin \theta) \qquad re^{i\theta}$$

The modulus, or absolute value, of z is

$$r = \sqrt{x^2 + y^2} = |x + iy|$$

The amplitude, or argument, of z is

$$\theta = \text{amp } z = \arg z = \arctan \frac{y}{x} = \text{amp } (x + iy)$$

A complex number can be represented geometrically in a plane, called the complex (Argand or Gauss) plane. In cartesian coordinates, the real

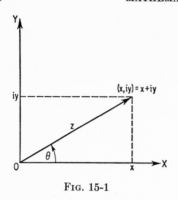

FIG. 15-1

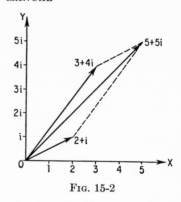

FIG. 15-2

component is plotted parallel to the x axis (real axis) and the imaginary component is plotted parallel to the y axis (Fig. 15-1). In polar coordinates, the length of the radius vector from the origin to the point representing a complex number is the modulus r. The angle the vector makes with the x axis is the amplitude θ (Fig. 15-1).

Two complex numbers can be added or subtracted geometrically by vector addition or subtraction (parallelogram law, Art. 14-7) of their radius vectors.

• Add $z_1 = 3 + 4i$ and $z_2 = 2 + i$.

$$z_1 + z_2 = (3 + 2) + i(4 + 1) = 5 + 5i$$

(See also Fig. 15-2.)

15-3. Complex Conjugate. The conjugate of a complex number

$$z = (x,iy) = x + iy$$

is the complex number

$$\bar{z} = (x,-iy) = x - iy$$

Geometrically, it can be represented by the reflection of point z in the x axis (Fig. 15-3). The polar form is

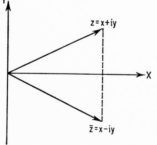

FIG. 15-3

$$\bar{z} = r[\cos(-\theta) + i \sin(-\theta)] = re^{-i\theta}$$

The sum of a complex number and its conjugate is a real number.

$$z + \bar{z} = x + iy + x - iy = 2x = 2 \operatorname{Re} z$$

The difference between a complex number and its conjugate is an imaginary number.

$$z - \bar{z} = x + iy - x + iy = 2iy = 2 \operatorname{Im} z$$

15-4. Products, Quotients, and Powers of Complex Numbers. In polar coordinates, the product of two complex numbers

$$z_1 = r_1(\cos \theta_1 + i \sin \theta_1) \qquad \text{and} \qquad z_2 = r_2(\cos \theta_2 + i \sin \theta_2)$$

is a complex number

$$z_1 z_2 = r_1 r_2 [\cos (\theta_1 + \theta_2) + i \sin (\theta_1 + \theta_2)]$$

Thus, the modulus of the product is the product of the moduli, $r_1 r_2$. The amplitude of the product is the sum of the amplitudes, $\theta_1 + \theta_2$ (Fig. 15-4).

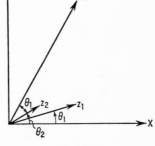

FIG. 15-4

• Multiply $z_1 = 3 + 4i$ by $z_2 = 12 + 5i$.
Convert these complex numbers to polar form. $r_1 = \sqrt{3^2 + 4^2} = 5$ and $r_2 = \sqrt{12^2 + 5^2} = 13$; amp $z_1 = \arctan \frac{4}{3} = 53°8'$; amp $z_2 = \arctan \frac{5}{12} = 22°37'$. Hence,

$$z_1 z_2 = 5 \times 13[\cos (53°8' + 22°37')$$
$$+ i \sin (53°8' + 22°37')]$$
$$= 65(\cos 75°45' + i \sin 75°45') = 16 + 63i$$

Multiplication by i rotates the radius vector of a complex number 90°; multiplication by $i^2 = -1$ rotates the radius vector 180°.

The quotient of the two complex numbers z_1 and z_2 is

$$\frac{z_1}{z_2} = \frac{r_1}{r_2} [\cos (\theta_1 - \theta_2) + i \sin (\theta_1 - \theta_2)]$$

• Divide $z_1 = 3 + 4i$ by $z_2 = 12 + 5i$.
In polar form, $z_1 = 5(\cos 53°8' + i \sin 53°8')$ and $z_2 = 13(\cos 22°37' + i \sin 22°37')$. Therefore,

$$\frac{z_1}{z_2} = \frac{5}{13}[\cos (53°8' - 22°37') + i \sin (53°8' - 22°37')]$$
$$= \frac{5}{13}(\cos 30°31' + i \sin 30°31') = 0.331 + 0.195i$$

If n is any real number,

$$z^n = r^n(\cos n\theta + i \sin n\theta) \qquad \text{De Moivre's formula}$$

• Cube $z = 3 + 4i$.
In polar form, $z = 5(\cos 53°8' + i \sin 53°8')$. Hence,

$$z^3 = 5^3[\cos 3(53°8') + i \sin 3(53°8')] = 125(\cos 159°24' + i \sin 159°24')$$
$$= 125(-0.936 + 0.352i) = -117 + 44i$$

• Compute the fourth roots of unity.
The amplitude of any real number is given by zero or a multiple of 2π. Thus,

in polar form, unity can be denoted by $z = \cos 2\pi k + i \sin 2\pi k$, where $k = 0,$ 1, 2, 3. Use of De Moivre's theorem gives, for the fourth roots,

$$w = z^{1/4} = \cos \frac{2\pi k}{4} + i \sin \frac{2\pi k}{4}$$

Substitution of 0, 1, 2, 3 for k in succession yields

$$w_1 = \cos 0 + i \sin 0 = +1$$
$$w_2 = \cos \frac{\pi}{2} + i \sin \frac{\pi}{2} = +i$$
$$w_3 = \cos \pi + i \sin \pi = -1$$
$$w_4 = \cos \frac{3\pi}{2} + i \sin \frac{3\pi}{2} = -i$$

The nth roots of unity are given in the complex plane by vectors to the vertices of a regular polygon of n sides inscribed in the circle $|z| = 1$, with one vertex at $z = 1$.

15-5. Analytic (Regular, Holomorphic) Functions. A function of $z = x + iy$ is analytic at a point z_0 if its derivative with respect to z exists at z_0 and every point in some neighborhood of z_0. The function is analytic in a domain of the z plane if it is analytic at every point in the domain.

Let $w = f(z) = f(x + iy)$. Then

$$w = u + iv$$

where u and v are functions of x and y. This function is analytic at $z_0 = x_0 + iy_0$ if

1. u and v are real, single-valued, continuous functions of x and y

2. The partial derivatives of u and v at z_0 and in its neighborhood are continuous and satisfy the Cauchy-Riemann equations:

$$\frac{\partial u}{\partial x} = \frac{\partial v}{\partial y} \quad \text{and} \quad \frac{\partial u}{\partial y} = -\frac{\partial v}{\partial x}$$

The Cauchy-Riemann equations in polar coordinates are

$$\frac{\partial u}{\partial r} = \frac{1}{r}\frac{\partial v}{\partial \theta} \quad \text{and} \quad \frac{1}{r}\frac{\partial u}{\partial \theta} = -\frac{\partial v}{\partial r}$$

Both u and v satisfy the Laplace equations:

$$\frac{\partial^2 u}{\partial x^2} + \frac{\partial^2 u}{\partial y^2} = 0 \quad \text{and} \quad \frac{\partial^2 v}{\partial x^2} + \frac{\partial^2 v}{\partial y^2} = 0$$

Hence, they are harmonic functions. They are called harmonic conjugates.

If $w = u + iv$ is analytic, the curves $u =$ constant and $v =$ constant are orthogonal.

Ordinary rules of differentiation and integration apply to analytic functions, and they can be represented by power series.

Points of a domain where a function is not analytic are called singular points of $f(z)$. For example, $f(z) = 1/z$ has a singular point at $z = 0$ because $f'(z)$ does not exist there.

• Show that $w = z^2$ is analytic at $z_0 = 0$.

Since $w = (x + iy)^2 = (x^2 - y^2) + 2xyi$,

$$u = x^2 - y^2 \qquad v = 2xyi$$

Substitution of $x = 0$ and $y = 0$ shows that u and v exist. It is evident also that they are continuous functions at $(0,0)$ and its neighborhood. Furthermore, their partial derivatives exist and are continuous, and they satisfy the Cauchy-Riemann equations.

$$\frac{\partial u}{\partial x} = 2x = \frac{\partial v}{\partial y} \qquad \text{and} \qquad \frac{\partial u}{\partial y} = -2y = -\frac{\partial v}{\partial x}$$

15-6. Derivatives of Analytic Functions. The derivative of $w = f(z)$ at a point z_0 is defined by

$$f'(z) = \frac{dw}{dz} = \lim \left[\frac{f(z_0 + \Delta z) - f(z_0)}{\Delta z} \right]_{\Delta z \to 0}$$

where $\Delta z = z - z_0$. Since z is a function of two variables, Δz may approach zero along different paths. If w is not analytic, the value of $f'(z)$ may vary with the path. The derivative of an analytic function has a single value regardless of the path taken when z approaches z_0.

If $w = f(z) = u + iv$ is analytic at z_0, then the derivative at z_0 is

$$f'(z) = \frac{\partial u}{\partial x} + i\frac{\partial v}{\partial x} = \frac{\partial v}{\partial y} - i\frac{\partial u}{\partial y}$$

In polar coordinates, with u and v as functions of the modulus r and the amplitude θ, the derivative is

$$f'(z) = (\cos \theta - i \sin \theta)\left(\frac{\partial u}{\partial r} + i\frac{\partial v}{\partial r} \right)$$

• Show that if $w = z^2$, $\dfrac{dw}{dz} = 2z$.

Since w is analytic (Art. 15-5), and $w = u + iv = (x^2 - y^2) + 2xyi$,

$$\frac{dw}{dz} = \frac{\partial u}{\partial x} + i\frac{\partial v}{\partial x} = 2x + 2yi = 2(x + yi) = 2z$$

The usual rules for differentiating sums, products, and quotients of real functions also hold for analytic complex functions, as do the formulas for differentiating elementary functions (Art. 15-7).

15-7. Elementary Complex Functions. Every polynomial

$$P(z) = a_0 + a_1 z + a_2 z^2 + \cdots + a_n z^n$$

is analytic at every point of the z plane. (Such functions are called entire.)

The exponential function is defined by

$$e^z = \exp z = e^x(\cos y + i \sin y)$$

It is analytic everywhere in the z plane. Its absolute value is e^x and its amplitude is y. If x is set equal to 0, then

$$e^{iy} = \exp iy = \cos y + i \sin y$$

Also, $e^{-iy} = \exp -iy = \cos y - i \sin y$

The exponential function is periodic with period $2\pi i$:

$$\exp (z + 2\pi i) = \exp z$$

The sine and cosine functions are defined by

$$\sin z = \frac{e^{iz} - e^{-iz}}{2i} \qquad \cos z = \frac{e^{iz} + e^{-iz}}{2}$$

They are analytic everywhere in the z plane. Other trigonometric functions and formulas can be obtained from these in the same way as for real variables. The sine and cosine functions also can be written as

$$\sin z = \sin (x + iy) = \sin x \cosh y + i \cos x \sinh y$$
$$\cos z = \cos (x + iy) = \cos x \cosh y - i \sin x \sinh y$$

These functions are not bounded in absolute value. Also, the real zeros of $\sin z$ and $\cos z$ are their only zeros.

The hyperbolic sine and cosine functions are defined by

$$\sinh z = \frac{e^z - e^{-z}}{2} \qquad \cosh z = \frac{e^z + e^{-z}}{2}$$

The other hyperbolic functions and formulas can be obtained from these in the same way as for real variables.

The logarithmic function is defined by

$$\log z = \log re^{i\theta} = \log r + i\theta \qquad r > 0$$

It is multivalued, for θ can be set equal to $\varphi \pm 2n\pi$, with $-\pi < \varphi \leq \pi$ and $n = 0, 1, 2, \ldots$.

The principal value of $\log z$ is obtained by setting $n = 0$. It is not defined at the origin or some points on the negative real axis, but it is analytic and single-valued where defined.

15-8. Mapping by Complex Functions. At least two complex planes are needed for graphical representation of $w = f(z)$, since this function represents a relationship between two pairs of variables u,v and x,y. Points in one plane are said to be mapped or transformed into corresponding points (images) in the other.

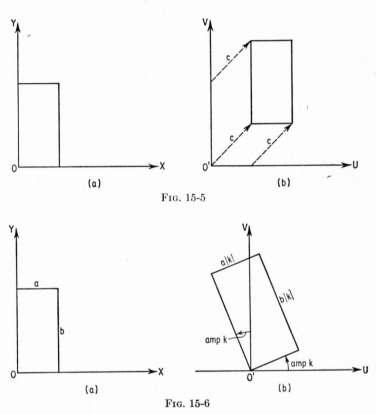

Fig. 15-5

Fig. 15-6

The linear function $w = z + c$, where c is a complex constant, maps a region of the z plane into one in the w plane of the same size, shape, and orientation, but each point is translated along the vector c (Fig. 15-5).

Mapping with the linear function $w = kz$, where k is a complex constant, consists of a rotation through the angle amp k and a magnification by a factor equal to the modulus of k, or $|k|$ (Fig. 15-6).

Mapping with the general linear function $w = kz + c$ is a combination of the transformations illustrated in Figs. 15-5 and 15-6, a translation along the vector c, a rotation amp k, and a magnification by $|k|$.

The transformation $w = z^n$, where n is a positive integer, maps the wedge-shaped region in Fig. 15-7a onto the upper half of the w plane (Fig. 15-7b). It transforms a circular arc $r = r_0$, $\theta_0 < \theta < (\theta_0 + 2\pi/n)$ into the circle $\rho = r_0{}^n$. Thus, the function $w = z^2$ maps the first quadrant of the z plane onto the upper half of the w plane and maps a semicircle in the z plane into a circle in the w plane.

The transformation $w = 1/z$ is equivalent to an inversion with respect to the unit circle $r = 1$ and a reflection in the real axis. (An inversion

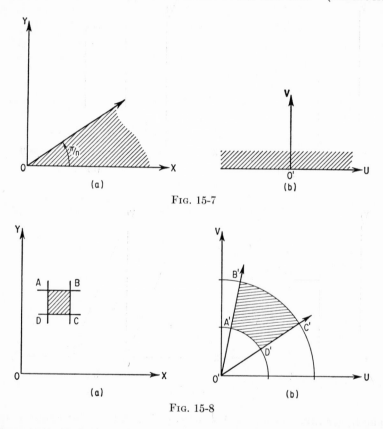

(a)   (b)

FIG. 15-7

(a)   (b)

FIG. 15-8

shifts a point z_0 along its radius vector to a point z_1 whose distance from the origin is $1/|z_0|$.) Points outside the unit circle in the z plane are mapped inside the unit circle $|w| = 1$ in the w plane, and those inside the unit circle in the z plane are mapped outside in the w plane. Circles not passing through $z = 0$ transform into circles not passing through $w = 0$. Every circle through $z = 0$ transforms into a straight line in the w plane. Lines in the z plane transform into circles through $w = 0$, except lines through $z = 0$, for which the images are lines through $w = 0$.

The transformation $w = e^z$ maps lines parallel to the y axis ($x = c$) into circles in the w plane ($\rho = e^c$). Lines parallel to the x axis ($y = k$) map into rays ($\varphi = k$). Thus, a rectangle with sides parallel to the x and y axes maps onto the portion of a circular sector cut off by two circles (Fig. 15-8). Also, the infinite strip $0 \leq y \leq \pi$ maps onto the upper half of the w plane. (See also Art. 15-9.)

15-9. Conformal Mapping. A transformation that preserves angles is conformal.

Mapping done with an analytic function $f(z)$ is conformal for all points z for which $f'(z) \neq 0$.

Every conformal transformation maps orthogonal curves into orthogonal curves. Thus, the lines $x = $ constant, $y = $ constant in the z plane transform into mutually orthogonal trajectories in the w plane. Similarly, the lines $u = $ constant, $v = $ constant in the w plane map into orthogonal trajectories in the z plane.

The angle through which the directed tangent to a curve at a point z_0 is turned by an analytic mapping function $f(z)$ is amp $f'(z_0)$.

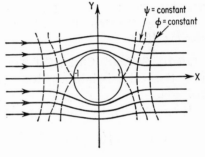

• Determine the streamlines for an ideal fluid flowing with uniform velocity around a long circular cylinder perpendicular to the direction of flow.

Fig. 15-9

Let the cylinder be represented by the circle $x^2 + y^2 = 1$ and let the flow be parallel to the x axis (Fig. 15-9). The streamlines are determined by the equation $\psi = $ constant, where ψ is the stream function, which is related to the velocity components v_x and v_y, parallel to the x and y axes respectively, by

$$\frac{\partial \psi}{\partial x} = -v_y \qquad \frac{\partial \psi}{\partial y} = v_x$$

(The streamlines are paths of particles of the fluid.) Both the stream function ψ and the velocity potential φ, whose derivatives with respect to x and y equal v_x and v_y respectively, satisfy Laplace's equation throughout any region containing no sources and sinks; they are related by the Cauchy-Riemann equations.

$$\frac{\partial \varphi}{\partial x} = \frac{\partial \psi}{\partial y} \qquad \frac{\partial \varphi}{\partial y} = -\frac{\partial \psi}{\partial x}$$

Hence, the complex potential $F(z) = \varphi + i\psi$ is an analytic function of the variable $z = x + iy$, and it can be shown that

$$\frac{dF}{dz} = v_x - iv_y$$

One of the streamlines $\psi = k$ must coincide with the boundary of the circular cylinder $|z| = r = 1$. All other streamlines are outside the circle. Take advantage of symmetry and consider only the upper half of the z plane. Now, a transformation that maps this whole region of flow in the z plane onto the upper half of the w plane, $v \geq 0$, is

$$w = z + \frac{1}{z}$$

(The x axis and circumference of the circle map onto the u axis; for $v = 0$ either when $y = 0$ or $x^2 + y^2 = 1$.) For uniform flow in this half plane, the complex potential is $F = cw$, where c is a real constant. Hence, for flow in the z plane $F = c(z + 1/z) = c[(r + 1/r) \cos \theta + i(r - 1/r) \sin \theta]$, from which

$$\psi = c \left(r - \frac{1}{r} \right) \sin \theta$$

The streamline $\psi = 0$ consists of the circle $r = 1$ and the x axis. For very large values of y, the streamline is given approximately by $r \sin \theta = y = \text{constant}$ (see Fig. 15-9).

15-10. Schwarz-Christoffel Transformation. An equation for mapping the real axis into a polygon or a half plane into the interior of a polygon.

If $z = x + iy$, the transformation

$$w = A \int (z - x_1)^{-k_1} (z - x_2)^{-k_2} \cdots (z - x_n)^{-k_n} \, dz + B$$

where $k_1\pi$, $k_2\pi$, ... represent the successive exterior angles of the polygon ($k_1 + k_2 + \cdots + k_n = 2$) and A and B are complex constants, maps the upper half plane $y > 0$ conformally into the interior of a polygon in the w plane. Vertices $w_1, w_2, \ldots, w_n$ correspond to points $x_1, x_2, \ldots, x_n$ on the x axis. For any given polygon, three of the x_j can be chosen arbitrarily.

If one of the x_j coincides with the point at infinity, the corresponding term is dropped from the transformation equation. For example, if x_n is at infinity,

$$w = A' \int (z - x_1)^{-k_1} (z - x_2)^{-k_2} \cdots (z - x_{n-1})^{-k_{n-1}} \, dz + B'$$

Application of the Schwarz-Christoffel transformation usually leads to integrals difficult to evaluate. Useful cases include so-called degenerate polygons, for which one or more angles $k_j\pi$ are zero.

• Map the half plane $y > 0$ into a right angle.

Here, $k_1\pi = \pi/2$; all other values of k_j equal zero. Let $x_1 = 0$; all other $x_j = \infty$. Hence,

$$w = A \int_0^z z^{-\frac{1}{2}} \, dz + B = 2Az^{\frac{1}{2}} + B$$

$B = 0$, since $w = 0$ when $z = 0$.   Hence, the required transformation is

$$w = 2Az^{1/2}$$

15-11. Integrals of a Complex Variable.   Let f be a complex-valued function of a real variable t, such that

$$f(t) = u(t) + iv(t)$$

where u and v are real-valued functions that are continuous over an interval (a,b), except possibly for a finite number of finite jumps.   The definite integral of f is defined by

$$\int_a^b f(t)\ dt = \int_a^b u(t)\ dt + i \int_a^b v(t)\ dt$$

One basic property of this integral is that its absolute value is equal to or less than the integral of the absolute value of f.

$$\left| \int_a^b f(t)\ dt \right| \leq \int_a^b |f(t)|\ dt$$

A contour is a continuous chain of a finite number of smooth arcs.   If a contour is closed and does not intersect itself, it is called a closed contour.

A line integral of a function f of a complex variable z from a point $z = a$ to a point $z = b$ is defined in terms of the values of $f(z)$ along a contour C extending from a to b.

$$\int_C f(z)\ dz = \int_a^b f[\varphi(t) + i\psi(t)] \left(\frac{d\varphi}{dt} + i\frac{d\psi}{dt} \right) dt$$

where $z = x + iy$

$x = \varphi(t)$

$y = \psi(t)$

If $f = u + iv$, the line integral can be written in terms of real line integrals.

$$\int_C f(z)\ dz = \int_C (u\ dx - v\ dy) + i \int_C (u\ dy + v\ dx)$$

Reversal of the sense of integration reverses the sign of the integral.

The length L of the contour C is given by

$$L = \int_C |dz|$$

If M is a constant such that $|f(z)| \leq M$ when z is on C, then

$$\left| \int_C f(z)\ dz \right| \leq ML$$

If C is a circle of radius r and center at any point z_0,

$$\int_C \frac{dz}{z - z_0} = 2\pi i$$

$$\int_C \frac{dz}{(z - z_0)^{n+1}} = 0 \qquad n = \pm 1, \pm 2, \ldots$$

15-12. Cauchy-Goursat Theorem. If a function f is analytic at all points interior to and on a closed contour C, then

$$\int_C f(z)\, dz = 0$$

(see also Art. 15-11). The theorem also holds if inside C there are a finite number of nonintersecting closed contours C_1, C_2, . . . and the integral is taken over C and all the C_j, keeping the interior of C not enclosed by the C_j on the left.

• Evaluate $\int_C dz/[z^2(z^2 + 9)]$ for the circles $|z| = 2$ described counterclockwise and $|z| = 1$ described clockwise (Fig. 15-10).

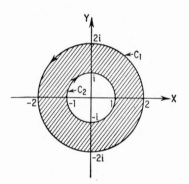

Fig. 15-10

The integrand is analytic except at $z = 0$ and $z = \pm 3i$, which lie outside the annular region between the two circles. Hence, by the Couchy-Goursat theorem, the integral is zero.

15-13. Cauchy's Integral Formula. If a function f is analytic everywhere inside and on a closed contour C, then

$$f(z_0) = \frac{1}{2\pi i} \int_C \frac{f(z)}{z - z_0}\, dz$$

where z_0 is any point inside C and the integral is taken around C with the interior on the left. This formula may also be written

$$f(z) = \frac{1}{2\pi i} \int_C \frac{f(t)}{t - z}\, dt$$

The formula indicates that every analytic function is completely determined in the interior of a closed region when the values of the function are given on the boundary.

• Evaluate $I = \int_C e^{-z}\, dz/(z + 1)$ over the circle $|z| = 2$.

Let $f(z) = e^{-z}$ and $z_0 = -1$ and apply Cauchy's integral formula. Note that e^{-z} is analytic and that -1 lies within the given circle. Then

$$I = 2\pi i e^{-z}\Big]_{z=-1} = 2\pi e i$$

Repeated differentiation of Cauchy's integral formula yields

$$f'(z) = \frac{1}{2\pi i} \int_C \frac{f(t)}{(t - z)^2}\, dt$$

$$f''(z) = \frac{1}{2\pi i} \int_C \frac{f(t)}{(t - z)^3}\, dt$$

and in general,

$$f^{(n)}(z) = \frac{1}{2\pi i} \int_C \frac{f(t)}{(t - z)^{n+1}}\, dt$$

If M is the maximum value of $|f(z)|$ on a circle of radius r,

$$|f^{(n)}(z)| \leq \frac{Mn!}{r^n}$$

15-14. Taylor's Series for a Complex Function.   If $f(z)$ is analytic at all points inside a circle C with center at z_0 and radius r, then at each point inside C,

$$f(z) = f(z_0) + f'(z_0)(z - z_0)$$
$$+ \frac{f''(z_0)}{2!}(z - z_0)^2 + \cdots + \frac{f^{(n)}(z_0)}{n!}(z - z_0)^n + \cdots$$

When $z_0 = 0$, this reduces to Maclaurin's series.

Both series are identical in form with the corresponding series for real functions.   For example, compare

$$e^x = 1 + \sum_{n=1}^{\infty} \frac{x^n}{n!} \qquad e^z = 1 + \sum_{n=1}^{\infty} \frac{z^n}{n!}$$

The maximum radius of the circle of convergence is the distance from z_0 to the nearest point at which $f(z)$ is not analytic (singular point).

15-15. Laurent's Series.   Consider an annular region R between two circles C_1 and C_2 about a point z_0 with radii r_1 and r_2, $r_2 < r_1$.   If $f(z)$ is analytic throughout R, then at each point in R

$$f(z) = a_0 + a_1(z - z_0) + a_2(z - z_0)^2 + \cdots + a_n(z - z_0)^n + \cdots$$
$$+ \frac{b_1}{z - z_0} + \frac{b_2}{(z - z_0)^2} + \cdots + \frac{b_n}{(z - z_0)^n} + \cdots$$

where $a_n = \dfrac{1}{2\pi i} \displaystyle\int_{C_1} \dfrac{f(z)\,dz}{(z - z_0)^{n+1}}$ $n = 0, 1, 2, \ldots$

$b_n = \dfrac{1}{2\pi i} \displaystyle\int_{C_2} \dfrac{f(z)\,dz}{(z - z_0)^{-n+1}}$ $n = 1, 2, \ldots$

each integral being taken counterclockwise.

15-16. Residues and Poles.   If $f(z)$ has an isolated singularity (point where it is not analytic) at $z = z_0$, the residue of $f(z)$ at z_0 is the coefficient of $(z - z_0)^{-1}$ in Laurent's series for $f(z)$ (Art. 15-15).

$$b_1 = \frac{1}{2\pi i} \int_C f(z)\,dz$$

Let $R_1, R_2, \ldots, R_n$ be the residues at singular points inside a closed contour C within and on which $f(z)$ is analytic except at those points. If C encloses a finite number of singular points, then

$$\int_C f(z)\,dz = 2\pi i(R_1 + R_2 + \cdots + R_n)$$

This is known as the residue theorem.

Let $f(z)$ be expanded in a Laurent series and $b_1, b_2, \ldots, b_m, b_{m+1}, \ldots, b_n$ be the coefficients of the negative powers of $z - z_0$. If coefficients $b_{m+1}, b_{m+2}, \ldots, b_n$ are all zero, but $b_m \neq 0$, the isolated singular point z_0 is called a pole of order m of the function $f(z)$. If $m = 1$, z_0 is called a simple pole.

If $\varphi(z) = (z - z_0)^m f(z)$ is analytic at z_0, $f(z)$ has a pole of order m at z_0. The residue of $\varphi(z)$ at z_0 is given by

$$b_1 = \frac{1}{(m-1)!} \left[\frac{d^{m-1}\varphi}{dz^{m-1}} \right]_{z=z_0} \qquad m > 1$$
$$b_1 = \lim \, [(z - z_0)f(z)]_{z \to z_0} \qquad m = 1$$

Residues are useful in evaluating integrals.

• Evaluate $I = \displaystyle\int_0^{2\pi} d\theta/(1 + \alpha \sin \theta)$ for $0 < \alpha < 1$.

Let $z = e^{i\theta}$, so $d\theta = dz/iz$ and $\sin \theta = (z - z^{-1})/2i$. Substitution for θ in I yields

$$I = \int_C \frac{dz}{iz[1 + \alpha(z - z^{-1})/2i]} = \frac{2}{\alpha} \int_C \frac{dz}{z^2 + 2iz/\alpha - 1}$$

where C is the circle $|z| = 1$. For $0 < \alpha < 1$, the integrand has only the singularity

$$z_0 = -\frac{i}{\alpha}(1 - \sqrt{1 - \alpha^2})$$

inside C, and z_0 is a simple pole.   Hence, the residue at z_0 is

$$\lim \left[\frac{1}{z^2 + 2iz/\alpha - 1}(z - z_0)\right]_{z \to z_0}$$

$$= \lim \left[\frac{1}{(z - z_0)(z - \bar{z}_0)}(z - z_0)\right]_{z \to z_0} = \frac{\alpha}{2i\sqrt{1 - \alpha^2}}$$

Application of the residue theorem to I gives

$$I = \frac{2}{\alpha}(2\pi i)\frac{\alpha}{2i\sqrt{1 - \alpha^2}} = \frac{2\pi}{\sqrt{1 - \alpha^2}}$$

BIBLIOGRAPHY

Churchill, R. V.: "Complex Variables and Applications," 2d ed., McGraw-Hill Book Company, Inc., New York, 1960.

Nehari, Zeev: "Conformal Mapping," McGraw-Hill Book Company, Inc., New York, 1952.

Sokolnikoff, I. S., and R. M. Redheffer: "Mathematics of Physics and Modern Engineering," McGraw-Hill Book Company, Inc., New York, 1958.

Chapter 16

PERMUTATIONS, COMBINATIONS, AND PROBABILITIES

16-1. Permutations. The ways in which a given set of items can be arranged if a specific number are taken at a time.

If, for example, three items are to be arranged taking two at a time, six arrangements or permutations are possible. The first position can be filled in three different ways; after that is done, the second position can be filled in only two different ways. Then both positions can be filled in 3×2, or six, different ways.

The number of permutations of n different items taken r at a time is

$$P(n,r) = n(n - 1)(n - 2) \cdots (n - r + 1)$$

(The symbol sometimes is written $_nP_r$ or P_r^n.)

• In how many ways can the first eight letters of the alphabet be arranged, taking them three at a time?

Here, $n = 8$, $r = 3$, and $n - r + 1 = 6$. Hence,

$$P(8,3) = 8 \times 7 \times 6 = 336$$

Permutations of n different things taken n at a time total

$$P(n,n) = n!$$

However, if p of them are alike, q others are alike, and also r others are alike,

$$P = \frac{n!}{p!q!r!}$$

16-2. Combinations. The ways in which a specific number of items can be selected from a given set of items in any order.

If, for example, A, B, and C are to be combined two at a time, the number of combinations is three (AB, BC, and AC) since BA, CB, and CA, though different permutations, are the same combinations.

The number of combinations of n different items taken r at a time is

$$C(n,r) = \frac{n(n - 1)(n - 2) \cdots (n - r + 1)}{r!} = \frac{n!}{r!(n - r)!} = C(n,n - r)$$

[The symbol $C(n,r)$ sometimes is written $_nC_r$ or $\binom{n}{r}$.] Note that

$$C(n,n) = C(n,0) = 1$$

• In how many ways can a committee of three be selected from a group of 21? Here, $n = 21$, $r = 3$, and $n - r + 1 = 19$.

$$C(21,3) = \frac{21 \times 20 \times 19}{3!} = 1,330$$

If n things are combined first one at a time, then two at a time, then three at a time, . . . , and finally n at a time, the total number of combinations is

$$\Sigma C = C(n,1) + C(n,2) + C(n,3) + \cdots + C(n,n) = 2^n - 1$$

• If a shop has ten different articles for sale, how many choices has a purchaser? The purchaser can buy one, or any two, or any three, etc., or all ten. So the number of choices is

$$\Sigma C = 2^{10} - 1 = 1,023$$

16-3. Simple Probability. Among several events equally likely to occur, the probability that a given event will occur is the ratio of the number of favorable cases to the total number of possible cases.

• What is the probability of rolling a ten on one throw with a pair of dice?

Each face of either die is equally likely to turn up (unless the dice are loaded), and each die, therefore, may show any one of six different numbers. Thus, the total number of ways the dice may fall is $6 \times 6 = 36$. But a ten can be rolled in only three ways, two fives, or a six on one die and a four on the other, or vice versa, so the probability of rolling a ten is

$$p = \tfrac{3}{36} = \tfrac{1}{12}$$

The probability that an event will not occur is the ratio of unfavorable cases to the total number of possible cases. If p is the probability of an event happening and q is the probability of it not happening, then

$$q = 1 - p$$

16-4. Compound Probabilities. The probability that two or more independent events will all happen is equal to the product of their probabilities.

$$p = p_1 p_2 p_3 p_4 \ldots$$

• What is the probability of rolling two tens in succession with two throws of a pair of dice?

The probability of rolling a ten on the first throw is $\tfrac{1}{12}$ (Art. 16-3). The probability of a ten on the second throw also is $\tfrac{1}{12}$. The probability of two tens in succession is

$$p = \frac{1}{12}\frac{1}{12} = \frac{1}{144}$$

If the probability of an event is p_1, and if, after this event has occurred, the probability of a second event is p_2, the probability that they will happen in this order is

$$p = p_1 p_2$$

This can be extended to apply to more events.

• A bag contains six black balls and three white balls. What is the probability of drawing two white balls in succession in two tries, taking one ball out at a time?

On the first try, the total number of ways of drawing a ball is $6 + 3 = 9$, and a white ball can be drawn in three ways. So the probability of a white ball on the first try is $\frac{3}{9} = \frac{1}{3}$. Now, eight balls remain in the bag, and only two are white. Hence, the probability of a white ball on the second try is $\frac{2}{8} = \frac{1}{4}$. Therefore, the probability of picking two white balls in succession is $\frac{1}{3} \times \frac{1}{4} = \frac{1}{12}$.

16-5. Probability of Mutually Exclusive Events. Two events are mutually exclusive if the occurrence of one excludes the occurrence of the other.

The probability that any one of a set of mutually exclusive events will occur is the sum of the probabilities of occurrence of each of the events.

$$p = p_1 + p_2 + p_3 + \cdots$$

p is called the total probability and p_i is called the partial probability.

• What is the probability of drawing either a black or a white ball on one try from a box containing three black, four white, and five red balls?

The total number of ways of drawing a ball is $3 + 4 + 5 = 12$. A black ball can be drawn in three ways, so the probability of drawing a black ball is $\frac{3}{12} = \frac{1}{4}$. The probability of drawing a white ball is $\frac{4}{12} = \frac{1}{3}$. The probability of drawing either a black ball or a white ball is $\frac{1}{4} + \frac{1}{3} = \frac{7}{12}$.

If the probability of an event A occurring is p_A and the probability of B is p_B, then the probability of either A or B or both is

$$p_{A+B} = p_A + p_B - p_{AB}$$

where p_{AB} is the probability of both A and B occurring.

• What is the probability that at least one four will appear when two dice are rolled?

The probability of a four on either die is $\frac{1}{6}$ and the probability of a four on both dies is $\frac{1}{6} \times \frac{1}{6} = \frac{1}{36}$. Thus the probability of at least one four is

$$p = \frac{1}{6} + \frac{1}{6} - \frac{1}{36} = \frac{11}{36}$$

The same result could have been obtained by noting that the total number of ways the dice can fall is 36. A four can be obtained on one die and any number on the other die in six ways. Also, a four can be obtained on the second die with any number on the first die in six ways. But if these results are added to get the total number of ways of rolling a four, the case of a pair of fours is counted twice. So actually, a four can be turned up in 11 ways. Hence, the probability of at least one four is $1\frac{1}{36}$.

The probability of A or B or C or any combination of these events (at least one of them) occurring is

$$p_{A+B+C} = p_A + p_B + p_C - p_{AB} - p_{AC} - p_{BC} + p_{ABC}$$

16-6. Probability of Repetitions. If p is the probability that an event will occur in any single trial, and $q = 1 - p$ is the probability that it will not occur, then the probability that the event will occur exactly r times in n trials is the binomial frequency function

$$p_r = C(n,r)p^r q^{n-r} = \frac{n!}{r!(n-r)!} p^r q^{n-r} = \frac{n(n-1)\cdots(n-r+1)}{r!} p^r q^{n-r}$$

• A's probability of winning a game is $\frac{1}{3}$. If he plays five times, what is the probability of his winning three games and losing two?

Here, $n = 5$, $r = 3$, and $q = 1 - \frac{1}{3} = \frac{2}{3}$.

$$C(5,3) = \frac{5 \times 4 \times 3}{3!} = 10$$

$$p_r = 10(\tfrac{1}{3})^3(\tfrac{2}{3})^2 = {}^{40}\!/_{243}$$

The probability that an event will happen at least r times in n trials is

$$p^n + C(n,n-1)p^{n-1}q + C(n,n-2)p^{n-2}q^2 + \cdots + C(n,r)p^r q^{n-r}$$

or $p^n + C(n,1)p^{n-1}q + C(n,2)p^{n-2}q^2 + \cdots + C(n,n-r)p^r q^{n-r}$

The probability that an event will happen at least once in n trials is $1 - q^n$.

• What is the probability of throwing at least three sixes in five throws with a single die?

The probability of rolling a six on a throw is $\frac{1}{6}$ and the probability of not succeeding is $\frac{5}{6}$. With $n = 5$ and $r = 3$, the probability of three sixes in five throws is

$$(\tfrac{1}{6})^5 + C(5,1)(\tfrac{1}{6})^4(\tfrac{5}{6}) + C(5,2)(\tfrac{1}{6})^3(\tfrac{5}{6})^2$$
$$= (1/7{,}776)(1 + 5 \times 5 + 10 \times 25) = 276/7{,}776$$

The probability that an event will happen at most r times in n trials is the binomial distribution

$$q^n + C(n,1)pq^{n-1} + C(n,2)p^2 q^{n-2} + \cdots + C(n,r)p^r q^{n-r}$$

• What is the probability of throwing no more than three sixes in five throws with a single die?

Here, $p = \frac{1}{6}$, $q = \frac{5}{6}$, $n = 5$, and $r = 3$, as in the previous example. The probability of at most three sixes is

$$(\tfrac{5}{6})^5 + C(5,1)(\tfrac{1}{6})(\tfrac{5}{6})^4 + C(5,2)(\tfrac{1}{6})^2(\tfrac{5}{6})^3 + C(5,3)(\tfrac{1}{6})^3(\tfrac{5}{6})^2$$
$$= (1/7{,}776)(3{,}125 + 5 \times 625 + 10 \times 125 + 10 \times 25) = 7{,}750/7{,}776$$

16-7. Expectation. The expected value of a random variable X is given by

$$E(X) = p_1x_1 + p_2x_2 + \cdots + p_rx_r$$

where $x_1, x_2, \ldots, x_r$ = specific values of X

$p_1, p_2, \ldots, p_r$ = probabilities that $X = x_1, x_2, \ldots, x_r$, respectively

If all values x_i are equally likely, $E(X)$ equals the arithmetic mean.

16-8. Distributions. The probabilities p_i associated with discrete values of a random variable x_i may be written in the form

$$p_i = f(x_i)$$

with $\Sigma f(x_i) = 1$, the summation extending over all values of i. The function $f(x_i)$ is called the frequency function and is proportional to the expected frequency of the event $X = x_i$ in a fixed number of observations.

The probability that x_i will not exceed a given value t is

$$F(t) = \Sigma f(x_i)$$

where the summation extends over the values of $x_i \leq t$. $F(t)$ is called the distribution function. If the variable may vary continuously, the distribution function becomes

$$F(t) = \int_{-\infty}^{t} f(x)\,dx$$

Here, $f(x)$ is the probability density. The probability that $a < x < b$, where a and b are given values, is

$$p_{a,b} = \int_{a}^{b} f(x)\,dx$$

Some important density functions are

Poisson: $\quad e^{-\mu x}\dfrac{(\mu x)^r}{r!} \qquad 0 \leq x < \infty \qquad \mu > 0 \qquad r$ = positive integer

Gauss:

$$\frac{1}{\sqrt{2\pi}\,\sigma}\,e^{-\frac{1}{2}[(x-\mu)/\sigma]^2} \qquad -\infty < x < \infty \qquad \sigma > 0 \qquad -\infty < \mu < \infty$$

Maxwell-Boltzmann: $\qquad 4a\sqrt{\dfrac{a}{\pi}}\,x^2e^{-ax^2}$

where x = random variable

μ, r, a, σ = constants

16-9. Probability Curve (Error Function). If, in a series of n observations, the errors are $x_1, x_2, \ldots, x_n$ and their sum is

$$\Sigma x_i = x_1 + x_2 + \cdots + x_n = 0$$

the arithmetic mean of the observed values is the best value that can be assigned to the measured quantity.

A normal (Gaussian) distribution of observed values is defined as a set of observations that satisfies

$$y = \frac{h}{\sqrt{\pi}} e^{-h^2 x^2}$$

where y = probability of an error x

h = measure of precision of observation

The curve represented by this equation (Fig. 16-1) is called an error curve or probability curve.

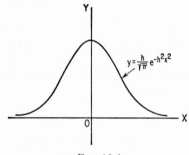

FIG. 16-1

The area under the curve between any two values of an error equals the probability that an error is between these values.

$$p_{a,b} = \frac{h}{\sqrt{\pi}} \int_a^b e^{-h^2 x^2} \, dx$$

If ρ is the residual or deviation of an observed value from the arithmetic mean of a set of n observations, then

$$h = \sqrt{\frac{n}{2\Sigma\rho^2}}$$

The probability that an error lies between $-X$ and X is

$$p = \frac{2}{\sqrt{\pi}} \int_0^x e^{-x^2} \, dx$$

where the upper limit $x = hX$. This function p is called the error function, erf x. Its value may be obtained from tables (see Bibliography).

• In a set of observations for which $h = 7$, what is the probability of an error lying in the interval (a) ± 0.2 (b) ± 0.1 (c) ± 0.05?

From a table, (a) for $x = 7 \times 0.2 = 1.4$, $p = \frac{2}{\sqrt{\pi}} \int_0^{1.4} e^{-x^2} \, dx = 0.95$; (b) for $x = 7 \times 0.1 = 0.7$, $p = 0.68$; (c) for $x = 7 \times 0.05 = 0.35$, $p = 0.38$.

BIBLIOGRAPHY

Burington, R. S., and D. C. May: "Handbook of Probability and Statistics with Tables," McGraw-Hill Book Company, Inc., New York, 1953.

Cramer, H.: "Elements of Probability Theory and Some of Its Applications," John Wiley & Sons, Inc., New York, 1955.

Dwight, H. B.: "Tables of Integrals and Other Mathematical Data," The Macmillan Company, New York, 1947.

Feller, W.: "An Introduction to Probability Theory and Its Applications," 2d ed., John Wiley & Sons, Inc., New York, 1957.

Jahnke, E., F. Emde, and F. Losch: "Tables of Higher Functions," McGraw-Hill Book Company, Inc., New York, 1960.

Munroe, M. E.: "The Theory of Probability," McGraw-Hill Book Company, Inc., New York, 1951.

Peirce, B. O., and R. M. Foster: "A Short Table of Integrals," Ginn & Company, Boston, 1956.

Sokolnikoff, I. S., and R. M. Redheffer: "Mathematics of Physics and Modern Engineering," McGraw-Hill Book Company, Inc., New York, 1958.

Uspensky, J. V.: "Introduction to Mathematical Probability," McGraw-Hill Book Company, Inc., New York, 1937.

Chapter 17

STATISTICS

Mathematical techniques are employed in statistics to interpret numerical data. These data may be obtained from experiments, measurements, counts of objects or events with specific characteristics, or from any observations providing a series of numbers. Statistics enables probability limits to be placed on the truth of statements made about a total system, or population, from analysis of a sample.

17-1. Index Number. A measure of the change that has occurred in a series of values arranged in a time sequence. Index numbers are usually based on a fixed period, either a specific year or an average of several years. A chain index, however, consists of index numbers each of which is linked to the previous number.

Since the items contributing to an index number may be measured in different units, the individual values must be weighted to reflect their relative importance. For example, to determine the weights to be applied to the price of two commodities comprising a price index, it would be reasonable to select two numbers in the same ratio as the total amount of money spent on each in the base period.

Suppose that $50 million worth of milk was sold in 1960 at an average of 25 cents per quart and $5 million worth of butter was sold at 80 cents per pound. Then any two numbers in the ratio 50:5 may be used to weight the prices. If 10 and 1 are chosen, the weighted value for milk in 1960 would be $25 \times 10 = 250$, and for butter it would be $80 \times 1 = 80$.

Let I_0 = the index number for the base period, I_1 = the index number for the next period, I_n = the index number for the nth period, m_0 = a measure of a contributing item in the base period, m_1 = the corresponding measure in the next period, m_n = the corresponding measure in the nth period, and w_0 = the weight, or quantity, to be applied to m_0 in the base period. Then the index numbers for successive periods are

$$I_0 = \frac{\Sigma m_0 w_0}{\Sigma m_0 w_0} 100 = 100$$

$$I_1 = \frac{\Sigma m_1 w_0}{\Sigma m_0 w_0} 100$$

$$I_n = \frac{\Sigma m_n w_0}{\Sigma m_0 w_0} 100$$

where the summation is to be extended over all the weighted measures contributing to the index number.

• Compute the milk-butter index number for 1965, taking the index number for 1960 as 100. Assume that milk sold for 25 cents per qt in 1960 and 30 cents per qt in 1965, and that butter sold for 80 cents per lb in 1960 and 90 cents per lb in 1965. Use a weight of 10 for milk and 1 for butter.

In 1960, the weighted value for milk was $25 \times 10 = 250$ and for butter was $80 \times 1 = 80$. Hence, for 1960, $\Sigma m_0 w_0 = 250 + 80 = 330$.

In 1965, the weighted value for milk was $30 \times 10 = 300$ and for butter was $90 \times 1 = 90$. Thus, for 1965, $\Sigma m_1 w_0 = 300 + 90 = 390$. With the 1960 index number as 100, the 1965 index number is

$$I = \frac{\Sigma m_1 w_0}{\Sigma m_0 w_0} 100 = \frac{390}{330} 100 = 118$$

17-2. Relative Frequency. Ratio of the number of times an event occurs to the total number of trials or observations. Thus, if an event is observed m times in N trials, its relative frequency is m/N.

The importance of relative frequency is that it is an approximation of the probability of occurrence of the event. As the number of trials are increased, the deviation of m/N from the probability p is likely to become small.

17-3. Histogram. A bar chart indicating how many items have the same measure or lie within a given range.

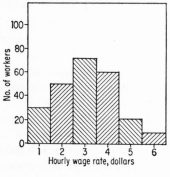

FIG. 17-1   FIG. 17-2

Figure 17-1 is a histogram showing the number of workers receiving various wage rates from a specific company. Note that each bar is centered on the wage for which the number of workers are counted.

17-4. Frequency Distributions and Curves (or Polygons). A frequency distribution is a collection of observations, counts, or measures listed, in order of size, with frequency of occurrence. If the measures and their frequencies are plotted, the curve or polygon through the points is called a frequency distribution curve or polygon.

The frequency polygon in Fig. 17-2 shows the same relationship between wage rates and number of workers as does Fig. 17-1. It could have been obtained directly from Fig. 17-1 by drawing lines between the midpoints of the upper bases of the rectangles (see also Arts. 17-11 and 17-12).

17-5. Cumulative Frequency Curve (Ogive). A curve in which each measure is plotted against the number of times it is (or is not) exceeded.

Figure 17-3 shows the cumulative frequency curve for a series of tests made to determine the compressive strength of a specific concrete mix.

The trend line is a straight line fitted closely to the majority of the points.

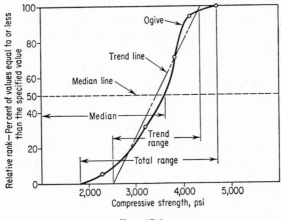

Fig. 17-3

Total range is the difference between the largest and smallest observations.

Trend range is the difference between the value at the intersection of the trend line and the 100 per cent relative-rank and the value at the intersection of the trend line and the 0 per cent line.

When an ogive has a single, clearly discernible trend line, the standard deviation of the data is approximately one-third the trend range. An approximation to the coefficient of variation, the ratio of the standard deviation to the arithmetic mean, can be obtained by dividing the standard deviation by the median value (intersection of the ogive with the 50 per cent relative-rank).

17-6. Measures of Central Tendency. Several measures are used in statistics to represent numerically such imprecise words as average, typical, normal, usual, and representative. Here are some common statistical measures of central tendency:

The arithmetic mean $\bar{X}$ of a group of numbers, or measures, is the sum of the measures ΣX divided by the number of items N.

$$\bar{X} = \frac{\Sigma X}{N} = \frac{\Sigma f_n X_n}{\Sigma f_n}$$

where f_n is the number of times a measure X_n occurs.   (f_n can also represent the weight to be applied to each item to obtain a weighted average.)

The arithmetic mean of a series of measures in which only random errors are present is the most probable value or best estimate.

When a series of measures consists of large numbers, computation of the arithmetic mean can be shortened by the following method:

Select an arbitrary reference value (as close to the mean as can be estimated). Subtract the reference number from each measure and compute the mean of these differences. Add this mean algebraically to the reference number to obtain the required arithmetic mean.

• Calculate the mean of 1,278.016, 1,280.324, and 1,274.841.

Take 1,278 as the reference value and subtract it from the given numbers. The results are 0.016, 2.324, and −3.159. The mean of these values is −0.273, so the required mean is 1,278 − 0.273 = 1,277.727.

The median M is the middle measure in a series in which all measures have been arranged according to size. Thus, equal numbers of items are above and below it. The median also is the 50th percentile.

If a series of measures are arranged in order of size, the p percentile is the measure below which p per cent of the measures lie. Thus, the 90th percentile is the item larger than 90 per cent of all the items in the series. The 25th percentile is called the first quartile, and the 75th percentile is called the third quartile. The multiple-of-ten percentiles are known as deciles; thus, the 30th percentile is the third decile.

The mode of a series of measures is the one that occurs most often. It corresponds to the highest point of a frequency curve.

• In a factory where one man earns $50,000, nine men earn $10,000 each, 40 men earn $6,000 each, and ten men earn $4,000 each, what are the arithmetic mean of their wages, the median wage, the first quartile, and the mode?

The arithmetic mean is

$$\bar{X} = \frac{\Sigma f_n X_n}{\Sigma f_n} = \frac{50,000 + 9 \times 10,000 + 40 \times 6,000 + 10 \times 4,000}{1 + 9 + 40 + 10} = 7,000$$

The median is $6,000, since as many men in this factory earn $6,000 or more as earn $6,000 or less. The first quartile also is $6,000, since 25 per cent, or 15, of the men earn that amount or less. The mode is also $6,000, since most men earn that amount.

The geometric mean is the nth root of the product of n measures.

$$GM = \sqrt[n]{X_1 X_2 X_3 \cdots X_n}$$

For example, 10 is the geometric mean of 5, 10, and 20.

$$\sqrt[3]{5 \times 10 \times 20} = 10$$

17-7. Measures of Variability. The compactness of a frequency distribution of a series of measures about a point of central tendency may be described by such terms as dispersion, spread, scatter, deviation, and variability. Several terms are used in statistics to measure variability.

The range, the difference between the largest and smallest values, is not a good quantitative description of compactness of data because it gives no indication of the form of the distribution between the extremes.

The semi-interquartile range or quartile deviation, half the difference between the first and third quartiles, has similar disadvantages.

The deviation from the mean $\bar{X}$ of each measure X in a series equals $X - \bar{X}$. The average deviation is the mean of the deviations from the mean.

The widely-used standard deviation is defined by

$$\sigma = \sqrt{\frac{\Sigma(X_i - \bar{X})^2}{N}} = \sqrt{\frac{\Sigma x_i^2}{N}}$$

where $X_i - \bar{X} = x_i$ = deviation of each measure from mean
N = number of items

For machine calculation, the following may be preferable to the use of x_i in calculating standard deviation:

$$\Sigma x_i^2 = \Sigma X_i^2 - \frac{(\Sigma X_i)^2}{N} = \Sigma X_i^2 - \bar{X}\Sigma X_i$$

• Determine the standard deviation of the series 3, 8, 9, 10, 12, 15.

The squares of these six numbers X_i^2 are 9, 64, 81, 100, 144, and 225. The sum of the numbers $\Sigma X_i = 57$ and the sum of their squares $\Sigma X_i^2 = 623$. Hence, the standard deviation is

$$\sigma = \sqrt{\frac{\Sigma X_i^2}{N} - \left(\frac{\Sigma X_i}{N}\right)^2} = \sqrt{623/6 - (57/6)^2} = 3.68$$

(See also Art. 17-8.)

17-8. Variance. The square of the standard deviation of a random variable X_i (Art. 17-7). It is defined by

$$\sigma^2 = \frac{\Sigma(X_i - \bar{X})^2}{N}$$

where $\bar{X}$ = mean
N = number of items

The variance of the means of a large number N of random samples is related to the variance of the population, or universe, by

$$\bar{\sigma}^2 = \frac{\sigma^2}{N}$$

(See also standard error, Art. 17-17.)

For small samples ($N < 25$), the variance of a population σ^2 can be estimated from the variance s^2 of a sample.

$$\sigma^2 = \frac{N}{N-1} s^2$$

The variance of the sum or difference of independently varying functions is equal to the sum of the separate variances.

The variance of a general function $X = f(x_1, x_2, \ldots, x_n)$ can be computed from the variances of the individual variables.

$$\sigma^2(X) = \left(\frac{\partial X}{\partial x_1}\right)^2 \sigma^2(x_1) + \left(\frac{\partial X}{\partial x_2}\right)^2 \sigma^2(x_2) + \cdots + \left(\frac{\partial X}{\partial x_n}\right)^2 \sigma^2(x_n)$$

17-9. Covariance.   This is defined for two measures X and Y by

$$\sigma_{XY}{}^2 = \frac{\Sigma(X - \bar{X})(Y - \bar{Y})}{N}$$

where N = number of paired measures
$\bar{X}$ = mean of X measures
$\bar{Y}$ = mean of Y measures
(See also Art. 17-21.)

17-10. Skewness and Kurtosis.   If a frequency curve is unsymmetric about a point of central tendency and the measures with greatest frequency of occurrence are clustered on the left, with a long spread, or tail,

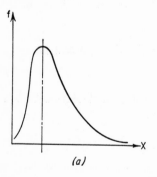 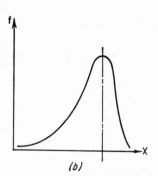

FIG. 17-4

on the right, the distribution is positively skewed (Fig. 17-4a).   If most of the cases are on the right, it is negatively skewed (Fig. 17-4b).

A distribution with high concentration in the tails as well as the center and with little between has high kurtosis (Fig. 17-5a).   One relatively low in the middle and tails has little kurtosis (Fig. 17-5b).

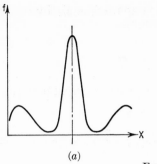 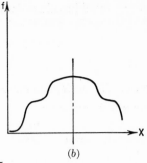

(a)   (b)

FIG. 17-5

These characteristics are measured by moments of the distribution about the arithmetic mean. These moments are defined by

$$\mu_k = \frac{\Sigma f_n x_n{}^k}{N}$$

where x_n = deviations from mean = $X_n - \bar{X}$
     f_n = frequency of occurrence of each measure X_n
     N = number of measures
     k = constant indicating degree of moment
The first moment $\mu_1 = 0$, since $\Sigma f_n x_n = 0$. The second moment

$$\mu_2 = \frac{\Sigma f_n x_n{}^2}{N} = \sigma^2$$

where σ^2 is the variance.
Skewness is measured by $\gamma_1 = \mu_3/\sigma^3$.
Kurtosis is measured by $\gamma_2 = \mu_4/\mu_2{}^2 - 3$ or by $\beta_2 = \mu_4/\mu_2{}^2$.
The normal distribution has zero kurtosis as measured by γ_2.

17-11. Binomial Distribution. The binomial frequency function

$$B(x) = C(N,x)p^x q^{N-x} = \frac{N!}{x!(N-x)!}\, p^x q^{N-x}$$
$$= \frac{N(N-1)\,\cdots\,(N-x+1)}{x!}\, p^x q^{N-x}$$

represents the probability of exactly x successes in N trials with constant probability p in each trial; $q = 1 - p$. If x is given successive values from 0 to N, x and $B(x)$ comprise a binomial distribution.

The expected number of successes in N trials $E(x) = Np$. The maximum value of $B(x)$ is the probability of the most probable number of successes. For the binomial distribution, the expected value and the most probable value are equal when Np is an integer; otherwise, the most probable value is within 1 of Np. For large values of N, the probability of the most probable value of x is given approximately by

$$B_{max} \approx \frac{1}{\sqrt{2\pi N pq}}$$

When both N and Np are large, $B(x)$ can be computed from the Laplace or normal approximation.

$$B(x) \approx \frac{1}{\sqrt{2\pi N pq}}\, e^{-X^2}$$

where $X^2 = (x - Np)^2/2Npq$.

When N is large but Np is moderate, $B(x)$ can be computed from Poisson's law (law of small numbers).

$$B(x) \approx \frac{(Np)^x}{x!}\, e^{-Np}$$

• What is the probability of five persons dying of cancer in one day in a certain large city when, on the average, two die daily there?

The expected number of deaths $Np = 2$. Hence, the probability of five deaths is approximately

$$B(5) \approx \frac{2^5}{5!}\, e^{-2} = {}^{32}\!/_{120}(0.135) = 0.036$$

17-12. Normal Distribution.　A series of numbers or measures which, when plotted with respect to their frequencies of occurrence, determine a

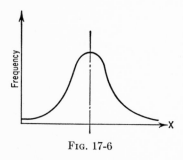

FIG. 17-6

bell-shaped, normal, or Gaussian curve (Art. 16-9; see Fig. 17-6). If y is the frequency of any measure X, the normal distribution relationship can be written

$$y = y_0 e^{-\frac{1}{2}S^2}$$

where y_0 = ordinate at mean $\bar{X}$

$e = 2.7183$

$S = (X - \bar{X})/\sigma$

σ = standard deviation (Art. 17-7)

For a normal curve, skewness as measured by γ_1 and kurtosis as measured by γ_2 are zero (Art. 17-10).

The ratio of the area under the normal curve between the mean and a given value of X to the total area under the curve equals the probability that a measure selected at random from the distribution will lie between X and $\bar{X}$. Most books on statistics have tables giving such probabilities for various values of the significance ratio $(X - \bar{X})/\sigma$. Here is a useful portion of such a table:

TABLE 17-1. PROBABILITY THAT $\dfrac{X - \bar{X}}{\sigma}$ OR $\dfrac{|X - \bar{X}|}{\sigma}$ WILL NOT BE EXCEEDED

$\dfrac{X - \bar{X}}{\sigma}$	Probability, %	$\dfrac{|X - \bar{X}|}{\sigma}$	Probability, %
1.00	34.13	±1.00	68.26
1.28	39.97	±1.28	80
1.64	44.95	±1.64	90
1.96	47.50	±1.96	95
2.33	49.01	±2.33	98
2.58	49.51	±2.58	99
3.29	49.95	±3.29	99.9

17-13. Level of Confidence. The probability, expressed in per cent, that a statement is true.

For example, we can say at the 95 per cent level of confidence that if a single measure is selected at random from a normal distribution, it will lie within 1.96 standard deviations σ of the mean (Art. 17-12). At the 98 per cent level we can say it will lie within 2.33σ of the mean, and at the 99 per cent level we can say it will lie within 2.58σ of the mean.

The confidence interval is the range within which the true mean lies at a given level of confidence. Thus, the 95 per cent confidence interval for the true mean of a population is $\bar{X} \pm 1.96\sigma$, where $\bar{X}$ is the mean of a random sample and σ is the standard error of the mean (Art. 17-17).

17-14. Level of Significance. When the null hypothesis (assumption that the true difference between two measures is zero) may be rejected, the difference between the measures is statistically significant. The level of significance is 100 minus the level of confidence (Art. 17-13).

Thus, a difference is significant at the 1 per cent level if we can say at a confidence level of 99 per cent that the null hypothesis is false.

17-15. Types of Errors. False rejection of a hypothesis is a type I error. False acceptance is a type II error.

If the significance level is set so low that there is small probability of a type I error, there is a high probability of a type II error. Usually, probabilities of 0.05 to 0.01 for type I errors give close to the minimum probabilities of both types of errors.

17-16. Power of a Test. Probability of rejecting a hypothesis when it is false. It equals one minus the probability of a type II error (Art. 17-15).

17-17. Standard Errors for Large Samples. A random sample of N items is one so selected that it is just as likely to be drawn from the whole population as any other set of N items.

Suppose a very large number of large random samples ($N \geq 25$) is chosen from a given population or system. Suppose also that the means of the measures in each sample have been computed to be $\bar{X}_i$. Then the distribution of these $\bar{X}_i$ will closely approximate the normal distribution, and the mean $\bar{X}$ of the $\bar{X}_i$ will be the mean of the individual measures comprising the population. The standard deviation of the distribution of $\bar{X}_i$ is the standard error of the mean $\bar{\sigma}$ of a given random sample.

The standard error of the mean can be computed when the standard deviation s of a sample is known.

$$\bar{\sigma} = \frac{s}{\sqrt{N-1}} \qquad s = \sqrt{\frac{\Sigma(X_i - \bar{X})^2}{N}}$$

where N = number of measures X_i in sample ($N \geq 25$)

  $\bar{X}$ = mean of sample

For small samples, see Art. 17-19.

• Determine the mean age at a confidence level of 99 per cent of all construction workers in New York City if a random sample of 50 workers has a mean age of 45 years and a standard deviation of 8.4 years.

The standard error of the mean of the sample is

$$\bar{\sigma} = \frac{8.4}{\sqrt{50-1}} = 1.2 \text{ years}$$

From Table 17-1 (Art. 17-12), the probability is 99 per cent that a random measure will deviate $\pm 2.58\bar{\sigma}$ or less from the mean. So at the 99 per cent confidence level, the mean age of all construction workers in the city is $45 \pm 2.58 \times 1.2 = 45 \pm 3.1$ years.

The standard error of any statistical measure obtained from a random sample is the standard deviation of its sampling distribution. Thus, the standard error σ_M of the median is the standard deviation of a distribution of medians for a very large number of random samples of the same size as the given sample. The standard error of the median can be determined from

$$\sigma_M = \tfrac{5}{4}\bar{\sigma}$$

where $\bar{\sigma}$ is the standard error of the mean. Similarly, the standard error of the standard deviation of a random sample equals $0.707\bar{\sigma}$.

If X and Y are two uncorrelated measures, the standard error of their difference $X - Y$ is

$$\bar{\sigma} = \sqrt{\bar{\sigma}_X^2 + \bar{\sigma}_Y^2}$$

where $\bar{\sigma}_X$ = standard error of X sample

  $\bar{\sigma}_Y$ = standard error of Y sample

• Two laboratories test a series of 50 concrete specimens in compression. One finds a mean strength of 3,000 psi with a standard deviation of 105 psi; the other finds a mean strength of 3,080 psi with a standard deviation of 84 psi. Is the difference of the means significant at the 5 per cent level?

The standard error of the mean for the specimens with the 3,000-psi mean is $105/\sqrt{50-1} = 15$ psi. The standard error for the specimens with the 3,080-psi mean is $84/\sqrt{50-1} = 12$ psi. The standard error of the difference of the means, therefore, is

$$\bar{\sigma} = \sqrt{(15)^2 + (12)^2} = 19.2 \text{ psi}$$

Now, from Table 17-1 (Art. 17-12), at a confidence level of 95 per cent, $1.96\bar{\sigma} = 1.96 \times 19.2 = 38$ psi will not be exceeded. Since the actual difference between the means is $3,080 - 3,000 = 80$, which is greater than 38, the difference is significant at the 5 per cent level. Thus, the difference between the means obtained by the two laboratories is not likely due to chance; their tests appear to disagree.

If X and Y are two related variables, the standard error of their difference is

$$\bar{\sigma} = \sqrt{\bar{\sigma}_X{}^2 + \bar{\sigma}_Y{}^2 - 2r\bar{\sigma}_X\bar{\sigma}_Y}$$

where $\bar{\sigma}_X, \bar{\sigma}_Y$ = standard errors of X, Y
$\qquad\quad r$ = coefficient of correlation for X, Y (Art. 17-21)

The standard error of a proportion is

$$\bar{\sigma} = \sqrt{\frac{p(1-p)}{N}}$$

where p is the hypothetical true proportion (not the observed proportion).

• A poll shows that 13 persons out of a sample of 100 use product Z. Can it be said that 10 per cent of the whole population uses it?

The standard error of the proportion is

$$\bar{\sigma} = \sqrt{\frac{0.10(1-0.10)}{100}} = 0.03$$

The difference between the observed and hypothetical true proportions is $0.13 - 0.10 = 0.03$. This is exactly equal to the standard error. Hence, the difference may be attributed to chance, and it is highly probable that 10 per cent of the population uses product Z.

17-18. Significance Ratios (t Test) for Large Samples. The t test is a means of estimating a true value from a sample and establishing confidence ranges within which the true value should lie.

The significance ratio t for a large sample (number of items $N \geq 25$) is the difference between $\bar{X}_s$, a mean obtained from the sample, and $\bar{X}$, a mean assumed to be the true mean, measured in units of the standard error of the mean $\bar{\sigma}$ (Art. 17-17).

$$t = \frac{|\bar{X}_s - \bar{X}|}{\bar{\sigma}}$$

If a large number of random samples of the same size are drawn from a population, the distribution of the significance ratios of the samples will be normal.

To test the hypothesis that the mean of a population has a specific value $\bar{X}$, find the mean and standard error of a large sample. Next, select an appropriate level of confidence on which the hypothesis may be accepted or rejected. (Usual levels are 95, 98, 99, and 99.9 per cent.) From Table 17-1 (Art. 17-12), pick for this confidence level the value of $|X - \bar{X}|/\sigma$ that should not be exceeded. Then, using the sample mean, hypothetical mean, and standard error of the sample, compute t. If it exceeds the value from Table 17-1, reject the hypothesis; if t is less, accept the hypothesis.

• A chemical company claims its product is 87.5 per cent pure. However, tests of 100 random samples from a shipment yield a mean purity of 87.1 per cent, with a standard error of 0.25 per cent. Can the shipment be accepted at a confidence level of 95 per cent on the assumption that the difference between measured purity and that claimed is due strictly to chance?

Table 17-1 indicates that at the 95 per cent confidence level, t should be equal to or less than 1.96. Now, with the sample mean 87.1 and the assumed true mean 87.5,

$$t = \frac{|87.1 - 87.5|}{0.25} = 1.60$$

Since $t < 1.96$, the shipment can be accepted, with the probability less than 5 chances in 100 that the decision is wrong.

17-19. Significance Ratios (t Test) for Small Samples. (See Art. 17-18 for large samples.) The assumption of a normal distribution of significance ratios for random samples is not valid for small samples (number of items $N < 25$). The distribution varies with the size of sample, and the minimum value of the significance ratio

$$t = \frac{|\bar{X}_s - \bar{X}|}{\bar{\sigma}}$$

required for a given significance level depends on the sample size. The effect of sample size is evaluated in terms of *degrees of freedom* in estimating the standard error.

The number of degrees of freedom equals the number of independent measurements that are available for estimating a statistical parameter. In the estimation of a standard deviation or standard error, for example, the number of degrees of freedom equals $N - 1$.

The minimum values of significance ratios required for significance at various levels for small samples are given in Table 17-2. Note that for infinite degrees of freedom the values of t are the same as for a normal distribution (Table 17-1, Art. 17-12).

TABLE 17-2. MINIMUM VALUE OF t FOR SIGNIFICANCE AT VARIOUS LEVELS

Degrees of freedom	Levels of significance			
	5%	2%	1%	0.1%
2	4.30	6.97	9.93	31.60
3	3.18	4.54	5.84	12.94
4	2.78	3.75	4.60	8.61
5	2.57	3.37	4.03	6.86
6	2.45	3.14	3.71	5.96
7	2.37	3.00	3.50	5.40
8	2.31	2.90	3.36	5.04
9	2.26	2.82	3.25	4.78
10	2.20	2.72	3.11	4.44
11	2.18	2.68	3.06	4.32
12	2.16	2.65	3.01	4.22
14	2.13	2.60	2.95	4.07
19	2.09	2.54	2.86	3.88
24	2.06	2.49	2.80	3.75
∞	1.96	2.33	2.58	3.29

• Determine the mean age at a confidence level of 99 per cent (significance level 1 per cent) of all construction workers in New York City if a random sample of 10 workers shows a mean age of 45 years and a standard deviation of 8.4 years. The standard error of the mean of the sample is

$$\bar{\sigma} = \frac{8.4}{\sqrt{10 - 1}} = 2.8 \text{ years}$$

The number of degrees of freedom is $10 - 1 = 9$. From Table 17-2, a t value of 3.25 will not be exceeded at the 1 per cent level of significance for 9 degrees of freedom. So at the 99 per cent confidence level, the mean age of all construction workers in the city is $45 \pm 3.25 \times 2.8 = 45 \pm 9.1$ years. (Comparison with the problem in Art. 17-17, where the sample consisted of 50 workers, indicates that a decrease in size of sample increased the range in which the true mean is likely to lie.)

The significance ratio for the difference in means $\bar{X} - \bar{Y}$ of two independent small samples is

$$t = \frac{\bar{X} - \bar{Y}}{\sqrt{[(N_X\sigma_X{}^2 + N_Y\sigma_Y{}^2)/(N_X + N_Y - 2)][(N_X + N_Y)/N_XN_Y]}}$$

where σ_X, σ_Y = standard deviations of samples of X, Y measures

N_X, N_Y = number of items in samples of X, Y measures

$N_X + N_Y - 2$ is the number of degrees of freedom.

• One laboratory tests 10 concrete specimens in compression. It finds a mean strength of 3,000 psi with a standard deviation of 105 psi. Another laboratory, making 17 tests, finds a mean strength of 3,080 psi with a standard deviation of 84 psi. Is the difference of the means significant at the 5 per cent level?

The number of degrees of freedom is $10 + 17 - 2 = 25$. From Table 17-2, for 25 degrees of freedom, the minimum value of the significance ratio for significance at the 5 per cent level is (interpolated) 2.06.

$$t = \frac{3,080 - 3,000}{\sqrt{[17(84)^2 + 10(105)^2/25] \times (17 + 10)/(17 \times 10)}} = 2.09$$

Since $t > 2.06$, the difference between the means is significant at the 5 per cent level.

17-20. Chi-squared Test.

χ^2 is a measure of the deviation of observed from expected values. For continuous data,

$$\chi^2 = \sum \frac{[X - E(X)]^2}{E(X)}$$

where the summation is carried out for all values of measure X, and $E(X)$ is the expected value of X (Art. 16-7). χ^2 varies with the number of degrees of freedom (number of independent measurements available for estimating it). For discrete measures, use

$$\chi^2 = \sum \frac{[|X - E(X)| - 0.5]^2}{E(X)}$$

The χ^2 test should not be applied to classes of data for which the expected value is less than 5; where possible, combine the data to exceed 5.

The probability of a larger value than χ^2 in a χ^2 test is given in Table 17-3 for degrees of freedom f up to 29. For larger values of f, χ^2 is approximated by

$$\chi^2 = \frac{(\sqrt{2f - 1} + z)^2}{2}$$

where z is the value of $(X - \bar{X})/\sigma$ in Table 17-1, Art. 17-12, corresponding to a probability equal to 50 per cent minus the given level of significance.

The χ^2 test is used to determine whether the null hypothesis should be rejected. The null hypothesis assumes that there is no significant difference between a value calculated from a sample and the corresponding value for the population from which the sample was drawn.

• A sample of 100 items contains 12 defectives. The manufacturer guarantees that not more than 10 per cent of delivered items will be defective. Is the sample result cause for rejection of the shipment?

The expected number of defectives is $0.10 \times 100 = 10$ and the expected number of nondefectives is $100 - 10 = 90$. The sample yielded 12 defectives and $100 - 12 = 88$ nondefectives. There is one degree of freedom, since the number of defectives also determines the number of nondefectives.

$$\chi^2 = \frac{(12 - 10 - 0.5)^2}{10} + \frac{(|88 - 90| - 0.5)^2}{90} = 0.225 + 0.025 = 0.250$$

From Table 17-3, with one degree of freedom, the 5 per cent level of significance corresponds to $\chi^2 = 3.84$. Hence, a decision to reject the shipment may be in error more than 5 times in 100 (in fact, more than 50 times in 100).

TABLE 17-3. MINIMUM VALUE OF χ^2 FOR SIGNIFICANCE AT VARIOUS LEVELS

Degrees of freedom	Levels of significance			
	5%	2%	1%	0.1%
1	3.84	5.41	6.64	10.83
2	5.99	7.82	9.21	13.82
3	7.82	9.84	11.35	16.27
4	9.49	11.67	13.28	18.47
5	11.07	13.39	15.09	20.52
6	12.59	15.03	16.81	22.46
7	14.07	16.62	18.48	24.32
8	15.51	18.17	20.09	26.13
9	16.92	19.68	21.67	27.88
10	18.31	21.16	23.21	29.59
11	19.68	22.62	24.73	31.26
12	21.03	24.05	26.22	32.91
13	22.36	25.47	27.69	34.53
14	23.69	26.87	29.14	36.12
19	30.14	33.69	36.19	43.82
24	36.15	40.27	42.98	51.18
29	42.56	46.69	49.59	58.30

• If the sample in the above example contains 18 defectives, should the shipment be rejected?

$$\chi^2 = \frac{(18 - 10 - 0.5)^2}{10} + \frac{(90 - 82 - 0.5)^2}{90} = 5.63 + 0.62 = 6.25$$

From Table 17-3, for one degree of freedom, this value is significant at close to the 1 per cent level; so the shipment can be rejected with a probability of only one chance in 100 of being wrong.

17-21. Correlation and Regression Coefficients. The coefficient of correlation is a measure of the degree of association between two variables. Let x be the deviation of a measure X from $\bar{X}$, the mean of its group $(x = X - \bar{X})$, and let y be the deviation of a measure Y from $\bar{Y}$, the mean of its group $(y = Y - \bar{Y})$. Then, the coefficient of correlation for the two groups is

$$r_{XY} = \frac{\sigma_{XY}^2}{\sigma_X \sigma_Y} = \frac{\Sigma xy}{N\sigma_X \sigma_Y} = \frac{\Sigma XY - N\bar{X}\bar{Y}}{N\sigma_X \sigma_Y}$$

where N = number of paired items

    σ_{XY}^2 = covariance of X and Y (Art. 17-9)

    σ_X = standard deviation of X group

    σ_Y = standard deviation of Y group

When X plotted against Y yields points scattered closely about a straight line, the coefficient of correlation will equal ± 1 if the relationship between X and Y is perfect and 0 if there is no relationship. If the line has a positive slope, r_{XY} will be positive; if it has a negative slope, r_{XY} will be negative. For other patterns of scatter than about a straight line, the coefficient may not be a reliable measure of relationship.

With data that can be correlated by a straight line, the least-squares line gives the best fit, because it minimizes the sum of the squares of the deviations of the dependent variable. The equation of the least-squares line is

$$Y = r_{XY} \frac{\sigma_Y}{\sigma_X} (X - \bar{X}) + \bar{Y}$$

This also is known as a regression equation. It can be used to estimate the probable value of Y corresponding to a specific X.

The standard error of an estimate of Y from X is

$$S_Y = \sigma_Y \sqrt{1 - r_{XY}^2}$$

It may be interpreted in the same way as standard error of the mean (Art. 17-17). It represents the standard deviation of Y about the least-square line.

$$S_Y^2 = \frac{\Sigma(Y - Y')^2}{N}$$

where Y = measured value

    Y' = value calculated from regression equation

    N = number of items summed

The regression coefficient b is the slope of the line.

$$b = \frac{\sigma_{XY}^2}{\sigma_X^2} = r_{XY} \frac{\sigma_Y}{\sigma_X}$$

• Ten tests are made of spot welds in steel. A plot of shear strength versus weld diameter (the independent variable because it can be determined with greater precision) yields points scattered about a straight line. If the mean strength $\bar{Y}$ is 975 lb, the mean diameter $\bar{X}$ is 224 mils, σ_Y = 182 lb, σ_X = 23 mils, and the sum of the products of the pairs X and Y, ΣXY = 2,220,000, determine the coefficients of correlation and regression, the regression equation, and the standard error of estimate.

The correlation coefficient is

$$r = \frac{\Sigma XY - N\bar{X}\bar{Y}}{N\sigma_X\sigma_Y} = \frac{2,220,000 - 10 \times 224 \times 975}{10 \times 23 \times 182} = 0.835$$

The regression coefficient is

$$b = r\frac{\sigma_Y}{\sigma_X} = 0.835\,\frac{182}{23} = 6.61$$

and the regression equation is

$$Y = 6.61(X - 224) + 975 = 6.61X - 507$$

The standard error of estimate is

$$S_Y = 182\,\sqrt{1 - (0.835)^2} = 99.9$$

17-22. Curve Fitting. Process of finding a curve that passes through or near a group of plotted points, indicating their general trend. The equation of the curve is an empirical equation.

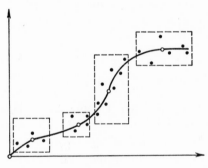

Fig. 17-7

An approximate method of fitting a curve to plotted points is to divide the group into several sets, find the arithmetic mean of the coordinates of each set (center of gravity), and draw the curve through these means. For fitting a straight line, use two sets; for a second degree curve, use three sets; for an nth degree curve, use $n + 1$ sets (Fig. 17-7).

If points are plotted on logarithmic coordinate paper, a straight line fitted to them represents $y = ax^n$, where n is the slope of the line and a is the intercept on the y axis.

If points are plotted on semilogarithmic paper (natural numbers along the x axis, $\log_{10}$ along the y axis), a straight line fitted to them represents $y = a(10)^{kx}$, where k is the slope of the line and a is the intercept on the y axis.

The best-fitting curve of a given type is one that minimizes the sum of the squares of the residuals or deviations of the dependent variable.

The residuals are the differences between ordinates of plotted points and corresponding ordinates to the curve (see also Art. 17-21).

If the equation of the least-square straight line fitted to a group of N points is assumed to be $y = mx + b$, where m is its slope and b is its intercept on the y axis, then m and b can be computed from the simultaneous solution of the equations:

$$m\Sigma x + Nb = \Sigma y$$
$$m\Sigma x^2 + b\Sigma x = \Sigma xy$$

• Fit a straight line to the following points: (1,3.4), (2,3.6), (3,4.6), and (4,6.4).

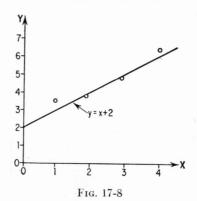

FIG. 17-8

Here, $N = 4$. $\Sigma x = 1 + 2 + 3 + 4 = 10$. $\Sigma x^2 = 30$, $\Sigma y = 18$, and $\Sigma xy = 1 \times 3.4 + 2 \times 3.6 + 3 \times 4.6 + 4 \times 6.4 = 50$. The equations are then

$$10m + 4b = 18$$
$$30m + 10b = 50$$

Their solution is $m = 1$, $b = 2$. Hence, the required line is (Fig. 17-8)

$$y = x + 2$$

To fit a second-degree least-square curve to a group of points requires the solution of three simultaneous equations. The procedure, which can be extended to higher-degree equations, is as follows:

Let $y = a_1 + a_2x + a_3x^2$ be the equation of the curve and (x_i, y_i), with $i = 1, 2, \ldots N$, be the given points. The residuals are represented by $v_i = a_1 + a_2x_i + a_3x_i{}^2 - y_i$.

If the curve is to make $\Sigma v_i{}^2$ a minimum, then the three equations to be solved are

$$\frac{\partial \Sigma v_i{}^2}{\partial a_k} = 2\,\Sigma v_i\,\frac{\partial v_i}{\partial a_k} = 0 \qquad k = 1, 2, 3$$

Since $\dfrac{\partial v_i}{\partial a_1} = 1$, $\dfrac{\partial v_i}{\partial a_2} = x_i$, and $\dfrac{\partial v_i}{\partial a_3} = x_i{}^2$, the equations, when $a_1 + a_2x_i + $

$a_3 x_i^2 - y_i$ is substituted for v_i, become

$$Na_1 + a_2\Sigma x_i + a_3\Sigma x_i^2 = \Sigma y_i$$
$$a_1\Sigma x_i + a_2\Sigma x_i^2 + a_3\Sigma x_i^3 = \Sigma x_i y_i$$
$$a_1\Sigma x_i^2 + a_2\Sigma x_i^3 + a_3\Sigma x_i^4 = \Sigma x_i^2 y_i$$

The summations are to be carried out from $i = 1$ to $i = N$, the number of points.

BIBLIOGRAPHY

Burington, R. S., and D. C. May: "Handbook of Probability and Statistics with Tables," McGraw-Hill Book Company, Inc., New York, 1953.

Duncan, A. J.: "Quality Control and Industrial Statistics," Richard D. Irwin, Inc., Homewood, Ill., 1959.

Fisher, R. A., and F. Yates: "Statistical Tables for Biological, Agricultural and Medical Research," Oliver & Boyd, Ltd., Edinburgh, 1953.

Fraser, D. A. S.: "Statistics: An Introduction," John Wiley & Sons, Inc., New York, 1958.

Grant, E. L.: "Statistical Quality Control," McGraw-Hill Book Company, Inc., New York, 1952.

Hoel, P. G.: "Introduction to Mathematical Statistics," John Wiley & Sons, Inc., New York, 1954.

Middlemiss, R. R.: "Analytic Geometry," 2d ed., McGraw-Hill Book Company, Inc., New York, 1955.

Paden, D. W., and E. F. Lindquist: "Statistics for Economics and Business," 2d ed., McGraw-Hill Book Company, Inc., New York, 1956.

Snedecor, G. W.: "Statistical Methods," 5th ed., Iowa State College Press, Ames, Iowa, 1956.

Sokolnikoff, I. S., and R. M. Redheffer: "Mathematics of Physics and Modern Engineering," McGraw-Hill Book Company, Inc., New York, 1958.

Volk, William: "Statistics for Chemical Engineers," McGraw-Hill Book Company, Inc., New York, 1958.

INDEX

Lines, parallel, 72–78, 80, 104, 157, 158, 306, 307
  unequal, 78, 79
Liquid measure, units of, 2, 4
Lituus, 187
Locus, 73, 102, 103
Logarithm, of e, 4
  series for, 279
  of ten, 4
Logarithmic equations, 64, 65, 179
  (*See also* Newton's method)
Logarithmic functions, 64, 65, 179, 198, 330, 331
Logarithmic series, 279
Logarithms, Briggs (common), 63, 179
  natural (base e), 63, 179
  operations with, 63, 64
Loss, 23
Lune, 99

Maclaurin's series, 279, 280, 337
Mantissa, 63
Mapping, 331–335
Maximum, 168, 200–202
Matrix, adjoint, 285
  algebra, 283, 284
  defined, 283
  determinants of, 294
  Hermitian, 285
  identity (unit), 286, 291, 294, 295
  inverse, 291
  order of, 283
  rank of, 294
  scalar, 285, 286
  skew-symmetric, 285
  submatrix of, 294
  symmetric, 285
  tranjugate, 285
  transformations of, 294, 295
  transpose, 285
Maxwell-Boltzmann density function, 344
Mean, arithmetic, 272, 344, 345, 349, 350
  geometric, 273, 350
  in proportion, 32
  theorem of, 203
Measurement systems, 2–4
Median, of observations, 349, 350
  of triangle, 77, 80, 83–85
Mega, 3
Menelaus theorem, 82
Meridian, 126–129, 139, 140
Metric coefficients, 305, 306, 314–317, 324
Metric system, 3
Micro, **3**

Micron, 3
Mil, 111
Milli, 3
Minimum, 168, 200–202
Minuend, 11, 13
Minute, 91–111
Mixed components, 305, 306, 319, 320
Mode, 350
Mollweide's equations, 122
Monomial, 26–28
Multiple, defined, 8
  least common, 8
Multiplicand, 15
Multiplication, algebraic, 27, 28
  of complex numbers, 325, 327
  of decimals, 15, 16, 18
  of determinants, 287, 289
  of fractions, 10–12
  graphical, 110
  logarithmic, 63, 64
  of matrices, 284
  by power of ten, 14, 15
  slide rule, 18, 19
  of surds, 37
  of tensors, 322
  of vectors, 301, 302, 306–309
Multiplier, 15

Nano, 3
Napier's analogies, 136
Nappes, 97
Nautical units, 3
Neighborhood, axiom of, 72
Newton's method, 52–54
Nomograms, 295–298
Normal, to curve, 165, 166, 310
  to plane, 159, 172
  to surface, 166, 317
  (*See also* Orthogonal curves; Perpendicular lines)
Normal distribution, 345, 353–358
Null hypothesis, 355, 356, 358, 360, 361
Numbers, absolute value of, 27, 251, 325
  approximate, addition, 14
    division, 15–19
    multiplication, 15, 16, 18
    rounding off, 13
    subtraction, 14
  in binary system, 6
  complex, 44, 325–339
    absolute value of, 325–327
    algebra of, 325–328
    amplitude of, 325–327
    argument of, 325–327
    conjugate of, 326
    geometry of, 325–327